T0215203

Lecture Notes in Computer Science　　8446

Commenced Publication in 1973
Founding and Former Series Editors:
Gerhard Goos, Juris Hartmanis, and Jan van Leeuwen

More information about this series at http://www.springer.com/series/7407

Marcello M. Bonsangue (Ed.)

Coalgebraic Methods in Computer Science

12th IFIP WG 1.3 International Workshop, CMCS 2014
Colocated with ETAPS 2014
Grenoble, France, April 5–6, 2014
Revised Selected Papers

 Springer

Editor
Marcello M. Bonsangue
Leiden Institute of Advanced
    Computer Science
Leiden University
Leiden
The Netherlands

ISSN 0302-9743   ISSN 1611-3349 (electronic)
ISBN 978-3-662-44123-7   ISBN 978-3-662-44124-4 (eBook)
DOI 10.1007/978-3-662-44124-4

Library of Congress Control Number: 2014946347

LNCS Sublibrary: SL1 – Theoretical Computer Science and General Issues

Springer Heidelberg New York Dordrecht London

Printed on acid-free paper

Springer is part of Springer Science+Business Media (www.springer.com)

Preface

The 12th International Workshop on Coalgebraic Methods in Computer Science, CMCS 2014, was held during April 5–6, 2014, in Grenoble, France, as a satellite event of the Joint Conference on Theory and Practice of Software, ETAPS 2014. In more than a decade of research, it has been established that a wide variety of state-based dynamical systems, such as transition systems, automata (including weighted and probabilistic variants), Markov chains, and game-based systems, can be treated uniformly as coalgebras. Coalgebra has developed into a field of its own interest presenting a deep mathematical foundation, a growing field of applications, and interactions with various other fields such as reactive and interactive system theory, object-oriented and concurrent programming, formal system specification, modal and description logics, artificial intelligence, dynamical systems, control systems, category theory, algebra, analysis, etc. The aim of the workshop is to bring together researchers with a common interest in the theory of coalgebras, their logics, and their applications.

Previous workshops of the CMCS series have been organized in Lisbon (1998), Amsterdam (1999), Berlin (2000), Genova (2001), Grenoble (2002), Warsaw (2003), Barcelona (2004), Vienna (2006), Budapest (2008), Paphos (2010), and Tallin (2012). Starting in 2004, CMCS has become a biennial workshop, alternating with the International Conference on Algebra and Coalgebra in Computer Science (CALCO), which, in odd-numbered years, has been formed by the union of CMCS with the International Workshop on Algebraic Development Techniques (WADT).

The CMCS 2014 program featured a keynote talk by Davide Sangiorgi (University of Bologna, Italy), an invited talk by Ichiro Hasuo (University of Tokyo, Japan), and an invited talk by Marina Lenisa (University of Udine, Italy). In addition, a special session on game theory and coalgebras was associated with Marina Lenisa's invited talk and featuring tutorials by Paul-Andre Mellies (Université Paris Denis Diderot, France) and Pierre Lescanne (Ecole Normale Superieure de Lyon, France).

This volume contains the revised contributions of the regular and invited papers presented at CMCS 2014. A special thanks goes to all the authors for the high quality of their contributions, and to the reviewers and Program Committee members for their help in improving the papers presented at CMCS 2014.

April 2014   Marcello M. Bonsangue

Organization

CMCS 2014 was organized as a satellite event of the Joint Conference on Theory and Practice of Software (ETAPS 2014), in close collaboration with the Leiden Institute of Advanced Computer Science (LIACS), Leiden University, The Netherlands.

Program Committee

Andreas Abel	Chalmers and Gothenburg University, Sweden
Davide Ancona	University of Genova, Italy
Adriana Balan	University Politehnica of Bucharest, Romania
Marta Bilkova	Charles University in Prague, Czech Republic
Filippo Bonchi	LIP ENS-Lyon, France
Marcello M. Bonsangue (Chair)	LIACS, Leiden University, The Netherlands
Joerg Endrullis	Free University of Amsterdam, The Netherlands
Rémy Haemmerlé	Universidad Politecnica de Madrid, Spain
Bart Jacobs	Radboud University Nijmegen, The Netherlands
Dexter Kozen	Cornell University, USA
Bartek Klin	University of Warsaw, Poland
Pierre Lescanne	ENS Lyon, France
Stefan Milius	Friedrich Alexander University of Erlangen-Nuernberg, Germany
Rob Myers	Technical University of Braunschweig, Germany
Dirk Pattinson	Australian National University, Australia
Daniela Petrisan	LIP ENS-Lyon, France
Grigore Rosu	University of Illinois at Urbana-Champaign, USA
Katsuhiko Sano	JAIST, Nomi, Japan
Monika Seisenberger	Swansea University, UK
Ana Sokolova	University of Salzburg, Austria

Publicity Chair

Alexandra Silva	Radboud University Nijmegen, The Netherlands

Sponsoring Institutions

IFIP WG 1.3
LIACS – Leiden University, The Netherlands

Contents

Invited Contributions

Higher-Order Languages: Bisimulation and Coinductive Equivalences (Extended Abstract)

Davide Sangiorgi[✉]

Università di Bologna and INRIA, Bologna, Italy
sangiorgi@gmail.com

1 Summary

Higher-order languages have been widely studied in functional programming, following the λ-calculus. In a higher-order calculus, variables may be instantiated with terms of the language. When multiple occurrences of the variable exist, this mechanism results in the possibility of copying the terms of the language.

Equivalence proof of computer programs is an important but challenging problem. Equivalence between two programs means that the programs should behave "in the same manner" under any context [Mor68]. Finding effective methods for equivalence proofs is particularly challenging in higher-order languages: pure functional languages like the λ-calculus, and richer languages including non-functional features such as non-determinism, information hiding mechanisms (e.g., generative names, store, data abstraction), concurrency, and so on.

Bisimulation [Par81a, Par81b, Mil89, San09, San12] has emerged as a very powerful operational method for proving equivalence of programs in various kinds of languages, due to the associated co-inductive proof method. Further, a number of enhancements of the bisimulation method have been studied, usually called *up-to techniques*. To be useful, the behavioral relation resulting from bisimulation—*bisimilarity*—should be a *congruence*. Bisimulation has been transplanted onto higher-order languages by Abramsky [Abr90]. This version of bisimulation, called *applicative bisimulations*, and variants of it, have received considerable attention [Gor93, GR96, Pit97, San98, Las98]. In short, two functions P and Q are applicatively bisimilar when their applications $P(M)$ and $Q(M)$ are applicatively bisimilar for any argument M.

Applicative bisimulations have some serious limitations. For instance, they are unsound under the presence of generative names [JR99] or data abstraction [SP05] because they apply bisimilar functions to an *identical* argument. Secondly, congruence proofs of applicative bisimulations are notoriously hard. Such proofs usually rely on Howe's method [How96]. The method appears however rather subtle and fragile, for instance under the presence of generative names [JR99], non-determinism [How96], or concurrency (e.g., [FHJ98]). Also, the method is very syntactical and lacks good intuition about when and why it works. Related to the problems with congruence are also the difficulties of

© IFIP International Federation for Information Processing 2014
M.M. Bonsangue (Ed.): CMCS 2014, LNCS 8446, pp. 3–9, 2014.
DOI: 10.1007/978-3-662-44124-4_1

applicative bisimulations with "up-to context" techniques (the usefulness of these techniques in higher-order languages and its problems with applicative bisimulations have been extensively studied by Lassen [Las98]; see also [San98, KW06]).

Congruence proofs for bisimulations usually exploit the bisimulation method itself to establish that the closure of the bisimilarity under contexts is again a bisimulation. To see why, intuitively, this proof does not work for applicative bisimulation, consider a pair of bisimilar functions P_1, Q_1 and another pair of bisimilar terms P_2, Q_2. In an application context they yield the terms $P_1 P_2$ and $Q_1 Q_2$ which, if bisimilarity is a congruence, should be bisimilar. However the argument for the functions P_1 and Q_1 are bisimilar, but not necessarily identical: hence we are unable to apply the bisimulation hypothesis on the functions.

Proposals for improving applicative bisimilarity include *environmental bisimulations* [SKS11, KLS11, PS12] and *logical bisimulations* [SKS07]. A key idea of environmental bisimulations is to make a clear distinction between the tested terms and the environment. An element of an environmental bisimulation has, in addition to the tested terms, a further component, the environment, which expresses the observer's current knowledge. (In languages richer than pure λ-calculi, there may be other components, for instance to keep track of generated names.) The bisimulation requirements for higher-order inputs and outputs naturally follow. For instance, in higher-order outputs, the values emitted by the tested terms are published to the environment, and are added to it, as part of the updated current knowledge. In contrast, when the tested terms perform a higher-order input (e.g., in λ-calculi the tested terms are functions that require an argument), the arguments supplied are terms that the observer can build using the current knowledge; that is, terms obtained by composing the values currently in the environment using the operators of the calculus.

A possible drawback of environmental bisimulations over, say, applicative bisimulations, is that the set of arguments to related functions that have to be considered in the bisimulation clause is larger (since it also includes non-identical arguments). As a remedy to this is offered by up-to techniques (in particular techniques involving up-to contexts), which are easier to establish for environmental bisimulations than for applicative bisimulations, and which allow us to considerably enhance the bisimulation proof method.

The difference between environmental bisimulations and logical bisimulations is that the latter does not make use of an explicit environment: the environment is implicitly taken to be the set of pairs forming the bisimulation. This simplifies the definition, but has the drawback of making the functional of bisimulation non-monotone. In λ-calculi one usually is able to show that the functional has nevertheless a greatest fixed-point which coincides with contextual equivalence. But in richer languages this does not appear to be possible.

For bisimulation and coinductive techniques, a non-trivial extension of higher-order languages concern probabilities. Probabilistic models are more and more pervasive. Not only they are a formidable tool when dealing with uncertainty and incomplete information, but they sometimes are a *necessity* rather than an alternative, like in computational cryptography (where, e.g., secure public key

encryption schemes need to be probabilistic [GM84]). A nice way to deal compu-
tationally with probabilistic models is to allow probabilistic choice as a primitive
when designing algorithms, this way switching from usual, deterministic computation to a new paradigm, called probabilistic computation. Examples of application areas in which probabilistic computation has proved to be useful include
natural language processing [MS99], robotics [Thr02], computer vision [CRM03],
and machine learning [Pea88].

This new form of computation, of course, needs to be available to program-
mers to be accessible. And indeed, various programming languages have been
introduced in the last years, spanning from abstract ones [JP89, RP02, PPT08]
to more concrete ones [Pfe01, Goo13], being inspired by various programming
paradigms like imperative, functional or even object oriented. A quite common
scheme consists in endowing any deterministic language with one or more primitives for probabilistic choice, like binary probabilistic choice or primitives for
distributions.

One class of languages which cope well with probabilistic computation are
functional languages. Indeed, viewing algorithms as functions allows a smooth
integration of distributions into the playground, itself nicely reflected at the level
of types through monads [GAB+13, RP02]. As a matter of fact, many exist-
ing probabilistic programming languages [Pfe01, Goo13] are designed around the
λ-calculus or one of its incarnations, like Scheme. All these allows to write higher-
order functions (programs can take functions as inputs and produce them as
outputs).

Bisimulation and context equivalence in a probabilistic λ-calculus have been
considered in [ALS14], where a technique is proposed for proving congruence of
probabilistic applicative bisimilarity. While the technique follows Howe's method,
some of the technicalities are quite different, relying on non-trivial "disentan-
gling" properties for sets of real numbers, these properties themselves proved
by tools from linear algebra. The bisimulation is proved to be sound for contextual equivalence. Completeness, however, fails: applicative bisimilarity is strictly
finer. A subtle aspect is also the late vs. early formulation of bisimilarity; with a
choice operator the two versions are semantically different; the congruence proof
of bisimilarity crucially relies on the late style.

Context equivalence and bisimilarity, however, coincidence on pure λ-terms.
The resulting equality is that induced by *Levy-Longo trees* (LLT), generally
accepted as the finest extensional equivalence on pure λ-terms under a lazy
regime. The proof follows Böhm-out techniques along the lines of [San94, SW01].
The result is in sharp contrast with what happens under a nondeterministic
interpretation of choice (or in the absence of choice), where context equivalence
is coarser than LLT equality.

A coinductive characterisation of context equivalence on the whole
probabilistic language is possible via an extension in which weighted formal
sums — terms akin to distributions — may appear in redex position. Thinking
of distributions as sets of terms, the construction reminds us of the reduction of
nondeterministic to deterministic automata. The technical details are however

quite different, because we are in a higher-order language and therefore — once more — we are faced with the congruence problem for bisimulation, and because formal sums may contain an *infinite* number of terms. The proof of congruence of bisimulation in this extended language uses the technique of logical bisimulation, therefore allowing bisimilar functions to be tested with *bisimilar* (rather than identical) arguments (more precisely, the arguments should be in the context closure of the bisimulation). In the probabilistic setting, however, the ordinary logical bisimulation game has to be modified substantially. For instance, formal sums represent possible evolutions of running terms, hence they should appear in redex position only (allowing them anywhere would complicate matters considerably). The obligation of redex position for certain terms is in contrast with the basic schema of logical bisimulation, in which related terms can be used as arguments to bisimilar functions and can therefore end up in arbitrary positions. This problem is solved by moving to *coupled logical bisimulations*, where a bisimulation is formed by a pair of relations, one on ordinary terms, the other on terms extended with formal sums. The bisimulation game is played on both relations, but only the first relation is used to assemble input arguments for functions.

In higher-order languages coinductive equivalences and techniques appear to be more fundamental than in first-order languages. Evidence of this are the above-mentioned results of correspondence between forms of bisimilarity and contextual equivalence in various λ-calculi. Contextual equivalence is a '*may*' of form of testing that, in first-order languages (e.g., CCS) is quite different from bisimilarity or even simulation equivalence. Indeed, in general, higher-order languages have a stronger discriminating power than first-order languages [BSV14]. For instance, if we use higher-order languages to test first-order languages, using (may-like) contextual equivalence, then the equivalences induced is often finer than the equivalences induced by first-order languages (usually trace equivalence); moreover, the natural definition of the former equivalences is coinductive, whereas that for the latter equivalences is inductive. In *distributed* higher-order languages, a construct that may strongly enhance the discriminating power is *passivation* [SS03, GH05a, LSS09a, LSS09b, LSS11, LPSS11, PS12, KH13]. Passivation offers the capability of capturing the content of a certain location into a variable, possibly copying it, and then restarting the execution in different contexts. The same discriminating power can also be obtained in call-by-value λ-calculi (that is, without concurrency or nondeterminism) extended with a location-like construct akin to a store of imperative λ-calculi, and operators for reading the content of this location, overriding it, and, if the location contains a process, for consuming such process (i.e., performing observations on the process actions). When the tested first-order processes are probabilistic, the difference in discriminating power between first-order and higher-order languages increases further: in higher-order languages equipped with passivation, or in a call-by-value λ-calculus, bisimilarity may be recovered [BSV14].

References

[Abr90] Abramsky, S.: The lazy lambda calculus. In: Turner, D.A. (ed.) Research Topics in Functional Programming, pp. 65–117. Addison-Wesley, Reading (1990)

[ALS14] Alberti, M., Lago, U.D., Sangiorgi, D.: On coinductive equivalences for higher-order probabilistic functional programs. In: Proceedings of POPL'14. ACM (2014)

[BSV14] Bernardo, M., Sangiorgi, D., Vignudelli, V.: On the discriminating power of passivation and higher-order interaction, Submitted (2014)

[CRM03] Comaniciu, D., Ramesh, V., Meer, P.: Kernel-based object tracking. IEEE Trans. Pattern Anal. Mach. Intell. **25**(5), 564–577 (2003)

[FHJ98] Ferreira, W., Hennessy, M., Jeffrey, A.: A theory of weak bisimulation for core CML. J. Funct. Program. **8**(5), 447–491 (1998)

[GAB+13] Gordon, A.D., Aizatulin, M., Borgström, J., Claret, G., Graepel, T., Nori, A.V., Rajamani, S.K., Russo, C.V.: A model-learner pattern for bayesian reasoning. In: POPL, pp. 403–416 (2013)

[GH05a] Godskesen, ChJ, Hildebrandt, T.: Extending Howe's method to early bisimulations for typed mobile embedded resources with local names. In: Sarukkai, S., Sen, S. (eds.) FSTTCS 2005. LNCS, vol. 3821, pp. 140–151. Springer, Heidelberg (2005)

[GM84] Goldwasser, S., Micali, S.: Probabilistic encryption. J. Comput. Syst. Sci. **28**(2), 270–299 (1984)

[Goo13] Goodman, N.D.: The principles and practice of probabilistic programming. In: POPL, pp. 399–402 (2013)

[Gor93] Gordon, A.D.: Functional programming and input/output. Ph.D. thesis, University of Cambridge (1993)

[GR96] Gordon, A.D., Rees, G.D.: Bisimilarity for a first-order calculus of objects with subtyping. In: Proceedings of the 23rd ACM SIGPLAN-SIGACT Symposium on Principles of Programming Languages, pp. 386–395 (1996)

[How96] Howe, D.J.: Proving congruence of bisimulation in functional programming languages. Inf. Comput. **124**(2), 103–112 (1996)

[JP89] Jones, C., Plotkin, G.D.: A probabilistic powerdomain of evaluations. In: LICS, pp. 186–195 (1989)

[JR99] Jeffrey, A., Rathke, J.: Towards a theory of bisimulation for local names. In: 14th Annual IEEE Symposium on Logic in Computer Science, pp. 56–66 (1999)

[KH13] Koutavas, V., Hennessy, M.: Symbolic bisimulation for a higher-order distributed language with passivation. In: D'Argenio, P.R., Melgratti, H. (eds.) CONCUR 2013 – Concurrency Theory. LNCS, vol. 8052, pp. 167–181. Springer, Heidelberg (2013)

[KLS11] Koutavas, V., Levy, P.B., Sumii, E.: From applicative to environmental bisimulation. Electr. Notes Theor. Comput. Sci. **276**, 215–235 (2011)

[KW06] Koutavas, V., Wand, M.: Small bisimulations for reasoning about higher-order imperative programs. In: Proceedings of the 33rd ACM SIGPLAN-SIGACT Symposium on Principles of Programming Languages, pp. 141–152 (2006)

[Las98] Lassen, S.B.: Relational reasoning about functions and nondeterminism. Ph.D. thesis, Department of Computer Science, University of Aarhus (1998)

[LPSS11] Lanese, I., Pérez, J.A., Sangiorgi, D., Schmitt, A.: On the expressiveness and decidability of higher-order process calculi. Inf. Comput. **209**(2), 198–226 (2011)

[LSS09a] Lenglet, S., Schmitt, A., Stefani, J.-B.: Howe's method for calculi with passivation. In: Bravetti, M., Zavattaro, G. (eds.) CONCUR 2009. LNCS, vol. 5710, pp. 448–462. Springer, Heidelberg (2009)

[LSS09b] Lenglet, S., Schmitt, A., Stefani, J.-B.: Normal bisimulations in calculi with passivation. In: de Alfaro, L. (ed.) FOSSACS 2009. LNCS, vol. 5504, pp. 257–271. Springer, Heidelberg (2009)

[LSS11] Lenglet, S., Schmitt, A., Stefani, J.-B.: Characterizing contextual equivalence in calculi with passivation. Inf. Comput. **209**(11), 1390–1433 (2011)

[Mil89] Milner, R.: Communication and Concurrency. Prentice Hall, Upper Saddle River (1989)

[Mor68] Morris, J.H. Jr.: Lambda-calculus models of programming languages. Ph.D. thesis, Massachusetts Institute of Technology (1968)

[MS99] Manning, C.D., Schütze, H.: Foundations of Statistical Natural Language Processing, vol. 999. MIT Press, Cambridge (1999)

[Par81a] Park, D.: A new equivalence notion for communicating systems. In: Maurer, G. (ed.) Bulletin EATCS, vol. 14, pp. 78–80 (1981). Abstract of the talk presented at the Second Workshop on the Semantics of Programming Languages, Bad Honnef, 16–20 March 1981. Abstracts collected in the Bulletin by B. Mayoh

[Par81b] Park, D.: Concurrency and automata on infinite sequences. In: Deussen, P. (ed.) Theoretical Computer Science. LNCS, vol. 104, pp. 167–183. Springer, Heidelberg (1981)

[Pea88] Pearl, J.: Probabilistic Reasoning in Intelligent Systems: Networks of Plausible Inference. Morgan Kaufmann, San Mateo (1988)

[Pfe01] Pfeffer, A.: IBAL: a probabilistic rational programming language. In: IJCAI, pp. 733–740. Morgan Kaufmann (2001)

[Pit97] Pitts, A.: Operationally-based theories of program equivalence. In: Pitts, A.M., Dybjer, P. (eds.) Semantics and Logics of Computation, pp. 241–298. Publications of the Newton Institute/Cambridge University Press, Cambridge (1997)

[PPT08] Park, S., Pfenning, F., Thrun, S.: A probabilistic language based on sampling functions. ACM Trans. Program. Lang. Syst. **31**(1), 1–46 (2008)

[PS12] Piérard, A., Sumii, E.: A higher-order distributed calculus with name creation. In: LICS, pp. 531–540. IEEE (2012)

[RP02] Ramsey, N., Pfeffer, A.: Stochastic lambda calculus and monads of probability distributions. In: POPL, pp. 154–165 (2002)

[San94] Sangiorgi, D.: The lazy lambda calculus in a concurrency scenario. Inf. Comp. **111**(1), 120–153 (1994)

[San98] Sands, D.: Improvement theory and its applications. In: Gordon, A.D., Pitts, A.M. (eds.) Higher Order Operational Techniques in Semantics, pp. 275–306. Publications of the Newton Institute/Cambridge University Press, Cambridge (1998)

[San09] Sangiorgi, D.: On the origins of bisimulation and coinduction. ACM Trans. Program. Lang. Syst. **31**(4), 15:1–15:41 (2009)

[San12] Sangiorgi, D.: Introduction to Bisimulation and Coinduction. Cambridge University Press, Cambridge (2012)

[SKS07] Sangiorgi, D., Kobayashi, N., Sumii, E.: Logical bisimulations and functional languages. In: Arbab, F., Sirjani, M. (eds.) FSEN 2007. LNCS, vol. 4767, pp. 364–379. Springer, Heidelberg (2007)

[SKS11] Sangiorgi, D., Kobayashi, N., Sumii, E.: Environmental bisimulations for higher-order languages. ACM Trans. Program. Lang. Syst. **33**(1), 5 (2011)

[SP05] Sumii, E., Pierce, B.C.: A bisimulation for type abstraction and recursion. In: Proceedings of the 32nd ACM SIGPLAN-SIGACT Symposium on Principles of Programming Languages, pp. 63–74 (2005)

[SS03] Schmitt, A., Stefani, J.-B.: The m-calculus: a higher-order distributed process calculus. In: Proceedings POPL'03, pp. 50–61. ACM (2003)

[SW01] Sangiorgi, D., Walker, D.: The Pi-Calculus - A Theory of Mobile Processes. Cambridge University Press, Cambridge (2001)

[Thr02] Thrun, S.: Robotic Mapping: A survey, pp. 1–35. Exploring Artificial Intelligence in the New Millennium, Schefferville (2002)

Generic Weakest Precondition Semantics from Monads Enriched with Order

Ichiro Hasuo[(✉)]

University of Tokyo, Tokyo, Japan
ichiro@is.s.u-tokyo.ac.jp

Abstract. We devise a generic framework where a weakest precondition semantics, in the form of indexed posets, is derived from a monad
whose Kleisli category is enriched by posets. It is inspired by Jacobs'
recent identification of a categorical structure that is common in various
predicate transformers, but adds generality in the following aspects: (1)
different notions of modality (such as "may" vs. "must") are captured
by Eilenberg-Moore algebras; (2) nested branching—like in games and in
probabilistic systems with nondeterministic environments—is modularly
modeled by a monad on the Eilenberg-Moore category of another.

1 Introduction

Among various styles of program semantics, the one by *predicate transformers* [5] is arguably the most intuitive. Its presentation is inherently logical, representing a program's behaviors by what properties (or *predicates*) hold before
and after its execution. Predicate transformer semantics therefore form a basis
of *program verification*, where specifications are given in the form of pre- and
post-conditions [14]. It has also been used for *refinement* of specifications into
programs (see e.g. [30]). Its success has driven extensions of the original nondeterministic framework, e.g. to the probabilistic one [18,24] and to the setting
with both nondeterministic and probabilistic branching [31].

A Categorical Picture. More recently, Jacobs in his series of papers [16,17]
has pushed forward a categorical view on predicate transformers. It starts with a
monad T that models a notion of branching. Then a program—henceforth called
a *(branching) computation*—is a Kleisli arrow $X \to TY$; and the the weakest
precondition semantics is given as a contravariant functor $\mathbb{P}^{\mathcal{K}\ell} : \mathcal{K}\ell(T)^{\mathrm{op}} \to \mathbb{A}$,
from the Kleisli category to the category $\mathbb{A}$ of suitable ordered algebras.

For example, in the basic nondeterministic setting, T is the powerset monad
$\mathcal{P}$ on **Sets** and $\mathbb{A}$ is the category $\mathbf{CL}_\wedge$ of complete lattices and $\wedge$-preserving
maps. The weakest precondition functor $\mathbb{P}^{\mathcal{K}\ell} : \mathcal{K}\ell(T)^{\mathrm{op}} \to \mathbf{CL}_\wedge$ then carries a
function $f : X \to \mathcal{P}Y$ to

$$\mathrm{wpre}(f) \; : \; \mathcal{P}Y \longrightarrow \mathcal{P}X \, , \qquad Q \longmapsto \{x \in X \mid f(x) \subseteq Q\} \, . \qquad (1)$$

Moreover it can be seen that: (1) the functor $\mathbb{P}^{\mathcal{K}\ell}$ factors through the comparison
functor $K : \mathcal{K}\ell(\mathcal{P}) \to \mathcal{EM}(\mathcal{P})$ to the Eilenberg-Moore category $\mathcal{EM}(\mathcal{P})$; and (2)
the extended functor $\mathbb{P}^{\mathcal{EM}}$ has a dual adjoint $\mathbb{S}$. The situation is as follows.

© IFIP International Federation for Information Processing 2014
M.M. Bonsangue (Ed.): CMCS 2014, LNCS 8446, pp. 10–32, 2014.
DOI: 10.1007/978-3-662-44124-4_2

$$\mathbf{CL}_\wedge \underset{\mathbb{P}^{\mathcal{E}\mathcal{M}}}{\overset{\mathbb{S}}{\underset{\perp}{\rightleftharpoons}}} (\mathbf{CL}_\vee)^{\mathrm{op}} \cong \mathcal{E}\mathcal{M}(\mathcal{P})^{\mathrm{op}} \tag{2}$$

$$\mathbb{P}^{\mathcal{K}\ell} = \mathbb{P}^{\mathcal{E}\mathcal{M}} \circ K^{\mathrm{op}} \qquad \mathcal{K}\ell(\mathcal{P})^{\mathrm{op}} \qquad K^{\mathrm{op}}$$

Here the functor K carries $f \colon X \to TY$ to $f^\dagger \colon \mathcal{P}X \to \mathcal{P}Y, P \mapsto \bigcup_{x \in P} f(x)$ and is naturally thought of as a strongest postcondition semantics. Therefore the picture (2)—understood as the one below—identifies a categorical structure that underlies predicate transformer semantics. The adjunction here—it is in fact an isomorphism in the specific instance of (2)—indicates a "duality" between forward and backward predicate transformers.

$$(\text{backward predicate transformers}) \underset{\perp}{\overset{\mathbb{S}}{\rightleftharpoons}} (\text{forward predicate transformers}) \tag{3}$$

$$\underset{\text{semantics}}{\text{weakest precondition}} \qquad \text{(computations)} \qquad \underset{\text{semantics}}{\text{strongest postcondition}}$$

Jacobs has identified other instances of (3) for: discrete probabilistic branching [16]; quantum logic [16]; and continuous probabilistic branching [17]. In all these instances the notion of *effect module*—originally from the study of quantum probability [6]—plays an essential role as algebras of "quantitative logics."

Towards Generic Weakest Precondition Semantics. In [16,17] the picture (3) is presented through examples and no categorical axiomatics—that induce the picture—have been explicitly introduced. Finding such is the current paper's aim. In doing so, moreover, we acquire additional generality in two aspects: *different modalities* and *nested branching*.

To motivate the first aspect, observe that the weakest precondition semantics in (1) is the *must* semantics. The *may* variant looks as interesting; it would carry a postcondition $Q \subseteq Y$ to $\{x \in X \mid f(x) \cap Q \neq \emptyset\}$. The difference between the two semantics is much like the one between the modal operators $\square$ and $\diamond$.

On the second aspect, situations are abound in computer science where a computation involves two layers of branching. Typically these layers correspond to two distinct *players* with conflicting interests. Examples are *games*, a two-player version of automata which are essential tools in various topics including model-checking; and *probabilistic systems* where it is common to include nondeterministic branching too for modeling the environment's choices. Further details will be discussed later in Sect. 3.

Predicates and Modalities from Monads. In this paper we present two categorical setups that are inspired by [4,23]—specifically by their use of $T1$ as a domain of *truth values* or *quantities*.

The first "one-player" setup is when we have only one layer of branching. Much like in [16,17] we start from a monad T. Assuming that T is *order-enriched*—in the sense that its Kleisli category $\mathcal{K}\ell(T)$ is **Posets**-enriched—we observe that:

- a natural notion of *truth value* arises from an object $T\Omega$ (typically $\Omega = 1$);
- and a modality (like "may" and "must") corresponds to a choice of an Eilenberg-Moore algebra $\tau \colon T(T\Omega) \to T\Omega$.

The required data set (T, Ω, τ) shall be called a *predicate transformer situation*. We prove that it induces a *weakest precondition semantics* functor $\mathcal{K}\ell(T)^{\mathrm{op}} \to$ **Posets**, and that it factors through $K \colon \mathcal{K}\ell(T) \to \mathcal{EM}(T)$, much like in (2). The general setup addresses common instances like the original nondeterministic one [5] and the probabilistic predicate transformers in [18,24]. Moreover it allows us to systematically search for different modalities, leading e.g. to a probabilistic notion of partial correctness guarantee (which does not seem well-known).

The other setup is the "two-player" one. It is much like a one-player setup built on another, with two monads T and R and two "modalities" τ and ρ. A potential novelty here is that R is a monad on $\mathcal{EM}(T)$; this way we manage some known complications in nested branching, such as the difficulty of combining probability and nondeterminism. We prove that the data set $(T, \Omega, \tau, R, \rho)$ gives rise to a weakest precondition semantics, as before. Its examples include: a logic of *forced predicates* in games; and the probabilistic predicate transformers in [31].

In this paper we focus on one side of predicate transformers, namely weakest precondition semantics. Many components in the picture of [16,17] are therefore left out. They include the adjoint $\mathbb{S}$ in (3), and the role of effect modules. Indeed, on the top-left corner of (3) we always have **Posets** which is less rich a structure than complete lattices or effect modules. Incorporating these is future work.

Organization of the Paper. In Sect. 2 we introduce our first "one-player" setup, and exhibit its examples. Our second "two-player" setup is first motivated in Sect. 3 through the examples of games and probabilistic systems, and is formally introduced in Sect. 4. Its examples are described in Sect. 5 in detail. In Sect. 6 we conclude.

Notations and Terminologies. For a monad T, a T-algebra $TX \xrightarrow{a} X$ shall always mean an *Eilenberg-Moore algebra* for T, making the diagrams on the right commute. For categorical backgrounds see e.g. [1,28].

Given a monad T on $\mathbb{C}$, an arrow in the Kleisli category $\mathcal{K}\ell(T)$ is denoted by $X \rightarrowtail Y$; an identity arrow is by $\mathrm{id}_X^{\mathcal{K}\ell(T)}$; and composition of arrows is by $g \odot f$. These are to be distinguished from $X \to Y$, id_X and $g \circ f$ in the base category $\mathbb{C}$.

$$
\begin{array}{c}
X \xrightarrow{\eta_X} TX \\
\mathrm{id} \searrow \quad \downarrow a \\
X
\end{array}
$$

$$
\begin{array}{ccc}
T(TX) & \xrightarrow{Ta} & TX \\
\mu_X \downarrow & & \downarrow a \\
TX & \xrightarrow{a} & X
\end{array}
$$

(4)

2   Generic Weakest Preconditions, One-Player Setting

2.1   Order-Enriched Monad

We use monads for representing various notions of "branching." These monads are assumed to have order-enrichment ($\sqsubseteq$ for, roughly speaking, "more options"); and this will be used for an entailment relation, an important element of logic.

The category **Posets** is that of posets and monotone functions.

Definition 2.1. An *order-enriched monad* T on a category $\mathbb{C}$ is a monad together with a **Posets**-enriched structure of the Kleisli category $\mathcal{K}\ell(T)$.

The latter means specifically: (1) each homset $\mathcal{K}\ell(T)(X,Y) = \mathbb{C}(X,TY)$ carries a prescribed poset structure; and (2) composition $\odot$ in $\mathcal{K}\ell(T)$ is monotone in each argument. Such order-enrichment typically arises from the poset structure of TY in the pointwise manner. In the specific setting of $\mathbb{C} =$ **Sets** such enrichment can be characterized by *substitutivity* and *congruence* of orders on TX; see [21].

Below are some examples of order-enriched monads. Our intuition about an order-enriched monad T is that it represents one possible branching type, where $\eta_X \colon X \to TX$ represents the trivial branching with a unique option and $\mu_X \colon T(TX) \to TX$ represents flattening 'branching twice' into 'branching once' (see [13]). In fact each of the examples below has the Kleisli category $\mathcal{K}\ell(T)$ enriched by the category **Cppo** of pointed cpo's and continuous maps—not just by **Posets**—and hence is suited for generic *coalgebraic trace semantics* [13].

Example 2.2.

1. The *lift monad* $\mathcal{L} = 1 + (\_)$—where the element of 1 is denoted by $\perp$— has a standard monad structure induced by coproducts. For example, the multiplication $\mu^{\mathcal{L}} \colon 1 + 1 + X \to 1 + X$ carries $x \in X$ to itself and both $\perp$'s to $\perp$. The set $\mathcal{L}X$ is a pointed dcpo with the flat order ($\perp \sqsubseteq x$ for each $x \in X$). The lift monad $\mathcal{L}$ models the "branching type" of potential nontermination.
2. The *powerset monad* $\mathcal{P}$ models (possibilistic) nondeterminism. Its action on arrows takes direct images: $(\mathcal{P}f)U = \{f(x) \mid x \in U\}$. Its unit is given by singletons: $\eta_X^{\mathcal{P}} = \{\_\} \colon X \to \mathcal{P}X$, and its multiplication is by unions: $\mu_X^{\mathcal{P}} = \bigcup \colon \mathcal{P}(\mathcal{P}X) \to \mathcal{P}X$.
3. The *subdistribution monad* $\mathcal{D}$ models probabilistic branching. It carries a set X to the set of (probability) subdistributions over X:

$$\mathcal{D}X := \left\{ d \colon X \to [0,1] \mid \sum_{x \in X} d(x) \le 1 \right\};$$

   such d is called a *sub*distribution since the values need not add to 1. Given an arrow $f \colon X \to Y$ in **Sets**, $\mathcal{D}f \colon \mathcal{D}X \to \mathcal{D}Y$ is defined by $(\mathcal{D}f)(d)(y) := \sum_{x \in f^{-1}(\{y\})} d(x)$. Its unit is the *Dirac* (or *pointmass*) *distribution*: $\eta_X^{\mathcal{D}}(x) = [x \mapsto 1; \ x' \mapsto 0 \ (\text{for } x' \ne x)]$; its multiplication is defined by $\mu_X^{\mathcal{D}}(\mathfrak{a}) = [x \mapsto \sum_{d \in \mathcal{D}X} \mathfrak{a}(d) \cdot d(x)]$ for $\mathfrak{a} \in \mathcal{D}(\mathcal{D}X)$.

Besides, the *quantum branching monad* $\mathcal{Q}$ is introduced in [12] for the purpose of modeling a quantum programming language that obeys the design principle of "quantum data, classical control." It comes with an order-enrichment, too, derived from the *Löwner partial order* between positive operators. Yet another example is the continuous variant of $\mathcal{D}$, namely the *Giry monad* on the category **Meas** of measurable spaces [7].

2.2 PT Situation and Generic Weakest Precondition Semantics

We introduce our first basic setup for our generic weakest precondition semantics. In our main examples we take $\mathbb{C} = \mathbf{Sets}$ and $\Omega = 1$ (a singleton).

Definition 2.2 (PT situation). A *predicate transformer situation* (a *PT situation* for short) over a category $\mathbb{C}$ is a triple (T, Ω, τ) of

- an order-enriched monad T on $\mathbb{C}$;
- an object $\Omega \in \mathbb{C}$; and
- an (Eilenberg-Moore) algebra $\tau \colon T(T\Omega) \to T\Omega$ that satisfies the following *monotonicity condition*: for each $X \in \mathbb{C}$, the correspondence

$$(\Phi_\tau)_X \,:\, \mathbb{C}(X, T\Omega) \longrightarrow \mathbb{C}(TX, T\Omega)\,, \text{ i.e. } \mathcal{K}\ell(T)(X, \Omega) \longrightarrow \mathcal{K}\ell(T)(TX, \Omega)\,,$$

$$\text{given by} \quad \left(X \xrightarrow{p} T\Omega\right) \longmapsto \left(TX \xrightarrow{Tp} T(T\Omega) \xrightarrow{\tau} T\Omega\right)$$

is monotone with respect to the order-enrichment of the Kleisli category $\mathcal{K}\ell(T)$ (Definition 2.1). Note here that $\Phi_\tau : \mathbb{C}(\_, T\Omega) \Rightarrow \mathbb{C}(T\_, T\Omega)$ is nothing but the natural transformation induced by the arrow τ via the Yoneda lemma.

The data τ is called a *modality*; see the introduction (Sect. 1) and also Sect. 2.3 below.

The following lemma gives a canonical (but not unique) modality for T.

Lemma 2.4. *If T is an order-enriched monad, (T, Ω, μ_Ω) is a PT situation.*

Proof. We have only to check the monotonicity condition of μ_Ω in Definition 2.2. It is easy to see that $(\Phi_{\mu_\Omega})_X = \mu_\Omega \circ T(\_) \colon \mathbb{C}(X, T\Omega) \to \mathbb{C}(TX, T\Omega)$ is equal to $(\_) \odot (\mathrm{id}_{TX})^\wedge \colon \mathcal{K}\ell(T)(X, \Omega) \to \mathcal{K}\ell(T)(TX, \Omega)$. Here $(\mathrm{id}_{TX})^\wedge \colon TX \nrightarrow X$ is the arrow that corresponds to the identity id_{TX} in $\mathbb{C}$. The claim follows from the monotonicity of $\odot$. □

We shall derive a weakest precondition semantics from a given PT situation (T, Ω, τ). The goal would consist of:

- a (po)set $\mathbb{P}^{\mathcal{K}\ell}(\tau)(X)$ of *predicates* for each object $X \in \mathbb{C}$, whose order $\sqsubseteq$ represents an *entailment relation* between predicates; and
- an assignment, to each *(branching) computation* $f \colon X \to TY$ in $\mathbb{C}$, a *predicate transformer*

$$\mathrm{wpre}(f) \,:\, \mathbb{P}^{\mathcal{K}\ell}(\tau)(Y) \longrightarrow \mathbb{P}^{\mathcal{K}\ell}(\tau)(X) \tag{5}$$

that is a monotone function.

Since a computation is an arrow $f \colon X \nrightarrow Y$ in $\mathcal{K}\ell(T)$, we are aiming at a functor

$$\mathbb{P}^{\mathcal{K}\ell}(\tau) \,:\, \mathcal{K}\ell(T)^{\mathrm{op}} \longrightarrow \mathbf{Posets}\,. \tag{6}$$

Such a functor is known as an *indexed poset*, a special case of *indexed categories*. These "indexed" structures are known to correspond to "fibered" structures (*poset fibrations* and *(split) fibrations*, respectively), and all these have been used as basic constructs in categorical logic (see e.g. [15]). An indexed poset like (6) therefore puts us on a firm footing.

Proposition 2.5 (the indexed poset $\mathbb{P}^{\mathcal{K}\ell}(\tau)$). *Given a PT situation (T, Ω, τ), the following defines an indexed poset $\mathbb{P}^{\mathcal{K}\ell}(\tau)\colon \mathcal{K}\ell(T)^{\mathrm{op}} \to \mathbf{Posets}.$*[1]

- *On an object $X \in \mathcal{K}\ell(T)$, $\mathbb{P}^{\mathcal{K}\ell}(\tau)(X) := \mathcal{K}\ell(T)(X, \Omega) = \mathbb{C}(X, T\Omega)$.*
- *On an arrow $f\colon X \nrightarrow Y$, $\mathbb{P}^{\mathcal{K}\ell}(\tau)(f)\colon \mathbb{C}(Y, T\Omega) \to \mathbb{C}(X, T\Omega)$ is defined by*

$$\left(Y \xrightarrow{q} T\Omega \right) \longmapsto \left(X \xrightarrow{f} TY \xrightarrow{Tq} T(T\Omega) \xrightarrow{\tau} T\Omega \right).$$

Proof. We need to check: the monotonicity of $\mathbb{P}^{\mathcal{K}\ell}(\tau)(f)$; and that the functor $\mathbb{P}^{\mathcal{K}\ell}(\tau)$ indeed preserves identities and composition of arrows. These will be proved later, altogether in the proof of Theorem 2.10. □

A consequence of the proposition—specifically the functoriality of $\mathbb{P}^{\mathcal{K}\ell}(\tau)$—is *compositionality* of the weakest precondition semantics: given two computations $f\colon X \to TY$, $g\colon Y \to TU$ and a postcondition $r\colon U \to T1$, we have

$$\mathbb{P}^{\mathcal{K}\ell}(\tau)(g \odot f)(r) \;=\; \mathbb{P}^{\mathcal{K}\ell}(\tau)(f)\big(\mathbb{P}^{\mathcal{K}\ell}(\tau)(g)(r) \big).$$

That is, the semantics of a sequential composition $g \odot f$ can be computed step by step.

2.3   Examples of PT Situations

For each of $T = \mathcal{L}, \mathcal{P}, \mathcal{D}$ in Example 2.2, we take $\Omega = 1$ and the set $T1$ is naturally understood as a set of "truth values" (an observation in [4, 23]):

$$\mathcal{L}1 \;=\; \left[\begin{array}{c} (\mathbf{tt} := *) \\ \sqcup \mathsf{I} \\ (\mathbf{ff} := \bot) \end{array} \right], \qquad \mathcal{P}1 \;=\; \left[\begin{array}{c} (\mathbf{tt} := 1) \\ \sqcup \mathsf{I} \\ (\mathbf{ff} := \emptyset) \end{array} \right], \quad \text{and} \quad \mathcal{D}1 \;=\; \big([0,1], \leq \big).$$

Here $*$ is the element of the argument 1 in $\mathcal{L}1$. Both $\mathcal{L}1$ and $\mathcal{P}1$ represent the Boolean truth values. In the $\mathcal{D}$ case a truth value is $r \in [0, 1]$; a predicate, being a function $X \to [0, 1]$, is hence a *random variable* that tells the certainty with which the predicate holds at each $x \in X$.

We shall introduce modalities for these monads T and $\Omega = 1$. The following observation (easy by diagram chasing) will be used.

Lemma 2.6. *The category $\mathcal{EM}(T)$ of Eilenberg-Moore algebra is iso-closed in the category of functor T-algebras. That is, given an Eilenberg-Moore algebra $a\colon TX \to X$, an arrow $b\colon TY \to Y$, and an isomorphism $f\colon X \xrightarrow{\cong} Y$ such that $f \circ a = b \circ Tf$, the arrow b is also an Eilenberg-Moore algebra.* □

The Lift Monad $\mathcal{L}$: τ_{total} and τ_{partial}. We have the following two modalities (and none other, as is easily seen).

---

[1] For brevity we favor the notation $\mathbb{P}^{\mathcal{K}\ell}(\tau)$ over more appropriate $\mathbb{P}^{\mathcal{K}\ell}(T, \Omega, \tau)$.

$$\tau_{\text{total}}, \tau_{\text{partial}} : \{\bot\} + \{\text{tt}, \text{ff}\} = \mathcal{L}(\mathcal{L}1) \quad \longrightarrow \quad \mathcal{L}1 = \{\text{tt}, \text{ff}\},$$
$$\tau_{\text{total}} : \bot \mapsto \text{ff}, \quad \text{tt} \mapsto \text{tt}, \quad \text{ff} \mapsto \text{ff},$$
$$\tau_{\text{partial}} : \bot \mapsto \text{tt}, \quad \text{tt} \mapsto \text{tt}, \quad \text{ff} \mapsto \text{ff}.$$

The one we obtain from multiplication $\mu_1^{\mathcal{L}}$ is τ_{total}; the other τ_{partial} is nonetheless important in program verification. Given $q \colon Y \to \mathcal{L}1$ and $f \colon X \to \mathcal{L}Y$ where f is understood as a possibly diverging computation from X to Y, the predicate

$$\mathbb{P}^{\mathcal{K}\ell}(\tau_{\text{partial}})(f)(q) = \tau_{\text{partial}} \circ \mathcal{L}q \circ f \quad : \quad X \longrightarrow \mathcal{L}1$$

carries $x \in X$ to tt in case $f(x) = \bot$, i.e., if the computation is diverging. This is therefore a *partial correctness* specification that is common in Floyd-Hoare logic (see e.g. [37]). In contrast, using τ_{total}, the logic is about *total correctness*.

The Powerset Monad $\mathcal{P}$: $\tau_\diamond$ and $\tau_\Box$. The monad multiplication $\mu_1^{\mathcal{P}}$ yields a modality which shall be denoted by $\tau_\diamond$. The other modality $\tau_\Box$ is given via the swapping $\sigma \colon \mathcal{P}1 \overset{\cong}{\to} \mathcal{P}1$:

$$\begin{array}{cc}
\mathcal{P}(\mathcal{P}1) \xrightarrow{\mathcal{P}\sigma} \mathcal{P}(\mathcal{P}1) & \text{explicitly,} \\
\tau_\Box \downarrow \quad \cong \quad \downarrow \tau_\diamond & \tau_\diamond\{\} = \text{ff}, \ \tau_\diamond\{\text{tt}\} = \text{tt}, \ \tau_\diamond\{\text{ff}\} = \text{ff}, \ \tau_\diamond\{\text{tt}, \text{ff}\} = \text{tt}; \quad (7) \\
\mathcal{P}1 \xleftarrow[\sigma]{\cong} \mathcal{P}1; & \tau_\Box\{\} = \text{tt}, \ \tau_\Box\{\text{tt}\} = \text{tt}, \ \tau_\Box\{\text{ff}\} = \text{ff}, \ \tau_\Box\{\text{tt}, \text{ff}\} = \text{ff}.
\end{array}$$

In view of Lemma 2.6, we have only to check that the map $\tau_\Box$ satisfies the monotonicity condition in Definition 2.2. We first observe that, for $h \colon X \to \mathcal{P}1$ and $U \in \mathcal{P}X$,

$$(\tau_\Box \circ \mathcal{P}h)(U) = \text{ff} \quad \Longleftrightarrow \quad \text{ff} \in (\mathcal{P}h)(U) \quad \Longleftrightarrow \quad \exists x \in U. \, h(x) = \text{ff},$$

where the first equivalence is by (7). Now assume that $f \sqsubseteq g \colon X \nrightarrow 1$ and $(\tau_\Box \circ \mathcal{P}g)(U) = \text{ff}$. For showing $\tau_\Box \circ \mathcal{P}f \sqsubseteq \tau_\Box \circ \mathcal{P}g$ it suffices to show that $(\tau_\Box \circ \mathcal{P}f)(U) = \text{ff}$; this follows from the above observation.

The modalities $\tau_\diamond$ and $\tau_\Box$ capture the *may* and *must* weakest preconditions, respectively. Indeed, given a postcondition $q \colon Y \to \mathcal{P}1$ and $f \colon X \to \mathcal{P}Y$, we have $\mathbb{P}^{\mathcal{K}\ell}(\tau_\diamond)(f)(q)(x) = \text{tt}$ if and only if there exists $y \in Y$ such that $y \in f(x)$ and $q(y) = \text{tt}$; and $\mathbb{P}^{\mathcal{K}\ell}(\tau_\Box)(f)(q)(x) = \text{tt}$ if and only if $y \in f(x)$ implies $q(y) = \text{tt}$.

Moreover, we can show that $\tau_\diamond$ and $\tau_\Box$ are the only modalities (in the sense of Definition 2.2) for $T = \mathcal{P}$ and $\Omega = 1$. Since the unit law in (4) forces $\tau\{\text{tt}\} = \text{tt}$ and $\tau\{\text{ff}\} = \text{ff}$, the only possible variations are the following τ_1 and τ_2 (cf. (7)):

$$\tau_1\{\} = \text{tt}, \ \tau_1\{\text{tt}, \text{ff}\} = \text{tt}; \quad \tau_2\{\} = \text{ff}, \ \tau_2\{\text{tt}, \text{ff}\} = \text{ff}.$$

Both of these, however, fail to satisfy the multiplication law in (4).

$$\begin{array}{ccc}
\{\{\}, \{\text{ff}\}\} \xrightarrow{\mathcal{P}\tau_1} \{\text{tt}, \text{ff}\} & \qquad & \{\{\}, \{\text{tt}\}\} \xrightarrow{\mathcal{P}\tau_2} \{\text{tt}, \text{ff}\} \\
\bigcup_{\mathcal{P}1} \downarrow \qquad\quad \downarrow \tau_1 & & \bigcup_{\mathcal{P}1} \downarrow \qquad\quad \downarrow \tau_2 \\
\{\text{ff}\} \xmapsto[\tau_1]{} \text{ff} \ \neq \ \text{tt} & & \{\text{tt}\} \xmapsto[\tau_2]{} \text{tt} \ \neq \ \text{ff}
\end{array}$$

The monotonicity condition in Definition 2.2, in the case of $T \in \{\mathcal{L}, \mathcal{P}\}$ (hence $T\Omega \cong 2$), coincides with monotonicity of a predicate lifting $2^{(-)} \Rightarrow 2^{T(-)}$. The latter is a condition commonly adopted in coalgebraic modal logic (see e.g. [25]).

The Subdistribution Monad $\mathcal{D}$: τ_{total} and τ_{partial}. The modality τ_{total}: $\mathcal{D}[0,1] \to [0,1]$ that arises from the multiplication $\mu_1^{\mathcal{D}}$ is such that: given $q\colon Y \to \mathcal{D}1$ and $f\colon X \to \mathcal{D}Y$, we have $\mathbb{P}^{\mathcal{K}\ell}(\tau_{\text{total}})(f)(q)(x) = \sum_{y \in Y} q(y) \cdot f(x)(y)$. This is precisely the expected value of the random variable q under the distribution $f(x)$; thus τ_{total} yields the *probabilistic predicate transformer* of [18,24].

In parallel to the powerset monad case, we have an isomorphism $\sigma\colon \mathcal{D}1 \xrightarrow{\cong} \mathcal{D}1, p \mapsto 1 - p$. Another modality $\tau_{\text{partial}}\colon \mathcal{D}[0,1] \to [0,1]$ then arises by $\tau_{\text{partial}} := \sigma \circ \tau_{\text{total}} \circ \mathcal{D}\sigma$ like in (7), for which we have

$$\tau_{\text{partial}}(d) \;=\; \left(1 - \sum_{r \in [0,1]} d(r)\right) + \sum_{r \in [0,1]} r \cdot d(r) \quad \text{and}$$

$$\mathbb{P}^{\mathcal{K}\ell}(\tau_{\text{partial}})(f)(q)(x) \;=\; \left(1 - \sum_{y \in Y} f(x)(y)\right) + \sum_{y \in Y} q(y) \cdot f(x)(y).$$

In the second line, the value $1 - \sum_{y \in Y} f(x)(y)$—the probability of f's divergence— is added to the τ_{total} case. Therefore the modalities τ_{partial} and τ_{total}, much like in the case of $T = \mathcal{L}$, carry the flavor of *partial* and *total* correctness guarantee.

To see that τ_{partial} is indeed a modality is easy: we use Lemma 2.6; and the monotonicity can be deduced from the following explicit presentation of $\tau_{\text{partial}} \circ \mathcal{D}p$ for $p\colon X \to \mathcal{D}1 = [0,1]$. For each $d \in \mathcal{D}X$,

$$(\tau_{\text{partial}} \circ \mathcal{D}p)(d) \;=\; \tau_{\text{partial}}\left[r \mapsto \sum_{x \in p^{-1}(\{r\})} d(x)\right]_{r \in [0,1]}$$

$$=\; \left(1 - \sum_{r \in [0,1]} \sum_{x \in p^{-1}(\{r\})} d(x)\right) + \sum_{r \in [0,1]} r \sum_{x \in p^{-1}(\{r\})} d(x)$$

$$=\; \left(1 - \sum_{x \in X} d(x)\right) + \sum_{x \in X} p(x) \cdot d(x).$$

We do not yet know if τ_{total} and τ_{partial} are the only modalities for $\mathcal{D}$ and $\Omega = 1$.

Remark 2.7. We note the difference between a subdistribution $d \in \mathcal{D}X$ and a predicate (i.e. a random variable) $p\colon X \to \mathcal{D}1$. An example of the latter is $p_\top$ that is everywhere 1—this is the *truth predicate*. In contrast, the former $d \in \mathcal{D}X$ is subject to the (sub)normalization condition $\sum_x d(x) \leq 1$. We understand it as one single "current state" whose whereabouts are known only probabilistically.

2.4   Factorization via the Eilenberg-Moore Category

The indexed poset $\mathbb{P}^{\mathcal{K}\ell}(\tau)\colon \mathcal{K}\ell(T)^{\mathrm{op}} \to$ **Posets** in Proposition 2.5 is shown here to factor through the comparison functor $K\colon \mathcal{K}\ell(T) \to \mathcal{E}\mathcal{M}(T)$, much like in (2). In fact it is possible to see K as a *strongest postcondition semantics*; see Remark 2.11.

We will be using the following result.

Lemma 2.8. *Let T be an order-enriched monad on $\mathbb{C}$, $X, Y, U \in \mathbb{C}$ and $f\colon X \to Y$ be an arrow in $\mathbb{C}$. Then $(\_) \circ f\colon \mathbb{C}(Y, TU) \to \mathbb{C}(X, TU)$ is monotone.*

Proof. Given $g\colon Y \to TU$ in $\mathbb{C}$,

$$g \circ f = \mu_U \circ \eta_{TU} \circ g \circ f = \mu_U \circ Tg \circ Tf \circ \eta_X = \mu_U \circ Tg \circ Tf \circ \mu_X \circ \eta_{TX} \circ \eta_X$$

$$= \mu_U \circ Tg \circ \mu_Y \circ T(Tf) \circ \eta_{TX} \circ \eta_X = \left(X \overset{J\eta_X}{\nrightarrow} TX \overset{Tf}{\nrightarrow} Y \overset{g}{\nrightarrow} U\right),$$

where $J: \mathbf{Sets} \to \mathcal{K}\ell(T)$ is the Kleisli inclusion that sends the arrow $\eta_X: X \to TX$ to $\eta_{TX} \circ \eta_X: X \nrightarrow TX$. In the calculation we used the monad laws as well as the naturality of η and μ. The correspondence $(\_) \odot (Tf \odot J\eta_X)$ is monotone by assumption (Definition 2.1); this proves the claim. □

Proposition 2.9. (the indexed poset $\mathbb{P}^{\mathcal{EM}}(\tau)$). *A PT situation (T, Ω, τ) induces an indexed poset $\mathbb{P}^{\mathcal{EM}}(\tau): \mathcal{EM}(T)^{\mathrm{op}} \to \mathbf{Posets}$ in the following way.*

– *On objects,*

$$\mathbb{P}^{\mathcal{EM}}(\tau)\begin{pmatrix} TX \\ \downarrow a \\ X \end{pmatrix} := \mathcal{EM}(T)\begin{pmatrix} TX & T(T\Omega) \\ \downarrow a, & \downarrow \tau \\ X & T\Omega \end{pmatrix}$$

where the order $\sqsubseteq$ on the set $\mathcal{EM}(T)(a, \tau)$ is inherited from $\mathbb{C}(X, T\Omega)$ via the forgetful functor $U^T: \mathcal{EM}(T) \to \mathbb{C}$.

– *On an arrow $f: (TX \xrightarrow{a} X) \to (TY \xrightarrow{b} Y)$,*

$$\mathbb{P}^{\mathcal{EM}}(\tau)(f): \mathcal{EM}(T)\begin{pmatrix} TY & T(T\Omega) \\ \downarrow b, & \downarrow \tau \\ Y & T\Omega \end{pmatrix} \longrightarrow \mathcal{EM}(T)\begin{pmatrix} TX & T(T\Omega) \\ \downarrow a, & \downarrow \tau \\ X & T\Omega \end{pmatrix}, \quad q \mapsto q \circ f.$$

Proof. The monotonicity of $\mathbb{P}^{\mathcal{EM}}(\tau)(f)$ follows from the order-enrichment of T via Lemma 2.8. The functoriality of $\mathbb{P}^{\mathcal{EM}}(\tau)$ is obvious. □

Theorem 2.10. *For a PT situation (T, Ω, τ), the following diagram commutes up-to a natural isomorphism. Here K is the comparison functor.*

$$
\begin{array}{c}
\mathbf{Posets} \xleftarrow{\quad \mathbb{P}^{\mathcal{EM}}(\tau) \quad} \mathcal{EM}(T)^{\mathrm{op}} \\
\mathbb{P}^{\mathcal{K}\ell}(\tau) \quad \Psi \Uparrow \cong \quad K^{\mathrm{op}} \\
\mathcal{K}\ell(T)^{\mathrm{op}}
\end{array}
\tag{8}
$$

Proof. (Also of Proposition 2.5) The natural isomorphism Ψ in question is of the type

$$\Psi_X: \quad \mathbb{P}^{\mathcal{K}\ell}(\tau)(X) = \mathbb{C}(X, T\Omega) \xrightarrow{\cong} \mathcal{EM}(T)\begin{pmatrix} T(TX) & T(T\Omega) \\ \downarrow \mu_X, & \downarrow \tau \\ TX & T\Omega \end{pmatrix} = \mathbb{P}^{\mathcal{EM}}(\tau)(KX)$$

and it is defined by the adjunction $\mathbb{C}(X, U^T \tau) \cong \mathcal{EM}(T)(\mu_X, \tau)$ where U^T is the forgetful functor. Explicitly: $\Psi_X(X \xrightarrow{p} T\Omega) = (TX \xrightarrow{Tp} T(T\Omega) \xrightarrow{\tau} T\Omega)$; and its inverse is $\Psi_X^{-1}(TX \xrightarrow{f} T\Omega) = (X \xrightarrow{\eta_X} TX \xrightarrow{f} T\Omega)$. The function Ψ_X is monotonic by the monotonicity of τ, see Definition 2.2; so is its inverse Ψ_X^{-1} by Lemma 2.8.

Let us turn to naturality of Ψ. Given $f: X \nrightarrow Y$ in $\mathcal{K}\ell(T)$, it requires

$$
\begin{array}{ccc}
\mathbb{C}(Y, T\Omega) & \xrightarrow{\quad \Psi_Y \quad}_{\cong} & \mathcal{EM}(T)(\mu_Y, \tau) \\
\mathbb{P}^{\mathcal{K}\ell}(\tau)(f) = \tau \circ T(\_) \circ f \downarrow & & \downarrow \mathbb{P}^{\mathcal{EM}}(\tau)(Kf) = (\_) \circ \mu_Y \circ Tf \\
\mathbb{C}(X, T\Omega) & \xrightarrow[\cong]{\quad \Psi_X \quad} & \mathcal{EM}(T)(\mu_X, \tau).
\end{array}
\tag{9}
$$

Indeed, given $q \colon Y \to T\Omega$,

$$\mathbb{P}^{\mathcal{EM}}(\tau)(Kf)(\Psi_Y q) = \mathbb{P}^{\mathcal{EM}}(\tau)(Kf)(\tau \circ Tq) = \tau \circ Tq \circ \mu_Y \circ Tf$$

$$= \tau \circ \mu_{T\Omega} \circ T(Tq) \circ Tf = \tau \circ T\tau \circ T(Tq) \circ Tf = \big(\Psi_X \circ \mathbb{P}^{\mathcal{K\ell}}(\tau)(f)\big)q \,,$$

where the third equality is naturality of μ and the fourth is the multiplication law of τ (see (4)). By this naturality, in particular, we have that $\mathbb{P}^{\mathcal{K\ell}}(\tau)(f)$ is monotone (since the other three arrows are monotone). This is one property needed in Proposition 2.5; the other—functoriality of $\mathbb{P}^{\mathcal{K\ell}}(\tau)$—also follows from naturality of Ψ, via the functoriality of K and $\mathbb{P}^{\mathcal{EM}}(\tau)$.   □

Remark 2.11. The comparison functor $K \colon \mathcal{K\ell}(T) \to \mathcal{EM}(T)$ can be seen as a strongest postcondition semantics. Given a (branching) computation $f \colon X \to TY$, we obtain

$$\mathrm{spost}(f) := \mu_Y \circ Tf \; : \; TX \longrightarrow TY$$

that is an algebra morphism Kf between free algebras. When $T = \mathcal{P}$ this indeed yields a natural notion: concretely it is given by $\mathrm{spost}^{\mathcal{P}}(f)(S) = \{y \mid \exists x \in S.\, y \in fx\}$; here we think of subsets as predicates. Some remarks are in order.

Firstly, note that the notion of predicate here diverges in general from the one in the weakest precondition semantics. This is manifest when $T = \mathcal{D}$: the former is a subdistribution $d \in \mathcal{D}X$ ("partial information on the current state's whereabouts", see Remark 2.7), while the latter is a random variable $p \colon X \to [0,1]$.

Secondly, there is no notion of modality involved here. This is unsatisfactory because, besides the above strongest postcondition semantics $\mathrm{spost}^{\mathcal{P}}(f)$ for $T = \mathcal{P}$ that carries the "may" flavor, the "must" variant $\mathrm{spost}^{\mathcal{P}}_{\mathrm{must}}(f)$ is also conceivable such that $\mathrm{spost}^{\mathcal{P}}_{\mathrm{must}}(f)(S) = \{y \mid \forall x \in S.\, y \in fx\}$. This does not arise from the comparison functor K.

3   The Two-Player Setting: Introduction

We extend the basic framework in the previous section by adding another layer of branching. This corresponds to adding another "player" in computations or systems. The additional player typically has an interest that conflicts with the original player's: the former shall be called *Opponent* and denoted by O, while the latter (the original player) is called *Player* P.[2]

The need for two players with conflicting interests is pervasive in computer science. One example is the (nowadays heavy) use of *games* in the automata-theoretic approach to model checking (see e.g. [8]). Games here can be understood as a two-player version of automata, where it is predetermined which player makes a move in each state. An example is above on the right, where P-states are x_0, x_3 and O-states are x_1, x_2. Typical questions asked here are about what Player P *can*

---

[2] Note that (capitalized) *Player* and *Opponent* are altogether called *players*.

force: can P force that x_3 be reached? (yes); can P force that x_0 be visited infinitely often? (no). In model checking, the dualities between $\wedge$ and $\vee$, ν and μ, etc. in the modal μ-calculus are conveniently expressed as the duality between P and O; and many algorithms and proofs rely on suitably formulated games and results on them (such as the algorithm in [19] that decides the winner of a parity game). Games have also been used in the coalgebraic study of fixed-point logics [26].

Another example of nested two-player branching is found in the process-theoretic study of *probabilistic systems*; see e.g. [3,33]. There it is common to include nondeterministic branching too: while probabilistic branching models the behavior of a *system* (such as a stochastic algorithm) that flips an internal coin, nondeterministic branching models the *environment*'s behavior (such as requests from users) on which no statistical information is available. In this context, probabilistic branching is often called *angelic* while nondeterministic one is *demonic*; and a common verification goal would be to ensure a property—with a certain minimal likelihood—whatever demonic choices are to be made.

3.1 Leading Example: Nondeterministic P and Nondeterministic O

Let us first focus on the simple setting where: P moves first and O moves second, in each round; and both P and O make nondeterministic choices. This is a setting suited e.g. for bipartite games where P plays first. A computation with such branching is modeled by a function

$$f : X \longrightarrow \mathcal{P}_\mathsf{P}(\mathcal{P}_\mathsf{O}Y), \tag{10}$$

where the occurrences of the powerset functor $\mathcal{P}$ are annotated to indicate which of the players it belongs to (hence $\mathcal{P}_\mathsf{P} = \mathcal{P}_\mathsf{O} = \mathcal{P}$). We are interested in what P can force; in this *logic of forced predicates*, the following notion of (pre)order seems suitable.

$$\mathfrak{a} \sqsubseteq \mathfrak{b} \quad \text{in } \mathcal{P}_\mathsf{P}(\mathcal{P}_\mathsf{O}Y) \quad \overset{\text{def.}}{\iff} \quad \forall S \in \mathfrak{a}.\, \exists S' \in \mathfrak{b}.\, S' \subseteq_{\mathcal{P}_\mathsf{O}Y} S \tag{11}$$

That is: if $\mathfrak{a}$ can force Opponent to $S \subseteq Y$, then $\mathfrak{b}$—that has a greater power—can force Opponent to better (i.e. smaller) $S' \subseteq Y$.

In fact, we shall now introduce a modeling alternative to (10) which uses up-closed families of subsets, and argue for its superiority, mathematical and conceptual. It paves the way to our general setup in Sect. 4.

For a set Y, we define $\mathcal{UP}Y$ to be the collection of *up-closed* families of subsets of Y, that is,

$$\mathcal{UP}Y := \{\mathfrak{a} \subseteq \mathcal{P}Y \mid \forall S, S' \subseteq Y.\, (S \in \mathfrak{a} \wedge S \subseteq S' \Rightarrow S' \in \mathfrak{a})\}. \tag{12}$$

On $\mathcal{UP}Y$ we define a relation $\sqsubseteq$ by: $\mathfrak{a} \sqsubseteq \mathfrak{b}$ if $\mathfrak{a} \subseteq \mathfrak{b}$. It is obviously a partial order.

Lemma 3.1. *1. For each set Y, the relation $\sqsubseteq$ in (11) on $\mathcal{P}_\mathsf{P}(\mathcal{P}_\mathsf{O}Y)$ is a pre-order. It is not a partial order.*

2. *For $\mathfrak{a} \in \mathcal{P}_\mathsf{P}(\mathcal{P}_\mathsf{O} Y)$, let $\uparrow\mathfrak{a} := \{S \mid \exists S' \in \mathfrak{a}.\, S' \subseteq S\}$ be its* upward closure. *Then the following is an equivalence of (preorders considered to be) categories; here ι is the obvious inclusion map.*

$$\mathcal{U}PY \underset{\iota}{\overset{\uparrow(\_)}{\underset{\simeq}{\rightleftarrows}}} \mathcal{P}_\mathsf{P}(\mathcal{P}_\mathsf{O} Y)$$

Proof. For 1., reflexivity and transitivity of $\sqsubseteq$ is obvious. To see it is not anti-symmetric consider $\{\emptyset, Y\}$ and $\{\emptyset\}$.

For 2., ι is obviously monotone. If $\mathfrak{a} \sqsubseteq \mathfrak{b}$ in $\mathcal{P}_\mathsf{P}(\mathcal{P}_\mathsf{O} Y)$, for any $S \in \uparrow\mathfrak{a}$ there exists $S' \in \mathfrak{b}$ such that $S' \subseteq S$, hence $S \in \uparrow\mathfrak{b}$. Therefore $\uparrow(\_)$ is monotone too. Obviously $\uparrow(\_) \circ \iota = \mathrm{id}$.

It must be checked that $\iota(\uparrow\mathfrak{a}) \simeq \mathfrak{a}$ for $\mathfrak{a} \in \mathcal{P}_\mathsf{P}(\mathcal{P}_\mathsf{O} Y)$, where $\simeq$ is the equivalence induced by $\sqsubseteq$. The $\sqsubseteq$ direction is immediate from the definition of $\uparrow\mathfrak{a}$; for the other direction, observe that in general $\mathfrak{a} \subseteq \mathfrak{b}$ implies $\mathfrak{a} \sqsubseteq \mathfrak{b}$ in $\mathcal{P}_\mathsf{P}(\mathcal{P}_\mathsf{O} Y)$. $\square$

Proposition 3.2. *For each set Y, $(\mathcal{U}PY, \sqsubseteq)$ is the poset induced by the preorder $\big(\mathcal{P}_\mathsf{P}(\mathcal{P}_\mathsf{O} Y), \sqsubseteq\big)$. Moreover $(\mathcal{U}PY, \sqsubseteq)$ is a complete lattice.*

Proof. The first half is immediate from Lemma 3.1. For the latter, observe that supremums are given by unions. $\square$

The constructions $\mathcal{P}_\mathsf{P}(\mathcal{P}_\mathsf{O}\_)$ and $\mathcal{U}P\_$ have been studied from a coalgebraic perspective in the context of *neighborhood frames* [9,10]. There a coalgebra for the former is a model of *non-normal* modal logic (meaning that axioms like $\Box p \wedge \Box q \to \Box(p \wedge q)$ and $\Box p \to \Box(p \vee q)$ can fail); one for the latter is a model of *monotone* modal logic (meaning that validity of $\Box p \to \Box(p \vee q)$ is retained). Proposition 3.2 shows that, as long as our interests are game-theoretic and are in the logical reasoning with respect to the preorder $\sqsubseteq$ in (11), we may just as well use $\mathcal{U}P\_$ which is mathematically better-behaved.

To argue further for the mathematical convenience of $\mathcal{U}P\_$, we look at its action on arrows. For $\mathcal{P}_\mathsf{P}(\mathcal{P}_\mathsf{O}\_)$ there are two obvious choices ($\mathcal{P}\mathcal{P}f$ and 2^{2^f}) of action on arrows, arising from the covariant and contravariant powerset functors, respectively. Given $f\colon X \to Y$ in **Sets**,

$$\mathcal{P}\mathcal{P}f, \ 2^{2^f} : \ \mathcal{P}_\mathsf{P}(\mathcal{P}_\mathsf{O} X) \longrightarrow \mathcal{P}_\mathsf{P}(\mathcal{P}_\mathsf{O} Y),$$
$$(\mathcal{P}\mathcal{P}f)\mathfrak{a} := \{\amalg_f S \mid S \in \mathfrak{a}\}, \quad 2^{2^f}\mathfrak{a} := \{T \subseteq Y \mid f^{-1}T \in \mathfrak{a}\}.$$

Here $\amalg_f S$ is the direct image of S by f.

These two choices are not equivalent with respect to $\sqsubseteq$ on $\mathcal{P}_\mathsf{P}(\mathcal{P}_\mathsf{O} Y)$. In general we have $2^{2^f}\mathfrak{a} \sqsubseteq (\mathcal{P}\mathcal{P}f)\mathfrak{a}$. To see that, assume $U \in 2^{2^f}\mathfrak{a}$, i.e. $f^{-1}U \in \mathfrak{a}$. Then $\amalg_f(f^{-1}U) \subseteq U$ (a general fact) and $\amalg_f(f^{-1}U) \in (\mathcal{P}\mathcal{P}f)\mathfrak{a}$. However the converse $2^{2^f}\mathfrak{a} \sqsupseteq (\mathcal{P}\mathcal{P}f)\mathfrak{a}$ can fail: consider $!\colon 2 \to 1$ (where $2 = \{0,1\}$) and $\mathfrak{a} = \{\{0\}\}$; then $2^{2^f}\mathfrak{a} = \emptyset$ while $(\mathcal{P}\mathcal{P}f)\mathfrak{a} = \{1\}$.

This discrepancy is absent with $\mathcal{UP}\_$. For a function $f\colon X \to Y$, the "covariant" action $\mathcal{UP}f$ and the "contravariant" action $\mathcal{UP}'f$ are defied as follows.

$$
\begin{array}{ccc}
\mathcal{UP}X \xrightarrow{\;\mathcal{UP}f\;} \mathcal{UP}Y \\
{\scriptstyle \iota}\downarrow \qquad\qquad \uparrow{\scriptstyle \uparrow(\_)} \\
\mathcal{P}_{\mathsf{P}}(\mathcal{P}_{\mathsf{O}}X) \xrightarrow[\;\mathcal{PP}f\;]{} \mathcal{P}_{\mathsf{P}}(\mathcal{P}_{\mathsf{O}}Y)
\end{array}
\qquad
\begin{array}{ccc}
\mathcal{UP}X \dashrightarrow^{\;\mathcal{UP}'f\;} \mathcal{UP}Y \\
\downarrow \qquad\qquad \downarrow \\
\mathcal{P}_{\mathsf{P}}(\mathcal{P}_{\mathsf{O}}X) \xrightarrow[\;2^{2^{f}}\;]{} \mathcal{P}_{\mathsf{P}}(\mathcal{P}_{\mathsf{O}}Y)
\end{array}
\qquad (13)
$$

On the left, ι and $\uparrow(\_)$ are as in Lemma 3.1. On the right $2^{2^{f}}$ restricts to $\mathcal{UP}X \to \mathcal{UP}Y$ (easy by the fact that f^{-1} is monotone); on the left such is not the case (consider $f\colon 1 \to 2$, $0 \mapsto 0$ and $\mathfrak{a} = \{1\}$) and we need explicit use of $\uparrow(\_)$.

Lemma 3.3. $\mathcal{UP}f = \mathcal{UP}'f$.

Proof. Let $\mathfrak{a} \in \mathcal{UP}X$ (hence up-closed). In view of Lemma 3.1, it suffices to show that $2^{2^{f}}\mathfrak{a} \simeq (\mathcal{PP}f)\mathfrak{a}$; we have already proved the $\sqsubseteq$ direction. For the other direction, let $S \in \mathfrak{a}$; proving $\mathrm{II}_f\, S \in 2^{2^{f}}\mathfrak{a}$ will prove $(\mathcal{PP}f)\mathfrak{a} \subseteq 2^{2^{f}}\mathfrak{a}$, hence $(\mathcal{PP}f)\mathfrak{a} \sqsubseteq 2^{2^{f}}\mathfrak{a}$. That $S \subseteq f^{-1}(\mathrm{II}_f\, S)$ is standard; since $\mathfrak{a}$ is up-closed we have $f^{-1}(\mathrm{II}_f\, S) \in \mathfrak{a}$. Therefore $\mathrm{II}_f\, S \in (2^{2^{f}})\mathfrak{a}$. $\qquad\square$

We therefore define $\mathcal{UP}\colon \mathbf{Sets} \to \mathbf{Sets}$ by (12) on objects and either of the actions in (13) on arrows. Its functoriality is obvious from (13) on the right.

3.2   Nondeterministic O, then Probabilistic P: Search for Modularity

We have argued for the convenience of the functor $\mathcal{UP}$, over $\mathcal{P}_{\mathsf{P}}(\mathcal{P}_{\mathsf{O}}\_)$, for modeling two-layer branching in games. A disadvantage, however, is that *modularity* is lost. Unlike $\mathcal{P}_{\mathsf{P}}(\mathcal{P}_{\mathsf{O}}\_)$, the functor $\mathcal{UP}\colon \mathbf{Sets} \to \mathbf{Sets}$ is not an obvious composite of two functors, each of which modeling each player's choice.

The same issue arises also in the systems with both probabilistic and nondeterministic branching (briefly discussed before). It is known (an observation by Gordon Plotkin; see e.g. [35]) that there is no distributive law $\mathcal{DP} \Rightarrow \mathcal{PD}$ of the subdistribution monad $\mathcal{D}$ over the powerset monad $\mathcal{P}$. This means we cannot compose them to obtain a new monad $\mathcal{PD}$. Two principal fixes have been proposed: one is to refine $\mathcal{D}$ into the *indexed valuation monad* that distinguishes e.g. $[x \mapsto 1/2, x \mapsto 1/2]$ from $[x \mapsto 1]$ (see [35]). The other way (see e.g. [34]) replaces $\mathcal{P}$ by the *convex powerset construction* and uses

$$
\mathcal{CD}X := \{\mathfrak{a} \subseteq \mathcal{D}X \mid p_i \in [0,1], \textstyle\sum_i p_i = 1, d_i \in \mathfrak{a} \Rightarrow \textstyle\sum_i p_i d_i \in \mathfrak{a}\}
$$

in place of $\mathcal{PD}$, an alternative we favor due to our process-theoretic interests (see Remark 5.8 later). However, much like with $\mathcal{UP}$, it is not immediate how to decompose $\mathcal{CD}$ into Player and Opponent parts.

We now introduce a categorical setup that addresses this issue of separating two players. It does so by identifying one layer of branching—like up-closed powerset and convex powerset—as a monad on an Eilenberg-Moore category.

4   Generic Two-Player Weakest Precondition Semantics

Definition 4.1 (2-player PT situation). A *2-player predicate transformer situation* over a category $\mathbb{C}$ is a quintuple $(T, \Omega, \tau, R, \rho)$ where:

- (T, Ω, τ) is a PT situation (Definition 2.2), where in particular $\tau \colon T(T\Omega) \to T\Omega$ is an Eilenberg-Moore algebra;
- R is a monad on the Eilenberg-Moore category $\mathcal{EM}(T)$; and
- $\rho \colon R\left(\begin{smallmatrix} T(T\Omega) \\ \downarrow\tau \\ T\Omega \end{smallmatrix} \right) \to \left(\begin{smallmatrix} T(T\Omega) \\ \downarrow\tau \\ T\Omega \end{smallmatrix} \right)$ is an Eilenberg-Moore R-algebra, that is also called a *modality*. It is further subject to the *monotonicity condition* that is much like in Definition 2.2: the map

$$\mathcal{EM}(T)\left(\begin{smallmatrix} TX \\ \downarrow a \\ X \end{smallmatrix}, \begin{smallmatrix} T(T\Omega) \\ \downarrow\tau \\ T\Omega \end{smallmatrix} \right) \longrightarrow \mathcal{EM}(T)\left(R\left(\begin{smallmatrix} TX \\ \downarrow a \\ X \end{smallmatrix} \right), \begin{smallmatrix} T(T\Omega) \\ \downarrow\tau \\ T\Omega \end{smallmatrix} \right), \quad f \longmapsto \rho \circ Rf$$

is monotone for each algebra a. Here the order of each homset is induced by the enrichment of $\mathcal{K\ell}(T)$ via $\mathcal{EM}(T)(b, \tau) \xrightarrow{U^T} \mathbb{C}(U^T b, T\Omega) = \mathcal{K\ell}(T)(U^T b, \Omega)$.

The situation is as in the following diagram.

$$ (14) $$

The composite adjunction yields a new monad $U^T U^R F^R F^T = U^T R F^T$ on $\mathbb{C}$; then from the Kleisli category $\mathcal{K\ell}(U^T R F^T)$ for the new monad we obtain a comparison functor to $\mathcal{EM}(R)$. It is denoted by K.

We have a monad R on $\mathcal{EM}(T)$ and an algebra (modality) ρ for it. This is much like in the original notion of PT situation, where $\tau \colon T(T\Omega) \to T\Omega$ is a modality from which we derived a weakest precondition semantics. Indeed, the following construction is parallel to Proposition 2.9.

Proposition 4.2 (the indexed poset $\mathbb{P}^{\mathcal{EM}}(\tau, \rho)$). *A 2-player PT situation* $(T, \Omega, \tau, R, \rho)$ *induces an indexed poset* $\mathbb{P}^{\mathcal{EM}}(\tau, \rho) \colon \mathcal{EM}(R)^{\mathrm{op}} \to$ **Posets** *over* $\mathcal{EM}(T)$ *by:*

- *on an object* $\alpha \in \mathcal{EM}(R)$,

$$\mathbb{P}^{\mathcal{EM}}(\tau, \rho)\left(\begin{smallmatrix} R(TX \xrightarrow{a} X) \\ \downarrow\alpha \\ (TX \xrightarrow{a} X) \end{smallmatrix} \right) := \mathcal{EM}(R)\left(\begin{smallmatrix} R(TX \xrightarrow{a} X) \\ \downarrow\alpha \\ (TX \xrightarrow{a} X) \end{smallmatrix}, \begin{smallmatrix} R(T(T\Omega) \xrightarrow{\tau} T\Omega) \\ \downarrow\rho \\ (T(T\Omega) \xrightarrow{\tau} T\Omega) \end{smallmatrix} \right)$$

where the order $\sqsubseteq$ on the set $\mathcal{EM}(R)(\alpha, \rho)$ is inherited from $\mathbb{C}(X, T\Omega)$ via the forgetful functors $\mathcal{EM}(R) \to \mathcal{EM}(T) \to \mathbb{C}$; and

– *on an arrow* $f\colon \left(\begin{smallmatrix} Ra \\ \downarrow\alpha \\ a \end{smallmatrix}\right) \to \left(\begin{smallmatrix} Rb \\ \downarrow\beta \\ b \end{smallmatrix}\right)$,

$$\mathbb{P}^{\mathcal{EM}}(\tau,\rho)(f)\colon \mathcal{EM}(R)\left(\begin{smallmatrix} Rb \\ \downarrow\beta \\ b \end{smallmatrix}, \begin{smallmatrix} R\tau \\ \downarrow\rho \\ R\tau \end{smallmatrix}\right) \longrightarrow \mathcal{EM}(R)\left(\begin{smallmatrix} Ra \\ \downarrow\alpha \\ a \end{smallmatrix}, \begin{smallmatrix} R\tau \\ \downarrow\rho \\ \tau \end{smallmatrix}\right), \quad q \longmapsto q \circ f.$$

Proof. The same as the proof of Proposition 2.9, relying on Lemma 2.8. $\qquad\square$

Much like in Theorem 2.10, composition of the indexed poset $\mathbb{P}^{\mathcal{EM}}(\tau,\rho)\colon$ $\mathcal{EM}(R)^{\mathrm{op}} \to \mathbf{Posets}$ and the comparison functor $K\colon \mathcal{K\ell}(U^T R F^T) \to \mathcal{EM}(R)$ will yield the weakest precondition calculus. The branching computations of our interest are therefore of the type $X \to U^T R F^T Y$. We will later see, through examples, that this is indeed what models the scenarios in Sect. 3.

Note that in what follows we rely heavily on the adjunction $F^T \dashv U^T$.

Proposition 4.3 (the indexed poset $\mathbb{P}^{\mathcal{K\ell}}(\tau,\rho)$). *A 2-player PT situation* (T,Ω,τ,R,ρ) *induces an indexed poset* $\mathbb{P}^{\mathcal{K\ell}}(\tau,\rho)\colon \mathcal{K\ell}(U^T R F^T)^{\mathrm{op}} \to \mathbf{Posets}$ *by:*

– *on an object* $X \in \mathcal{K\ell}(U^T R F^T)$, $\mathbb{P}^{\mathcal{K\ell}}(\tau,\rho)(X) := \mathcal{K\ell}(T)(X,\Omega) = \mathbb{C}(X,T\Omega)$;
– *given an arrow* $f\colon X \nrightarrow Y$ *in* $\mathcal{K\ell}(U^T R F^T)$, *it induces an arrow* $f^{\wedge}\colon F^T X \to R(F^T Y)$ *in* $\mathcal{EM}(T)$; *this is used in*

$$\mathcal{EM}(T)(F^T Y, \tau) \to \mathcal{EM}(T)(F^T X, \tau), \quad q \longmapsto \left(F^T X \xrightarrow{f^{\wedge}} R(F^T Y) \xrightarrow{Rq} R\tau \xrightarrow{\rho} \tau\right).$$

The last map defines an arrow $\mathbb{P}^{\mathcal{K\ell}}(\tau,\rho)(f)\colon \mathbb{P}^{\mathcal{K\ell}}(\tau,\rho)(Y) \to \mathbb{P}^{\mathcal{K\ell}}(\tau,\rho)(X)$ *since we have* $\mathbb{P}^{\mathcal{K\ell}}(\tau,\rho)(U) = \mathbb{C}(U,T\Omega) \cong \mathcal{EM}(T)(F^T U, \tau)$.

We have the following natural isomorphism, where K is the comparison in (14).

$$\mathbf{Posets} \xleftarrow{\;\;\mathbb{P}^{\mathcal{EM}}(\tau,\rho)\;\;} \mathcal{EM}(R)^{\mathrm{op}} \qquad (15)$$
$$\underset{\mathbb{P}^{\mathcal{K\ell}}(\tau,\rho)}{\swarrow} \underset{\Xi\bullet\Psi\Uparrow\cong}{\quad} \mathcal{K\ell}(U^T R F^T)^{\mathrm{op}} \underset{K^{\mathrm{op}}}{\nearrow}$$

Proof. Note here that the comparison functor K is concretely described as follows: $KX = F^R(F^T X)$ on objects, and use the correspondence

$$\mathcal{K\ell}(U^T R F^T)(X,Y) = \mathbb{C}(X, U^T U^R F^R F^T Y) \cong \mathcal{EM}(T)(F^T X, U^R F^R F^T Y)$$
$$\cong \mathcal{EM}(R)(F^R F^T X, F^R F^T Y) = \mathcal{EM}(R)(KX, KY)$$

for its action on arrows. We claim that the desired natural isomorphism $\Xi \bullet \Psi$ is the (vertical) composite

$$\mathbb{P}^{\mathcal{K\ell}}(\tau,\rho)(X) = \mathbb{C}(X,T\Omega) \xrightarrow{\Psi_X} \mathcal{EM}(T)(F^T X, \tau)$$
$$\xrightarrow{\Xi_X} \mathcal{EM}(R)(F^R F^T X, \rho) = \mathbb{P}^{\mathcal{EM}}(\tau,\rho)(KX)$$

where Ψ and Ξ are isomorphisms induced by adjunctions.

We have to check that Ψ_X and Ξ_X are order isomorphisms. The map Ψ_X is monotone due to the monotonicity condition on τ (Definition 2.2); so is Ψ_X^{-1}

by Lemma 2.8. Similarly, Ξ_X is monotone by the monotonicity condition on ρ (Definition 4.1); so is Ξ_X^{-1} by Lemma 2.8.

We turn to the naturality: the following diagram must be shown to commute, for each $f\colon X \twoheadrightarrow Y$ in $\mathcal{K}\ell(U^T R F^T)$.

$$\begin{array}{ccccc}
\mathbb{C}(Y, T\Omega) & \xrightarrow[\cong]{\Psi_Y} & \mathcal{EM}(T)(F^T Y, \tau) & \xrightarrow[\cong]{\Xi_Y} & \mathcal{EM}(R)(F^R(F^T Y), \rho) \\
{\scriptstyle \mathbb{P}^{\mathcal{K}\ell}(\tau, \rho)(f)}\Big\downarrow & & \Big\downarrow {\scriptstyle \rho \circ R(\_) \circ f^\wedge} & & \Big\downarrow {\scriptstyle \mathbb{P}^{\mathcal{EM}}(\tau, \rho)(Kf) = (\_) \circ Kf} \\
\mathbb{C}(X, T\Omega) & \xrightarrow[\cong]{\Psi_X} & \mathcal{EM}(T)(F^T X, \tau) & \xrightarrow[\cong]{\Xi_X} & \mathcal{EM}(R)(F^R(F^T Y), \rho)\,.
\end{array}$$
$$(16)$$

The square on the left commutes by the definition of $\mathbb{P}^{\mathcal{K}\ell}(\tau, \rho)(f)$ (Proposition 4.3); the one on the right is much like the one in (9) and its commutativity can be proved in the same way. Note here that $Kf = \mu_{F^T Y}^R \circ R(f^\wedge)$.

Since the diagram (16) commutes, and since Ψ and Ξ are order isomorphisms and $\mathbb{P}^{\mathcal{EM}}(\tau, \rho)(Kf)$ is monotone (Proposition 4.2), we have that $\mathbb{P}^{\mathcal{K}\ell}(\tau, \rho)f$ is monotone. The functoriality of $\mathbb{P}^{\mathcal{K}\ell}(\tau, \rho)$ is easy, too. This concludes the proof. □

5 Examples of 2-Player PT Situations

5.1 Nondeterministic Player and then Nondeterministic Opponent

We continue Sect. 3 and locate the monad $\mathcal{UP}$—and the logic of forced predicates— in the general setup of Sect. 4. We identify a suitable 2-player PT situation $(\mathcal{P}, 1, \tau_\square, \mathsf{R_G}, \rho_\mathsf{P})$, in which $T = \mathcal{P}$, $\Omega = 1$ and $\tau = \tau_\square$ that is from Sect. 2.3. The choice of $\tau_\square$ corresponds to the demonic nature of Opponent's choices: Player can force those properties that hold *whatever choices* Opponent makes.

To introduce the monad $\mathsf{R_G}$ on $\mathcal{EM}(\mathcal{P})$—corresponding to the up-closed powerset construction—we go via the following standard isomorphism.

Lemma 5.1. *Let* $C\colon \mathcal{EM}(\mathcal{P}) \to \mathbf{CL}_\wedge$ *be the functor such that* $C\big(\begin{smallmatrix} \mathcal{P}X \\ \downarrow a \\ X \end{smallmatrix}\big) := (X, \sqsubseteq_a)$, *where the order is defined by* $x \sqsubseteq_a y$ *if* $x = a\{x, y\}$. *Conversely, let* $D\colon \mathbf{CL}_\wedge \to \mathcal{EM}(\mathcal{P})$ *be such that* $D(X, \sqsubseteq) := \big(\begin{smallmatrix} \mathcal{P}X \\ \downarrow \wedge \\ X \end{smallmatrix}\big)$. *Both act on arrows as identities.*

Then C *and* D *constitute an isomorphism* $\mathcal{EM}(\mathcal{P}) \stackrel{\cong}{\to} \mathbf{CL}_\wedge$. □

The monad $\mathsf{R_G}$ is then defined to be the composite $\mathsf{R_G} := D \circ \mathsf{Dw} \circ C$, using the *down-closed powerset monad* Dw on $\mathbf{CL}_\wedge$.

$$\mathsf{R_G} \,\big(\!\!\curvearrowleft\, \mathcal{EM}(\mathcal{P}) \underset{C}{\overset{D}{\underset{\cong}{\rightleftarrows}}} \mathbf{CL}_\wedge \,\curvearrowright\!\!\big)\, \mathsf{Dw} \qquad (17)$$

The switch between *up-closed* subsets in $\mathcal{UP}$ and *down-closed* subsets Dw may seem confusing. Later in Proposition 5.3 it is shown that everything is in harmony; and after all it is a matter of presentation since there is an isomorphism

$\mathbf{CL}_\wedge \xrightarrow{\cong} \mathbf{CL}_\vee$ that reverses the order in each complete lattice. The switch here between up- and down-closed is essentially because: the bigger the set of Opponent's options is, the smaller the power of Player (to force Opponent to somewhere) is.

Concretely, the monad $\mathsf{Dw}\colon \mathbf{CL}_\wedge \to \mathbf{CL}_\wedge$ carries a complete lattice $(X, \sqsubseteq)$ to the set $\mathsf{Dw}X := \{S \subseteq X \mid x \sqsubseteq x',\, x' \in S \Rightarrow x \in S\}$. We equip $\mathsf{Dw}X$ with the inclusion order; this makes $\mathsf{Dw}X$ a complete lattice, with sups and infs given by unions and intersections, respectively. An arrow $f\colon X \to Y$ is carried to $\mathsf{Dw}f\colon \mathsf{Dw}X \to \mathsf{Dw}Y$ defined by $S \mapsto {\downarrow}(\mathrm{II}_f S)$. Here ${\downarrow}(\_)$ denotes the downward closure and it is needed to ensure down-closedness (consider a $\bigwedge$-preserving map $f\colon 1 \to 2$, $0 \mapsto 1$ where $0 \sqsubseteq 1$ in 2). The monad structure of Dw is given by: $\eta^{\mathsf{Dw}}_X\colon X \to \mathsf{Dw}X$, $x \mapsto {\downarrow}\{x\}$; and $\mu^{\mathsf{Dw}}_X\colon \mathsf{Dw}(\mathsf{Dw}X) \to \mathsf{Dw}X$, $\mathfrak{a} \mapsto \bigcup \mathfrak{a}$. Note in particular that η^{Dw}_X is $\bigwedge$-preserving. As in (17) we define $\mathsf{R}_{\mathsf{G}} := D \circ \mathsf{Dw} \circ C$.

Finally, let us define the data $\rho_{\mathsf{P}}\colon \mathsf{R}_{\mathsf{G}}(\tau_\square) \to \tau_\square$ in the 2-player PT situation. Via the isomorphism (17) we shall think of it as an Dw-algebra, where the $\mathcal{P}$-algebra $\tau_\square$ is identified with the 2-element complete lattice $[\mathsf{ff} \sqsubseteq \mathsf{tt}]$ (the order is because $\tau_\square\{\mathsf{tt}, \mathsf{ff}\} = \mathsf{ff}$). Therefore we are looking for a $\bigwedge$-preserving map

$$\mathsf{Dw}[\mathsf{ff} \sqsubseteq \mathsf{tt}] = \big[\emptyset \sqsubseteq \{\mathsf{ff}\} \sqsubseteq \{\mathsf{ff}, \mathsf{tt}\}\big] \xrightarrow{C\rho_{\mathsf{P}}} [\mathsf{ff} \sqsubseteq \mathsf{tt}]$$

subject to the conditions of an Eilenberg-Moore algebra in (4). In fact such $C(\rho_{\mathsf{P}})$ is uniquely determined: preservation of $\top$ forces $(C\rho_{\mathsf{P}})\{\mathsf{ff}, \mathsf{tt}\} = \mathsf{tt}$; the unit law forces $(C\rho_{\mathsf{P}})\{\mathsf{ff}\} = \mathsf{ff}$ and monotonicity of $C\rho_{\mathsf{P}}$ then forces $(C\rho_{\mathsf{P}})\emptyset = \mathsf{ff}$.

Lemma 5.2. $(\mathcal{P}, 1, \tau_\square, \mathsf{R}_{\mathsf{G}}, \rho_{\mathsf{P}})$ *thus obtained is a 2-player PT situation.*

Proof. It remains to check the monotonicity condition (Definition 4.1) for ρ_{P}. We shall again think in terms of complete lattices and $\bigwedge$-preserving maps; then the requirement is that the map $\big(X \xrightarrow{f} [\mathsf{ff} \sqsubseteq \mathsf{tt}]\big) \mapsto \big(\mathsf{Dw}X \xrightarrow{\rho_{\mathsf{P}} \circ \mathsf{Dw}f} [\mathsf{ff} \sqsubseteq \mathsf{tt}]\big)$ is monotone. Assume $g \sqsubseteq f$, $S \in \mathsf{Dw}X$ and $(\rho_{\mathsf{P}} \circ \mathsf{Dw}f)(S) = \mathsf{ff}$. It suffices to show that $(\rho_{\mathsf{P}} \circ \mathsf{Dw}g)(S) = \mathsf{ff}$; this follows from the observation that, for $h = f$ or g,

$$(\rho_{\mathsf{P}} \circ \mathsf{Dw}h)(S) = \mathsf{ff} \iff (\mathsf{Dw}h)S \subseteq \{\mathsf{ff}\} \iff \forall x \in S.\, hx = \mathsf{ff}.\qquad\square$$

Let us check that the logic $\mathbb{P}^{\mathcal{K}\ell}(\tau_\square, \rho_{\mathsf{P}})$ associated with this 2-player PT situation is indeed the logic of forced predicates in Sect. 3.1. For instance, we want "computations" $X \to U^{\mathcal{P}}\mathsf{R}_{\mathsf{G}}F^{\mathcal{P}}Y$ to coincide with "computations" $X \to \mathcal{U}\mathcal{P}Y$.

Proposition 5.3. *For any set X we have $U^{\mathcal{P}}\mathsf{R}_{\mathsf{G}}F^{\mathcal{P}} = \mathcal{U}\mathcal{P}X$. In fact they are isomorphic as complete lattices, that is, $\mathsf{Dw} \circ C \circ F^{\mathcal{P}} = \mathcal{U}\mathcal{P}\colon \mathbf{Sets} \to \mathbf{CL}_\wedge$ where the functor $\mathcal{U}\mathcal{P}$ is equipped with the inclusion order.*

Proof. Given $X \in \mathbf{Sets}$, the definition of C dictates that $C(F^{\mathcal{P}}X) = (\mathcal{P}X, \supseteq)$ and its order be given by the reverse inclusion order. Hence $\mathsf{Dw}(C(F^{\mathcal{P}}X))$ is the collection of families $\mathfrak{a} \subseteq \mathcal{P}X$ that are $\supseteq$-down-closed, i.e. $\subseteq$-up-closed. It is easily checked that the two functors coincide on arrows, too, using the characterization on the left in (13). $\qquad\square$

Next we describe the logic $\mathbb{P}^{\mathcal{K}\ell}(\tau_\square, \rho_\mathsf{P})$ (Proposition 4.3) in concrete terms. We base ourselves again in $\mathbf{CL}_\wedge$ via the isomorphism $\mathcal{EM}(\mathcal{P}) \cong \mathbf{CL}_\wedge$ in (17). Consider a postcondition $q \colon Y \to \mathcal{P}1$ and a branching computation $f \colon X \to \mathcal{UP}Y$. These are in one-to-one correspondences with the following arrows in $\mathbf{CL}_\wedge$:

$$q^\wedge \colon C(F^\mathcal{P}Y) = (\mathcal{P}Y, \supseteq) \longrightarrow [\mathbf{ff} \sqsubseteq \mathbf{tt}] = C(\tau_\square),$$
$$f^\wedge \colon C(F^\mathcal{P}X) = (\mathcal{P}X, \supseteq) \longrightarrow \mathsf{Dw}(\mathcal{P}Y, \supseteq) = C(\mathsf{R}_\mathsf{G}(F^\mathcal{P}Y)),$$

where we used Proposition 5.3. Since $q^\wedge$ are $f^\wedge$ are $\bigwedge$-preserving, we have

$$q^\wedge W = q^\wedge\left(\bigcup_{y \in W}\{y\}\right) = q^\wedge\left(\bigwedge_{y \in W}\{y\}\right) = \bigwedge_{y \in W} q^\wedge\{y\} = \bigwedge_{y \in W} qy;$$

and similarly $f^\wedge S = \bigcap_{x \in S} fx$. Recall that $\mathsf{Dw}(\mathcal{P}Y, \supseteq)$ has the inclusion order.

Now Proposition 4.3 states that the weakest precondition $\mathbb{P}^{\mathcal{K}\ell}(\tau_\square, \rho_\mathsf{P})(f)(q)$ is the arrow $X \to \mathcal{P}1$ that corresponds, via the adjunction $C \circ F^\mathcal{P} \dashv U^\mathcal{P} \circ D$, to

$$(\mathcal{P}X, \supseteq) \xrightarrow{f^\wedge} \mathsf{Dw}(\mathcal{P}Y, \supseteq) \xrightarrow{\mathsf{Dw}(q^\wedge)} \mathsf{Dw}[\mathbf{ff} \sqsubseteq \mathbf{tt}] \xrightarrow{\rho_\mathsf{P}} [\mathbf{ff} \sqsubseteq \mathbf{tt}] \quad \text{in} \mathbf{CL}_\wedge.$$

Unweaving definitions it is straightforward to see that, for $S \subseteq X$,

$$\left(\rho_\mathsf{P} \circ \mathsf{Dw}(q^\wedge) \circ f^\wedge\right)S = \mathbf{tt} \iff \exists W \subseteq Y.\,(\forall x \in S.\,W \in fx \wedge \forall y \in W.\,qy = \mathbf{tt});$$
$$\text{therefore } \mathbb{P}^{\mathcal{K}\ell}(\tau_\square, \rho_\mathsf{P})(f)(q)(x) = \mathbf{tt} \iff \exists W \subseteq Y.\,(W \in fx \wedge \forall y \in W.\,qy = \mathbf{tt}).$$
$$(18)$$

The last condition reads: among the set fx of possible moves of Player, there exists a move W, from which q holds no matter what Opponent's move y is. Therefore $\mathbb{P}^{\mathcal{K}\ell}(\tau_\square, \rho_\mathsf{P})(f)(q)(x) = \mathbf{tt}$ if Player can *force* the predicate q from x after the (two-layer branching) computation f.

5.2   Nondeterministic Opponent and then Nondeterministic Player

We change the order of Player and Opponent: O moves first and then P moves. The general setup in Sect. 4 successfully models this situation too, with a choice of a 2-player PT situation $(\mathcal{P}, 1, \tau_\diamond, \mathsf{R}_\mathsf{G}, \rho_\mathsf{O})$ that is dual to the previous one.

The modality $\tau_\diamond$ is from Sect. 2.3. Although the monad R_G is the same as in Sect. 5.1, we now prefer to present it in terms of $\mathcal{EM}(\mathcal{P}) \cong \mathbf{CL}_\vee$ instead of $\mathbf{CL}_\wedge$. The reason is that this way the algebra $\tau_\diamond$ gets identified with $[\mathbf{ff} \sqsubseteq \mathbf{tt}]$, which is intuitive. The situation is as follows.

$$(19)$$

The functors C'' and D'' carries a complete lattice $(X, \sqsubseteq)$ to $(X, \sqsupseteq)$, reversing the order. The monad Up is defined by $\mathsf{Up} := C'' \circ \mathsf{Dw} \circ D''$; concretely it carries $(X, \sqsubseteq)$ to the set of its up-closed subsets, equipped with the *reverse* inclusion order $\sqsupseteq$. That is,

$$\mathsf{Up}(X, \sqsubseteq) := \left(\{ S \subseteq X \mid S \ni x \sqsubseteq x' \Rightarrow x' \in S \}, \supseteq \right).$$

We have $\mathsf{R_G} = D \circ \mathsf{Dw} \circ C = D' \circ \mathsf{Up} \circ C'$.

The modality $\rho_{\mathsf{O}} \colon \mathsf{R_G}(\tau_\diamond) \to \tau_\diamond$ is identified, via the isomorphism C' in (19), with an Up-algebra on $[\mathsf{ff} \sqsubseteq \mathsf{tt}]$. The latter is a $\bigvee$-preserving map

$$\mathsf{Up}[\mathsf{ff} \sqsubseteq \mathsf{tt}] = \left[\{\mathsf{ff}, \mathsf{tt}\} \sqsubseteq \{\mathsf{tt}\} \sqsubseteq \emptyset \right] \xrightarrow{C'\rho_{\mathsf{O}}} [\mathsf{ff} \sqsubseteq \mathsf{tt}];$$

note here that the order in $\mathsf{Up}(X, \sqsubseteq)$ is the reverse inclusion $\supseteq$. Such $C'\rho_{\mathsf{O}}$ is uniquely determined (as before): the unit law forces $(C'\rho_{\mathsf{O}})\{\mathsf{tt}\} = \mathsf{tt}$; preservation of $\bot$ forces $(C'\rho_{\mathsf{O}})\{\mathsf{tt}, \mathsf{ff}\} = \mathsf{ff}$; and then by monotonicity $(C'\rho_{\mathsf{O}})\emptyset = \mathsf{tt}$.

It is straightforward to see that $(\mathcal{P}, 1, \tau_\diamond, \mathsf{R_G}, \rho_{\mathsf{O}})$ is indeed a 2-player PT situation; the proof is symmetric to the one in Sect. 5.1. Also symmetrically, the weakest precondition semantics $\mathbb{P}^{\mathcal{K}\ell}(\tau_\diamond, \rho_{\mathsf{O}})$ is concretely described as follows: given a postcondition $q \colon Y \to \mathcal{P}1$ and a branching computation $f \colon X \to \mathcal{UP}Y$,

$$\mathbb{P}^{\mathcal{K}\ell}(\tau_\diamond, \rho_{\mathsf{O}})(f)(q)(x) = \mathsf{tt} \iff \forall W \subseteq Y. \left(W \in fx \Rightarrow \exists y \in W. \, qy = \mathsf{tt} \right).$$

This is dual to (18) and reads: whatever move W Opponent takes, there exists Player's move $y \in W$ so that q holds afterwards.

We note that the analogue of Proposition 5.3 becomes: $\mathsf{Up} \circ C' \circ F^{\mathcal{P}} = \mathcal{UP} \colon \mathbf{Sets} \to \mathbf{CL}_{\bigvee}$, where each $\mathcal{UP}X$ is equipped with the *reverse* inclusion order. This order $(\mathfrak{a} \sqsubseteq \mathfrak{b}$ in $\mathcal{UP}X$ if $\mathfrak{a} \supseteq \mathfrak{b})$ is intuitive if we think of $\sqsubseteq$ as the power of Player.

Remark 5.4. The constructions have been described in concrete terms; this is for intuition. An abstract view is possible too: the modality $\tau_\diamond$ is the dual of $\tau_\Box$ via the swapping σ (see (7)); and the other modality ρ_{O} is also the dual of ρ_{P} by $\rho_{\mathsf{O}} = \left(\mathsf{R_G}(\tau_\diamond) \xrightarrow{\mathsf{R_G}\sigma} \mathsf{R_G}(\tau_\Box) \xrightarrow{\rho_{\mathsf{P}}} \tau_\Box \xrightarrow{\sigma} \tau_\diamond \right)$.

5.3 Nondeterministic Opponent and then Probabilistic Player

In our last example Opponent O moves nondeterministically first, and then Player P moves probabilistically. Such nested branching is in many process-theoretic models of probabilistic systems (see Sect. 3, in particular Sect. 3.2), most notably in Segala's *probabilistic automata* [27]. We identify a 2-player PT situation $(\mathcal{D}, 1, \tau_{\text{total}}, \mathcal{C}v, \rho_{\inf})$ for this situation; then the associated logic $\mathbb{P}^{\mathcal{K}\ell}(\tau_{\text{total}}, \rho_{\inf})$ is that of the probabilistic predicate transformers in [31]. The modality τ_{total} is from Sect. 2.3. The other components $(\mathcal{C}v, \rho_{\inf})$ are to be described in terms of *convex cones* and their *convex subsets*.

In what follows a $\mathcal{D}$-algebra is referred to as a *convex cone*, adopting the notation $\sum_{i \in I} w_i x_i$ to denote $a\left([x_i \mapsto w_i]_{i \in I} \right) \in X$ in a convex cone $a \colon \mathcal{D}X \to$

X. Here I is a countable index set,[3] $w_i \in [0,1]$, and $\sum_{i \in I} w_i \leq 1$. Note that, since $\mathcal{D}$ is the *subdistribution* monad, the zero distribution $\mathbf{0}$ is allowed in $\mathcal{D}X$ and therefore a convex cone $a \colon \mathcal{D}X \to X$ has its *apex* $a\mathbf{0} \in X$. Similarly, a morphism of $\mathcal{D}$-algebras is referred to as a *convex linear map*.

Definition 5.5 (convex subset). A subset $S \subseteq X$ of a convex cone $a \colon \mathcal{D}X \to X$ is said to be *convex* if, for any $p_i \in [0,1]$ such that $\sum_{i \in I} p_i = 1$ and any $x_i \in S$, the convex combination $\sum_{i \in I} p_i x_i$ belongs to S.

We emphasize that in the last definition $\sum_i p_i$ is required to be $= 1$. This is unlike $\sum_i w_i \leq 1$ in the definition of convex cone. Therefore a convex subset S need not include the apex $a\mathbf{0}$; one can think of the base of a 3-dimensional cone as an example. This variation in the definitions is also found in [34, Sect. 2.1.2]; one reason is technical: if we allow $\sum_i p_i \leq 1$ then it is hard to find the monad unit of $\mathcal{C}v$ (see below). Another process-theoretic reason is described later in Remark 5.8.

Definition 5.6 (the monad $\mathcal{C}v$). The functor $\mathcal{C}v \colon \mathcal{EM}(\mathcal{D}) \to \mathcal{EM}(\mathcal{D})$ carries a convex cone $a \colon \mathcal{D}X \to X$ to $\mathcal{C}vX := \{ S \subseteq X \mid S \text{ is convex} \}$; the latter is a convex cone by
$$\sum_i w_i S_i := \{ \textstyle\sum_i w_i x_i \mid x_i \in S_i \}.$$
It is easy to see that $\sum_i w_i S_i$ is indeed a convex subset of X. Given a convex linear map $f \colon X \to Y$, $\mathcal{C}vf \colon \mathcal{C}vX \to \mathcal{C}vY$ is defined by $(\mathcal{C}vf)S := \amalg_f S$, which is obviously convex in Y, too.

The monad structure of $\mathcal{C}v$ is as follows. Its unit is $\eta_X^{\mathcal{C}v} := \{\_\} \colon X \to \mathcal{C}vX$; note that a singleton $\{x\}$ is a convex subset of X (Definition 5.5). The monad multiplication is $\mu_X^{\mathcal{C}v} := \bigcup \colon \mathcal{C}v(\mathcal{C}vX) \to \mathcal{C}vX$. It is easy to see that $\eta_X^{\mathcal{C}v}$ and $\mu_X^{\mathcal{C}v}$ are convex linear maps, and that they satisfy the monad axioms.

We introduce the last component, namely the modality $\rho_{\inf} \colon \mathcal{C}v(\tau_{\text{total}}) \to \tau_{\text{total}}$. A convex subset S of the carrier $\mathcal{D}1 = [0,1]$ of τ_{total} is nothing but an interval (its endpoints may or may not be included); $\rho_{\inf}$ then carries such S to its infimum $\inf S \in [0,1]$. That $\rho_{\inf}$ is convex linear, and that it satisfies the Eilenberg-Moore axioms, are obvious.

Much like in Lemma 5.2, we obtain:

Lemma 5.7. $(\mathcal{D}, 1, \tau_{\text{total}}, \mathcal{C}v, \rho_{\inf})$ *thus obtained is a 2-player PT situation.* □

The resulting logic $\mathbb{P}^{\mathcal{K}\ell}(\tau_{\text{total}}, \rho_{\inf})$ is as follows. Given a postcondition $q \colon Y \to \mathcal{D}1$ and a computation $f \colon X \to U^{\mathcal{D}}\mathcal{C}vF^{\mathcal{D}}Y$, the weakest precondition is

$$\mathbb{P}^{\mathcal{K}\ell}(\tau_{\text{total}}, \rho_{\inf})(f)(q)(x) = \inf \{ \textstyle\sum_{y \in Y} d(y) \cdot q(y) \mid d \in f(x) \}. \tag{20}$$

Here d is a subdistribution chosen by Opponent; and the value $\sum_{y \in Y} d(y) \cdot q(y)$ is the expected value of the random variable q under the distribution d. Therefore the weakest precondition computed above is the least expected value of q when Opponent picks a distribution in harm's way. This is the same as in [31].

---

[3] The countability requirement is superfluous since, if $\sum_{i \in I} p_i = 1$, then only countably many p_i's are nonzero.

Remark 5.8. The use of the convex powerset construction, instead of (plain) powersets, was motivated in Sect. 3.2 through the technical difficulty in getting a monad. Convex powersets are commonly used in the process-theoretic study of probabilistic systems, also because they model a *probabilistic scheduler*: Opponent (called a *scheduler* in this context) can not only pick one distribution but also use randomization in doing so. See e.g. [2].

The definition of convex subset (Definition 5.5)—where we insist on $\sum_i p_i = 1$ instead of ≤ 1—is natural in view of the logic $\mathbb{P}^{\mathcal{K}\ell}(\tau_{\text{total}}, \rho_{\text{inf}})$ described above. Relaxing this definition entails that the zero distribution $\mathbf{0}$ is always included in a "convex subset," and hence always in Opponent's options. This way, however, the weakest precondition in (20) can always be forced to 0 and the logic gets trivial.

We can also model the situation where the roles of Player and Opponent are swapped: we can follow the same path as in Remark 5.4 and obtain a 2-player PT situation $(\mathcal{D}, 1, \tau_{\text{partial}}, \mathcal{C}v, \rho_{\text{sup}})$; the resulting modality ρ_{sup} carries an interval to its supremum.

6   Conclusions and Future Work

Inspired by Jacobs' recent work [16,17] we pursued a foundation of predicate transformers (more specifically weakest precondition semantics) based on an order-enriched monad. There different notions of modality (such as "may" vs. "must") are captured by Eilenberg-Moore algebras. Nested branching with two conflicting players can be modeled in a modular way, too, by a monad R on an Eilenberg-Moore category $\mathcal{EM}(T)$. Instances of this generic framework include probabilistic weakest preconditions, those augmented with nondeterminism, and the logic of forced predicates in games.

As future work we wish to address the components in the picture (2–3) that are missing in the current framework. A generic weakest precondition calculus presented in a *syntactic* form is another direction. Most probably relationships between monads and algebraic theories (see e.g. [32]) will be exploited there. So-called *healthiness conditions*—i.e. characterization of the image of $\mathbb{P}^{\mathcal{K}\ell}(\tau)$ in (8), to be precise its action on arrows—are yet another topic, generalizing [5,31].

The current work is hopefully a step forward towards a coalgebraic theory of automata (on infinite trees), games and fixed-point logics. For example, we suspect that our categorical formulation of the logic of forced predicates be useful in putting *game (bi)simulation* (studied e.g. in [22,36]) in coalgebraic terms. Possibly related, we plan to work on the relationship to the coalgebraic theory of traces and simulations formulated in a Kleisli category [11,13] since all the monads in Example 2.2 fit in this trace framework.

In this paper we relied on an order-enrichment of a monad to obtain the entailment order. We are nevertheless interested in what our current framework brings for other monads, like the ones that model computational effects [29] (global state, I/O, continuation, etc.). Also interesting is a higher-order extension of the current work, where the logic will probably take the form of dependent types. Related work in this direction is [20].

Acknowledgments. Thanks are due to Kazuyuki Asada, Kenta Cho, Corina Cîrstea and Tetsuri Moriya for useful discussions. The author is supported by Grants-in-Aid for Young Scientists (A) No. 24680001, JSPS, and by Aihara Innovative Mathematical Modelling Project, FIRST Program, JSPS/CSTP.

References

1. Barr, M., Wells, C.: Toposes, Triples and Theories. Springer, Berlin (1985)
2. Cattani, S., Segala, R.: Decision algorithms for probabilistic bisimulation. In: Brim, L., Jančar, P., Křetínský, M., Kučera, A. (eds.) CONCUR 2002. LNCS, vol. 2421, pp. 371–385. Springer, Heidelberg (2002)
3. Cheung, L.: Reconciling nondeterministic and probabilistic choices. Ph.D. thesis, Radboud Univ. Nijmegen (2006)
4. Cîrstea, C.: A coalgebraic approach to linear-time logics. In: Muscholl, A. (ed.) FOSSACS 2014. LNCS, vol. 8412, pp. 426–440. Springer, Heidelberg (2014)
5. Dijkstra, E.W.: A Discipline of Programming. Prentice Hall, Upper Saddle River (1976)
6. Foulis, D.J., Bennett, M.K.: Effect algebras and unsharp quantum logics. Found. Physics **24**(10), 1331–1352 (1994)
7. Giry, M.: A categorical approach to probability theory. In: Proceedings of the Categorical Aspects of Topology and Analysis. Lecture Notes in Mathematics, vol. 915, pp. 68–85 (1982)
8. Grädel, E., Thomas, W., Wilke, T. (eds.): Automata, Logics, and Infinite Games: A Guide to Current Research. LNCS, vol. 2500. Springer, Heidelberg (2002)
9. Hansen, H.H., Kupke, C.: A coalgebraic perspective on monotone modal logic. Electr. Notes Theor. Comput. Sci. **106**, 121–143 (2004)
10. Hansen, H.H., Kupke, C., Pacuit, E.: Neighbourhood structures: Bisimilarity and basic model theory. Log. Methods Comput. Sci. **5**(2), 1–32 (2009)
11. Hasuo, I.: Generic forward and backward simulations II: probabilistic simulation. In: Gastin, P., Laroussinie, F. (eds.) CONCUR 2010. LNCS, vol. 6269, pp. 447–461. Springer, Heidelberg (2010)
12. Hasuo, I., Hoshino, N.: Semantics of higher-order quantum computation via geometry of interaction. In: LICS. pp. 237–246. IEEE Computer Society (2011)
13. Hasuo, I., Jacobs, B., Sokolova, A.: Generic trace semantics via coinduction. Log. Methods Comput. Sci. **3**(4:11), 1–36 (2007)
14. Hoare, C.A.R.: An axiomatic basis for computer programming. Commun. ACM, **12**, 576–580, 583 (1969)
15. Jacobs, B.: Categorical Logic and Type Theory. North Holland, Amsterdam (1999)
16. Jacobs, B.: New directions in categorical logic, for classical, probabilistic and quantum logic. http://arxiv.org/abs/1205.3940
17. Jacobs, B.: Measurable spaces and their effect logic. In: LICS. pp. 83–92. IEEE Computer Society (2013)
18. Jones, C.: Probabilistic non-determinism. Ph.D. thesis, Univ. Edinburgh (1990)
19. Jurdziński, M.: Small progress measures for solving parity games. In: Reichel, H., Tison, S. (eds.) STACS 2000. LNCS, vol. 1770, pp. 290–301. Springer, Heidelberg (2000)
20. Katsumata, S.: Relating computational effects by ⊤⊤-lifting. Inf. Comput. **222**, 228–246 (2013)

21. Katsumata, S., Sato, T.: Preorders on monads and coalgebraic simulations. In: Pfenning, F. (ed.) FOSSACS 2013. LNCS, vol. 7794, pp. 145–160. Springer, Heidelberg (2013)
22. Kissig, C., Venema, Y.: Complementation of coalgebra automata. In: Kurz, A., Lenisa, M., Tarlecki, A. (eds.) CALCO 2009. LNCS, vol. 5728, pp. 81–96. Springer, Heidelberg (2009)
23. Kock, A.: Monads and extensive quantities (2011). arXiv:1103.6009
24. Kozen, D.: Semantics of probabilistic programs. J. Comput. Syst. Sci. **22**(3), 328–350 (1981)
25. Kupke, C., Pattinson, D.: Coalgebraic semantics of modal logics: an overview. Theor. Comput. Sci. **412**(38), 5070–5094 (2011)
26. Kupke, C., Venema, Y.: Coalgebraic automata theory: basic results. Log. Methods Comput. Sci. **4**(4), 1–43 (2008)
27. Lynch, N.A., Segala, R., Vaandrager, F.W.: Compositionality for probabilistic automata. In: Amadio, R.M., Lugiez, D. (eds.) CONCUR 2003. LNCS, vol. 2761, pp. 208–221. Springer, Heidelberg (2003)
28. Lane, S.: Categories for the Working Mathematician, 2nd edn. Springer, Berlin (1998)
29. Moggi, E.: Notions of computation and monads. Inf. Comp. **93**(1), 55–92 (1991)
30. Morgan, C.: Programming from Specifications. Prentice-Hall, London (1990)
31. Morgan, C., McIver, A., Seidel, K.: Probabilistic predicate transformers. ACM Trans. Program. Lang. Syst. **18**(3), 325–353 (1996)
32. Plotkin, G., Power, J.: Adequacy for algebraic effects. In: Honsell, F., Miculan, M. (eds.) FOSSACS 2001. LNCS, vol. 2030, pp. 1–24. Springer, Heidelberg (2001)
33. Sokolova, A.: Coalgebraic analysis of probabilistic systems. Ph.D. thesis, Techn. Univ. Eindhoven (2005)
34. Tix, R., Keimel, K., Plotkin, G.D.: Semantic domains for combining probability and non-determinism. Elect. Notes Theor. Comput. Sci. **129**, 1–104 (2005)
35. Varacca, D., Winskel, G.: Distributing probabililty over nondeterminism. Math. Struct. Comp. Sci. **16**(1), 87–113 (2006)
36. Venema, Y., Kissig, C.: Game bisimulations between basic positions. Talk at Highlights of Logic, Games and Automata, Paris (2013)
37. Winskel, G.: The Formal Semantics of Programming Languages. MIT Press, Cambridge (1993)

Coalgebraic Multigames

Marina Lenisa$^{(\boxtimes)}$

Dipartimento di Matematica e Informatica, Università di Udine, Udine, Italy
marina.lenisa@uniud.it

Abstract. *Coalgebraic games* have been recently introduced as a generalization of Conway games and other notions of games arising in different contexts. Using coalgebraic methods, games can be viewed as elements of a *final coalgebra* for a suitable functor, and operations on games can be analyzed in terms of (generalized) *coiteration schemata*. Coalgebraic games are *sequential* in nature, *i.e.* at each step either the Left (L) or the Right (R) player moves (*global polarization*), moreover only a *single* move can be performed at each step. Recently, in the context of Game Semantics, *concurrent games* have been introduced, where global polarization is abandoned, and multiple moves are allowed. In this paper, we introduce *coalgebraic multigames*, which are situated half-way between traditional sequential games and concurrent games: global polarization is still present, however *multiple* moves are possible at each step, *i.e.* a team of L/R players moves in parallel. Coalgebraic operations, such as sum and negation, can be naturally defined on multigames. Interestingly, sum on coalgebraic multigames turns out to be related to Conway's *selective sum* on games, rather than the usual (sequential) *disjoint sum*. Selective sum has a parallel nature, in that at each step the current player performs a move in *at least* one component of the sum game, while on disjoint sum the current player performs a move in *exactly* one component at each step. A monoidal closed category of coalgebraic multigames in the vein of a Joyal category of Conway games is then built. The relationship between coalgebraic multigames and games is then formalized via an equivalence of the multigame category and a monoidal closed category of coalgebraic games where tensor is selective sum.

1 Introduction

In [14], *coalgebraic games* have been introduced as a generalization of Conway games [11] to possibly non-terminating games. Coalgebras offer an elementary but sufficiently abstract framework, where games are represented as elements of a *final coalgebra* for a suitable functor, by abstracting away superficial features of positions, and operations are smoothly defined as *final morphisms* via *(generalized) coiteration schemata*. Coalgebraic games have been further studied and generalized in [15–18]. In particular, in [17], coalgebraic games have been shown to subsume also games arising in the context of Game Semantics. In [3] a similar

---

Work supported by MIUR PRIN Project CINA 2010LHT4KM.

M.M. Bonsangue (Ed.): CMCS 2014, LNCS 8446, pp. 33–49, 2014.
DOI: 10.1007/978-3-662-44124-4_3

notion of coalgebraic game has been used to model games arising in Economics (see also [23, 24]).

Coalgebraic games are 2-player games of perfect information, the two players being *Left* (L) and *Right* (R). A game is identified with its *initial* position. At any position, there are *moves* for L and R taking to new positions of the game. Contrary to other approaches in the literature, where games are defined as graphs, we view possibly non-wellfounded games as points of a *final coalgebra* of graphs, *i.e. minimal* graphs w.r.t. bisimilarity. This coalgebraic representation is more in the spirit of Conway's original presentation, and it is motivated by the fact that the existence of winning/non-losing strategies is invariant w.r.t. graph bisimilarity.

Coalgebraic games are *sequential* in nature, *i.e.* at each step either L or R moves (*global polarization*), moreover only a *single* move can be performed at each step. Recently, in the context of Game Semantics, *concurrent games* have been introduced [2, 10, 13, 30], where global polarization is abandoned, and multiple moves are allowed.

In this paper, we introduce *coalgebraic multigames*, which are situated half-way between traditional sequential games and concurrent games: global polarization is still present, however *multiple* moves are possible at each step, *i.e.* a team of L/R players moves in parallel. Coalgebraic operations, such as sum and negation, can be naturally defined on multigames via (generalized) coiteration schemata.

The notion of coalgebraic multigame introduced in the present paper is inspired by that of multigame recently defined by the same authors in [19], in the context of Game Semantics. The approach in [19] is slightly different, in that (multi)games are defined via trees of positions, rather than coalgebraically as sets/minimal graphs.

The main difference between coalgebraic multigames and games lies in the fact that in the *sum* of games, at each step, the current player can move in exactly one component, while in the *multigame sum*, by exploiting the parallel nature of multigames, the current player can perform a *multimove* consisting of atomic moves on *both* components. Sum on games amounts to Conway's *disjoint sum*, which corresponds to interleaving semantics and standard tensor product in Game Semantics (see *e.g.* [1, 20]), while multigame sum is related to Conway's *selective sum*, a form of parallel sum on games, where the current player can possibly move in *both* components.

We formalize the relationship between coalgebraic games and multigames in categorical terms. In particular, inspired by Joyal's categorical construction of Conway games [22], we build a symmetric monoidal closed category of coalgebraic multigames, where tensor is multigame sum. Namely, in [22], Joyal showed how to endow (well-founded) Conway games and winning strategies with a structure of compact closed category. This construction is based on the *disjunctive sum* of games, which induces a tensor product, and, in combination with *negation*, yields *linear implication*. Recently, the above categorical construction has been generalized to non-wellfounded games with disjoint or selective sum

[16–18], while, in the context of Linear Logic and Semantics, various categories of possibly non-wellfounded Conway games have been introduced and studied in [19,25,26,28].

In the present paper, we build a category of coalgebraic multigames and strategies in the line of the above constructions. Moreover, we show that this is *equivalent* to a category of coalgebraic games with a *parallel tensor product* inspired by Conway's selective sum. In particular, we carry out these categorical constructions in the context of *polarized (multi)games*, *i.e.* (multi)games where each position is marked as L or R, that is only L or R can move from that position, R starts, and L/R positions strictly alternate. Polarized games typically arise in Game Semantics, see *e.g.* [1,19,20].

Technically, the main difficulty in defining the above categories of (multi)games with parallel tensor product lies in the definition of strategy composition, which is not a straightforward adaptation of usual composition, but it requires a non-standard *parallel application* of strategies.

Our categorical constructions are related in particular to those carried out in [19] in the context of Game Semantics. In [19], the usefulness of multigames for modeling parallel languages is shown by providing a (universal) model of a simple parallel language, *i.e.* unary PCF with *parallel or*.

The interest of coalgebraic (multi)games is manifold. Multigames help in clarifying/factorizing the steps taking from sequential games (with global polarization and single moves) to concurrent games (no global polarization, multiple moves), offering a model of parallelism with a low level of complexity but still of a set-theoretic nature, compared to more complex concurrent games. Notably, the coalgebraic approach, both in the game and multigame version, appears significantly simpler than the traditional Game Semantics approach, where definitions of games and strategies require complex additional structures, such as equivalences on plays and strategies in the style of [1], or pointers in the arena style [20]. Such extra structure is not needed in the coalgebraic framework.

Related Work. Coalgebraic methods for modeling games have been used also in [5], where the notion of *membership game* has been introduced. This corresponds to a subclass of our coalgebraic games, where at any position L and R have the same moves, and all infinite plays are deemed winning for player II (the player who does not start). However, no operations on games are considered in that setting. In the literature, various notions of bisimilarity equivalences have been considered on games, see *e.g.* [6,29]. But, contrary to our approach, such games are defined as *graphs* of positions, and equivalences on graphs, such as trace equivalences or various bisimilarities are considered. By defining games as the elements of a final coalgebra, we directly work up to bisimilarity of game graphs.

Summary. In Sect. 2, we introduce the notions of coalgebraic multigame, play and strategy. In Sect. 3, we define operations on the final coalgebra of multigames via suitable (generalized) coiteration schemata. In Sect. 4, we build a monoidal closed category of polarized coalgebraic multigames and strategies, where tensor is sum. In Sect. 5, the relationship between coalgebraic multigames and games is expressed in categorical terms via an equivalence between the category of multigames and a monoidal closed category of coalgebraic games where tensor is selective sum. Conclusions and directions for future work appear in Sect. 6.

We assume the reader familiar with basic notions of coalgebras, see [21], and basic categorical definitions.

2 Coalgebraic Multigames and Strategies

In this section, we introduce the definitions of *coalgebraic multigame*, *play*, and *strategy*.

We consider a general notion of 2-player multigame of perfect information, where the two players are called *Left* (L) and *Right* (R). On a multigame, at each step, players can perform a *multimove*, *i.e.* a non-empty *finite set* of *atomic moves*. A multigame X is identified with its *initial position*; at any position, there are (multi)moves for L and R, taking to new *positions* of the game. By abstracting superficial features of positions, multigames can be viewed as elements of the final coalgebra for the functor $F_{\mathcal{M}_\mathcal{A}}(A) = \mathcal{P}_{<\kappa}(\mathcal{M}_\mathcal{A} \times A)$, where $\mathcal{A}$ is a set of *atomic moves*, each atomic move is marked with the name of the player who performs the move, $\mathcal{M}_\mathcal{A}$ is the set of multimoves, *i.e.* the finite powerset of atomic moves with the same polarity, and $\mathcal{P}_{<\kappa}$ is the set of all subsets of cardinality $< \kappa$, where κ can be ω if only games with finitely many moves are considered, or it can be an inaccessible cardinal if we are interested in more general games. The coalgebra structure captures, for any position, the moves of the players and the corresponding next positions.

We work in the category Set^* of sets belonging to a universe satisfying the *Antifoundation Axiom*, see [4,12], where the objects are the sets with *hereditary cardinal* less than κ, and whose morphisms are the functions with hereditary cardinal less than κ[1]. Of course, we could work in the category *Set* of well-founded sets, but we prefer to use Set^* so as to be able to use identities, *i.e.* extentional equalities in formal set theory, rather than isomorphisms in some naive set theory. Formally, we define:

Definition 1 (Coalgebraic Multigames)

– *Let $\mathcal{A}$ be a set of atoms with functions:*
  (i) $\mu : \mathcal{A} \to \mathcal{N}$ yielding the name of the move (for a set $\mathcal{N}$ of names),
  (ii) $\lambda : \mathcal{A} \to \{L, R\}$ yielding the player who has moved.
  We assume that $\mathcal{A}$ is closed under complementation, i.e. $a \in \mathcal{A} \Rightarrow \bar{a} \in \mathcal{A}$, *where $\mu\bar{a} = \mu a$ and $\lambda\bar{a} = \overline{\lambda a}$, with $\overline{R} = L$ and $\overline{L} = R$.*
– *Let $\mathcal{M}_\mathcal{A}$ be the powerset of all finite sets of atomic moves with the same polarity, i.e.*

$$\mathcal{M}_\mathcal{A} = \{\alpha \in \mathcal{P}_f(\mathcal{A}) \mid \forall a, a' \in \alpha. \ \lambda a = \lambda a'\} .$$

---

[1] We recall that the *hereditary cardinal* of a set is the cardinality of its transitive closure, namely the cardinality of the downward membership tree which has the given set as its root.

– Let $F_{\mathcal{M}_\mathcal{A}} : Set^* \to Set^*$ be the functor defined by

$$F_{\mathcal{M}_\mathcal{A}}(A) = \mathcal{P}_{<\kappa}(\mathcal{M}_\mathcal{A} \times A)$$

with the usual definition on morphisms, and let $(\mathbf{M}_\mathcal{A}, id)$ be the final $F_{\mathcal{M}_\mathcal{A}}$-coalgebra[2].

A coalgebraic multigame *is an element X of the carrier $\mathbf{M}_\mathcal{A}$ of the final coalgebra.*

The elements of the final coalgebra $\mathbf{M}_\mathcal{A}$ are the *minimal* graphs up to bisimilarity.

Clearly, every multigame can be viewed as a game whose moves are the multimoves. However, as we will see, multigame sum (as well as other constructions) is not preserved under this mapping.

We call *player I* the player who starts the multigame (who can be L or R in general), and *player II* the other. Once a player has moved on a multigame X, this leads to a new multigame/position X'. We define the *plays* on X as the sequences of pairs *move-position* from X; moves in a play are alternating:

Definition 2 (Plays). *A play on a coalgebraic game X_0 is a possibly empty finite or infinite sequence of pairs in $\mathcal{M}_\mathcal{A} \times \mathbf{M}_\mathcal{A}$, $s = \langle \alpha_1, X_1 \rangle \dots$ such that*

$$\forall n > 0. \ (\langle \alpha_n, X_n \rangle \in X_{n-1} \ \& \ \lambda \alpha_{n+1} = \overline{\lambda \alpha_n}) \ .$$

We denote by $Play_X$ the set of plays on X and by $FPlay_X$ the set of finite plays.

Here we focus on a general notion of (deterministic) *partial* strategy. Formally, strategies in our framework are partial functions on finite plays ending with a position where the player is next to move, and yielding (if any) a pair in $\mathcal{M}_\mathcal{A} \times \mathbf{M}_\mathcal{A}$, consisting of "a move of the given player together with a next position" on the game X. In what follows, we denote by

– $FPlay_X^{LI}$ ($FPlay_X^{RI}$) the set of possibly empty finite plays on the game X on which L (R) acts as player I, and ending with a position where the turn is L (R), *i.e.* $s = \epsilon$ or $s = \langle \alpha_1, X_1 \rangle \dots \langle \alpha_n, X_n \rangle$, where $\lambda \alpha_1 = L$ ($\lambda \alpha_1 = R$) and $\lambda \alpha_n = R$ ($\lambda \alpha_n = L$).
– $FPlay_X^{LII}$ ($FPlay_X^{RII}$) the set of finite plays on the game X on which L (R) acts as player II, and ending with a position where R (L) was last to move, *i.e.* $s = \langle \alpha_1, X_1 \rangle \dots \langle \alpha_n, X_n \rangle$, where $\lambda \alpha_1 = R$ ($\lambda \alpha_1 = L$) and $\lambda \alpha_n = R$ ($\lambda \alpha_n = L$).

Formally, we define:

Definition 3 (Strategies). *Let X be a coalgebraic multigame. A strategy σ for LI (i.e. L acting as player I) is a partial function $\sigma : FPlay_X^{LI} \to \mathcal{M}_\mathcal{A} \times \mathbf{M}_\mathcal{A}$ such that, for any $s \in FPlay_X^{LI}$,*

$$\sigma(s) = \langle \alpha, X' \rangle \implies \lambda \alpha = L \ \& \ s \langle \alpha, X' \rangle \in FPlay_X \ .$$

---

[2] The final coalgebra of the powerset functor exists since the powerset functor is bounded by κ.

Similarly, one can define strategies for players LII, RI, RII.
For any player LI, LII, RI, RII, we define the opponent player as RII, RI, LII, LI, respectively.
A counterstrategy of a strategy for a player in {LI,LII,RI,RII} is a strategy for the opponent player.

We are interested in studying the interactions of a strategy for a given player with the (counter)strategies of the opponent player. When a player plays on a game according to a strategy σ, against an opponent player who follows a (counter)strategy σ', a play arises. Formally, we define:

Definition 4 (Product of Strategies). *Let X be a coalgebraic multigame.*
(i) Let s be a play on X, and σ a strategy for a player in {LI,LII,RI,RII}. Then s is coherent with σ if, for any proper prefix s' of s, ending with a position where the player is next to move, $\sigma(s') = \langle \alpha, X' \rangle \implies s'\langle \alpha, X' \rangle$ is a prefix of s.
(ii) Given a strategy σ on X and a counterstrategy σ', we define the product of *σ and σ', $\sigma * \sigma'$, as the unique play coherent with both σ and σ'.*

Notice that a play arising from the product of strategies is alternating.

We call *well-founded multigames* those multigames which correspond to well-founded sets as elements of the final coalgebra $\mathbf{M}_{\mathcal{A}}$, and *non-wellfounded multigames* the non-wellfounded sets in $\mathbf{M}_{\mathcal{A}}$. Clearly, strategies on well-founded multigames generate only finite plays, while strategies on non-wellfounded multigames can generate infinite plays.

A special subclass of multigames on which we focus on in the sequel is that of *polarized multigames*. Such multigames have the following special structure: R starts, at any non-ending position only moves either for R or for L are available and along any path in the game graph R/L moves strictly alternate. Polarized multigames play a central rôle in the construction of our categories of multigames. Such games arise in traditional Game Semantics of Linear Logic and Programming Languages, see *e.g.* [1,20].

3   Multigame Operations

In this section, we show how to define various operations on multigames, including *sum, negation, linear implication*, and *infinite sum*. The crucial operation on multigames is sum, which, as we will see, is related to Conway's *selective sum* on traditional games, and it will give rise to a tensor product on the category of multigames defined in Sect. 4 below. In our coalgebraic framework, operations can be conveniently defined via *final morphisms*, using (some generalizations of) the standard *coiteration schema*. Before defining multigame operations, we start by recalling a useful generalized coiteration schema introduced in [9].

3.1   Guarded Coiteration Schema

Here we recall a generalized coiteration schema based on λ-*bialgebras*, for λ a distributive law, which will be used in the sequel for defining multigame operations.

Definition 5 (Guarded Coiteration). *Let* (Ω, α_Ω) *be a final coalgebra for a functor* $F : Set^* \to Set^*$, *and let* A *be a set. A* guarded specification *for a morphism* $h : A \to \Omega$ *is of the form:*

$$\alpha_\Omega \circ h = F(g \circ G(h)) \circ \delta_A \ , \tag{1}$$

where $\delta_A : A \to F(GA)$ *and* $g : G(\Omega) \to \Omega$ *are given,* g *is the* guard. *I.e.* h *makes the following diagram commutes:*

$$\begin{array}{ccc}
A & \xrightarrow{\ h\ } & \Omega \\
{\scriptstyle \delta_A}\downarrow & & \downarrow{\scriptstyle \alpha_\Omega} \\
FGA & \xrightarrow[F(g\circ Gh)]{} & F\Omega
\end{array}$$

Under suitable conditions, the above schema admits a unique solution, see [9] for more details. In particular, a functor $G : Set^* \to Set^*$ is required and a generalized distributive law $\lambda : GF \xrightarrow{.} FG$ for which $(\Omega, g, \alpha_\Omega)$ is a λ-bialgebra.

3.2 Operations on Multigames

Sum. In the context of multigames, the following notion of sum arises naturally. On the sum multigame, at each step, the next player selects either one (non-ended) or *both* component multigames, and makes a legal move in each of the selected components, while the component which has not been chosen (if any) remains unchanged.

Definition 6 (Sum). *The* sum *of two multigames* $\triangledown : \mathbf{M}_{\mathcal{A}} \times \mathbf{M}_{\mathcal{A}} \longrightarrow \mathbf{M}_{\mathcal{A}}$ *is defined by:*

$$X \triangledown Y = \{\langle \alpha', X' \triangledown Y \rangle \mid \langle \alpha, X' \rangle \in X\} \cup \{\langle \beta', X \triangledown Y' \rangle \mid \langle \beta, Y' \rangle \in Y\} \cup$$
$$\{\langle \alpha + \beta, X' \triangledown Y' \rangle \mid \langle \alpha, X' \rangle \in X \ \& \ \langle \beta, Y' \rangle \in Y\} \ ,$$

where α', β' *are the sets obtained from* α, β *by adding tags to atomic moves.*

The above definition corresponds to a *standard coiteration schema*, namely the function $\triangledown$ is the *final morphism* from the coalgebra $(\mathbf{M}_{\mathcal{A}} \times \mathbf{M}_{\mathcal{A}}, f_\triangledown)$ to the final coalgebra $(\mathbf{M}_{\mathcal{A}}, \mathrm{id})$, where the coalgebra morphism $f_\triangledown : \mathbf{M}_{\mathcal{A}} \times \mathbf{M}_{\mathcal{A}} \longrightarrow F_{\mathcal{M}_{\mathcal{A}}}(\mathbf{M}_{\mathcal{A}} \times \mathbf{M}_{\mathcal{A}})$ is defined by:

$$f_\triangledown(X, Y) = \{\langle \alpha', \langle X', Y \rangle \rangle \mid \langle \alpha, X' \rangle \in X\} \cup \{\langle \beta', \langle X, Y' \rangle \rangle \mid \langle \beta, Y' \rangle \in Y\} \cup$$
$$\{\langle \alpha + \beta, \langle X', Y' \rangle \rangle \mid \langle \alpha, X' \rangle \in X \ \& \ \langle \beta, Y' \rangle \in Y\}$$

Remark. *Notice that, in composing multigames via the sum, we keep track of the moves coming from the two different components by using tags. This definition is different from original Conway's sum on games, which is a purely set-theoretic* *extensional operation, possibly allowing for identifications between the two components. Our sum definition, where we keep track of the component in which each move has been performed, is necessary* e.g. *for extending sum to a bifunctor in categories of multigames and strategies such as that defined in Sect. 4 below, or even to define strategy composition in this category. Nonetheless, notice that, from a* determinacy *point of view, this sum and Conway's original one behave in the same way, i.e. they are equivalent w.r.t. the existence of (winning) strategies.*

Negation. The *negation* is a unary multigame operation, which allows us to build a new game, where the rôles of L and R are exchanged. For $\alpha \in \mathcal{M}_\mathcal{A}$, we define

$$\overline{\alpha} = \{\overline{a} \mid a \in \alpha\} \ .$$

The definition of multigame negation is as follows:

Definition 7 (Negation). *The* negation $^{-} : \mathbf{M}_\mathcal{A} \longrightarrow \mathbf{M}_\mathcal{A}$ *is defined by:*

$$\overline{X} = \{\langle \overline{\alpha}, \overline{X'} \rangle \mid \langle \alpha, X' \rangle \in X\} \ .$$

Also negation is an instance of the coiteration schema. It is the final morphism from the coalgebra $(\mathbf{M}_\mathcal{A}, f_-)$ to the final coalgebra $(\mathbf{M}_\mathcal{A}, \mathrm{id})$, where the coalgebra morphism $f_- : \mathbf{M}_\mathcal{A} \longrightarrow F_{\mathcal{M}_\mathcal{A}}(\mathbf{M}_\mathcal{A})$ is defined by:

$$f_-(X) = \{\langle \overline{\alpha}, X' \rangle \mid \langle \alpha, X' \rangle \in X\} \ .$$

Linear Implication. Using the above notions of sum and negation, we can now define the following *linear implication*, which corresponds to the notion of linear implication in Linear Logic:

Definition 8 (Linear Implication). *The* linear implication *of the multigames* X, Y, $X \multimap Y$, *is defined by*

$$X \multimap Y = \overline{X} \triangledown Y \ .$$

Infinite Sum. We can enrich multigames with a further interesting unary coalgebraic operation, $\triangledown^\infty$, an *infinite sum*: on the game $\triangledown^\infty x$, at each step, the current player can perform a move in *finitely many* of the infinite components of X. Our infinite sum is related to the exponential modality defined on a category of games in [28], and it will induce a comonad on the categories of multigames that we will consider in Sect. 4.

Definition 9 (Infinite Sum). *We define the* infinite sum $\triangledown^\infty : \mathbf{M}_\mathcal{A} \longrightarrow \mathbf{M}_\mathcal{A}$ *by:*

$$\triangledown^\infty X = \{\langle \alpha_1 + \ldots + \alpha_n, X_1' \triangledown \ldots \triangledown X_n' \triangledown (\triangledown^\infty X) \rangle \mid n \geq 1 \ \&$$
$$\langle \alpha_1, X_1' \rangle, \ldots, \langle \alpha_n, X_n' \rangle \in X \ \& \ \lambda \alpha_1 = \ldots = \lambda \alpha_n\} \ .$$

The above specification defines a unique function $\triangledown^\infty$, since it is an instance of the guarded coiteration of Definition 5 above, where

– the functor $G : Set^* \to Set^*$ is defined by

$$G(A) = \coprod_{n \geq 1} (\mathbf{M}_{\mathcal{A}}^n \times A)$$

– the guard $g : G(\mathbf{M}_{\mathcal{A}}) \longrightarrow \mathbf{M}_{\mathcal{A}}$ is defined by

$$g(X_1, \dots, X_{n+1}) = X_1 \triangledown \dots \triangledown X_{n+1}$$

– the function $\delta_{\mathbf{M}_{\mathcal{A}}} : \mathbf{M}_{\mathcal{A}} \to \mathcal{P}_{<\kappa}(\mathcal{M}_{\mathcal{A}} \times \coprod_{n \geq 1}(\mathbf{M}_{\mathcal{A}}^n \times \mathbf{M}_{\mathcal{A}}))$ is defined by

$$\delta_{\mathbf{M}_{\mathcal{A}}}(X) = \{(\alpha_1 + \dots + \alpha_n, X_1', \dots, X_n', X) \mid n \geq 1 \,\&\, \lambda\alpha_1 = \dots = \lambda\alpha_n \,\&\,$$
$$(\alpha_1, X_1'), \dots, (\alpha_n, X_n') \in X\}$$

– $(\mathbf{M}_{\mathcal{A}}, g, id)$ is a λ-bialgebra for $\lambda : GF \to FG$ distributive law defined by

$$\lambda_A : \coprod_{n \geq 1}(\mathbf{M}_{\mathcal{A}}^n \times \mathcal{P}_{<\kappa}(\mathcal{M}_{\mathcal{A}} \times A)) \to \mathcal{P}_{<\kappa}(\mathcal{M}_{\mathcal{A}} \times \coprod_{n \geq 1}(\mathbf{M}_{\mathcal{A}}^n \times A))$$

$$\lambda_A(X_1, \dots, X_{n+1}) = \{(\alpha_1 + \dots + \alpha_k, X_1', \dots, X_{n+1}') \mid k \geq 1 \,\&\,$$
$$\langle \alpha_1, X_{i_1}' \rangle \in X_{i_1}, \dots, \langle \alpha_k, X_{i_k}' \rangle \in X_{i_k} \,\&\, 1 \leq i_1, \dots, i_k \leq n \,\&\,$$
$$\forall j \in \{1, \dots, n\} \setminus \{i_1, \dots, i_k\}.\, X_j' = X_j\} \,.$$

3.3   Operations on Polarized Multigames

Notice that polarized multigames are *not* closed under the operations defined above. However, one can define corresponding sums and linear implications on polarized multigames, simply by "pruning" the graphs of the resulting multigames. A similar construction has been considered also in [18], in the case of games. In our setting, the *pruning* operation of a multigame into a polarized one can be defined as a coalgebraic operation, using a definition by *mutual recursion*:

Definition 10 (Pruning). *Let* $(\ )_+, (\ )_- : \mathbf{M}_{\mathcal{A}} \longrightarrow \mathbf{M}_{\mathcal{A}}$ *be the mutually recursive functions defined as:*

$$\begin{cases} (X)_+ = \{\langle \alpha, (X')_- \rangle \mid \langle \alpha, X' \rangle \in X \,\&\, \lambda\alpha = L\} \\ (X)_- = \{\langle \alpha, (X')_+ \rangle \mid \langle \alpha, X' \rangle \in X \,\&\, \lambda\alpha = R\} \,. \end{cases}$$

We define the pruning *operation as* $(\ )_-$.

Once we have the pruning operation, we can define *polarized sums* and *linear implication*:

Definition 11 (Polarized Operations). *Let* X, Y *be polarized multigames. We define:*

– *the* polarized sum *as the multigame* $(X \triangledown Y)_-$;
– *the* polarized linear implication *as the multigame* $(X \multimap Y)_-$;
– *the* polarized infinite sum *as the multigame* $(\triangledown^\infty X)_-$.

In the following, by abuse of notation, we will use the same symbols for polarized operations and the corresponding operations on all multigames.

4   Categories of Multigames and Strategies

We define a monoidal closed category $\mathcal{Y}_{M_A}$, whose *objects* are polarized multi-games and whose *morphisms* are strategies. We work with *polarized* multigames, since the whole class of multigames fails to give a category, because of lack of identities, as we will see.

The main difficulty in defining this category is the definition of composition, which is based on a non-standard *parallel composition of strategies*. The difficulty arises from the fact that a move in a strategy between X and Y can include atomic moves on both X and Y.

A Monoidal Closed Category of Multigames. Let $\mathcal{Y}_{M_A}$ be the category defined by:

Objects: polarized multigames.

Morphisms: a morphism between multigames X and Y, $\sigma : X \multimap Y$, is a strategy for LII on $X \multimap Y$. Notice that in a strategy on $X \multimap Y$, Player R can only open in Y, but then the Player L can move in X or Y or in both components, and so on.

Identity: the identity $id_X : X \multimap X$ is the *copy-cat strategy*. This definition works thanks to the fact that multigames are polarized, so as, on the multigame $X \multimap X = (\overline{X} \triangledown X)$, R can only open on X, then L proceeds by copying the move on $\overline{X}$ and so on, at each step R has exactly one component to move in. Notice that, if the games were *not* polarized, then R could play on both components X and $\overline{X}$, preventing L to apply the copy-cat strategy.

Composition: strategy composition is defined as follows.

Given strategies $\sigma : X \multimap Y$ and $\tau : Y \multimap Z$, $\tau \circ \sigma : X \multimap Z$ is obtained via the *swivel-chair strategy*, using the terminology of [7] (or the copy-cat strategy, in Game Semantics terminology), and a non-standard *parallel application* of strategies as follows.

The opening move by R on $X \multimap Z$ must be on Z, since multigames are polarized. Then consider the L reply given by the strategy τ on $Y \multimap Z$, if it exists, otherwise the whole composition is undefined. If L moves in Z, then we take this as the L move in the strategy $\tau \circ \sigma$. If the L move according to τ is in the Y component of $Y \multimap Z$ or in both components Y and Z, then we use the swivel chair to view the L move in the Y component as a R move in the Y component of $X \multimap Y$. Now, if L has a reply in $X \multimap Y$ according to σ, then L moves in X or in Y or in both X and Y. In the first case, the L move in X together with the possible previous move by L in Z form the L reply to the opening R move. In the latter two cases, using the swivel chair, the move in Y can be viewed as a R move in the Y component of $Y \multimap Z$, and we go on in this way: three cases can arise. Eventually, the L multimove is all in X or in Z, or σ or τ is undefined, or the dialogue between the Y components does not stop. In the first case, the last move on X or Z, together with a possible previous move on Z or X, form the answer to the opening R move, in the latter two cases the composition is undefined.

Now, in case of convergence, in order to understand how the strategy $\tau \circ \sigma$ behaves after the first pair of RL moves, it is convenient to list all situations which can arise after these initial moves, according to which player is next to move in each component. Namely, by case analysis, one can show that, after the first RL moves, the following four cases can arise:

1. $X^L \multimap Y^R \quad Y^L \multimap Z^R$
2. $X^R \multimap Y^L \quad Y^R \multimap Z^R$
3. $X^R \multimap Y^L \quad Y^R \multimap Z^R$
4. $X^R \multimap Y^R \quad Y^L \multimap Z^R$

where X^L (X^R) denotes that L (R) is next to move in that component. Notice that case 1 above corresponds to the initial situation. Thus we are left to discuss the behavior of $\tau \circ \sigma$ in the other cases. In case 2, R can only open in X; this case can be dealt with similarly as the initial case 1, and after a pair of RL moves, it takes again in one of the four situations above. Cases 3 and 4 are the interesting ones, where we need to apply the strategies σ and τ in *parallel*, by exploiting the parallelism of multigames. These two cases are dealt with similarly. Let us consider case 3. Then R can open in Z or in X or in both components.

- If R opens in Z, then the reply of L via τ must be in Z, since, by definition of configuration 3, L cannot play in the Y component of $Y \multimap Z$. This will be also the reply of L in the composition $\tau \circ \sigma$, and the final configuration coincides with configuration 3.
- If R opens in X, then the L reply given by σ can be in X, or in Y, or both in X and Y. In the first case, *i.e.* if the L reply is in X, this will be the reply of $\tau \circ \sigma$, and the new configuration coincides with configuration 3. In the latter two cases, the L move in the Y component of $X \multimap Y$ can be viewed, via the swivel chair, as a R move in $Y \multimap Z$. By definition of configuration 3, L can only reply in Y via τ, and, either after finitely many applications of the swivel chair the L reply via σ ends up in X, or the dialogue between the Y components goes on indefinitely, or σ or τ are undefined. In the latter two cases, the overall composition is undefined, while in the first case the L move in X will be the reply in $\tau \circ \sigma$ to the R move, and the final configuration still coincides with configuration 3.
- Finally, if R opens in X and Z, we apply the two strategies σ and τ in parallel: by the form of configuration 3, the L answer of τ must be in Z, while the L answer via σ can be either in X or in Y or both in X and Y. In this latter case, an infinite dialogue between the Y components arises (or σ or τ is undefined at some point), and hence the overall composition is undefined.
  If the L reply via σ is in X, then this, together with the L reply via τ in Z, will form the L move in $\tau \circ \sigma$, and the final configuration coincides with configuration 3 itself. If the L reply via σ is in Y, then again, either the dialogue between the Y components goes on indefinitely, or σ or τ is undefined at some point, or, after finitely many applications of the swivel chair, σ will finally provide a L move in X. This, together with the L reply via τ in Z, will form the L move in $\tau \circ \sigma$, and the final configuration coincides with configuration 3 again.

Similarly, one can deal with case 4. This proves closure under composition of the category $\mathcal{Y}_{\mathrm{M}_A}$.

Associativity of composition can also be proven by case analysis on the polarity of the current player in the various components.

Assume strategies $\sigma : X \multimap Y$, $\tau : Y \multimap Z$, $\theta : Z \multimap W$. We have to prove that $\theta \circ (\tau \circ \sigma) = (\theta \circ \tau) \circ \sigma$. Since multigames are polarized, in any of the two compositions, R can only open in W. Now, in any of the two compositions, one should consider the possible replies by L. We only discuss one case, the remaining being dealt with similarly. Assume the L reply via θ is in Z. In both compositions, we proceed to apply the swivel chair, by viewing this latter move as a R move in the Z component of $Y \multimap Z$. Then, in both compositions, we consider the L reply via τ. Assume L replies both in Y and in Z. At this point, the two compositions proceed differently, since in $\theta \circ (\tau \circ \sigma)$ we first apply the swivel chair to the move in Z and we go on until we get a L answer in W, and then we apply the swivel chair to the L move in Y. In $(\theta \circ \tau) \circ \sigma$, these two steps are reversed, first we apply the swivel chair to the L move in Y until we get a L reply in X, then we apply the swivel chair to the L move in Z until we get a L reply in W. The point is that these two steps, working on separate parts of the board (*i.e.* different components), are independent and can be exchanged. As a consequence, the behavior of $\theta \circ (\tau \circ \sigma)$ and $(\theta \circ \tau) \circ \sigma$ is the same.

The multigame constructions of tensor product and linear implication can be made functorial, determining a structure of a symmetric monoidal closed category on $\mathcal{Y}_{\mathrm{M}_A}$, with the empty multigame as tensor unit. In particular, in defining the bifunctor $\triangledown$, we proceed as follows. Let $\sigma_1 : X_1 \to Y_1$ and $\sigma_2 : X_2 \to Y_2$ be strategies. In order to define the strategy $\sigma_1 + \sigma_2 : X_1 + X_2 \to Y_1 + Y_2$, we let the two strategies σ_1 and σ_2 play in parallel. *I.e.* we consider R opening move: if it is in the Y_1 (Y_2) component, then the L answer will be given by the strategy σ_1 (σ_2); if the R move is both in Y_1 and Y_2, then we apply the two strategies in parallel. We proceed in a similar way for the next moves. Clearly, for the above definition of $\sigma_1 + \sigma_2$, it is essential that in the sum multigame we keep track of the components from which each move comes from. Otherwise, without tags in the moves of the sum, when *e.g.* $Y_1 = Y_2$, we would not know on which component the R opening move on $Y_1 + Y_2$ comes, and then we do not know whether applying σ_1 or σ_2. This justifies our definition of sum in Sect. 3.2.

Summarizing, we have:

Theorem 1. *The category* $\mathcal{Y}_{\mathrm{M}_A}$ *is symmetric monoidal closed.*

Finally, one could show that the infinite sum operation $\triangledown^\infty$ induces a symmetric monoidal comonad, determining on $\mathcal{Y}_{\mathrm{M}_A}$ a structure of linear category in the sense of [8]. We omit the details of this construction. Notice that, contrary to traditional Game Semantics, in our coalgebraic framework this construction does not require the definition of an equivalence on strategies and a quotient operation on them in the style of [1].

5   Relating Multigames to Games

In this section, we show that coalgebraic multigames are related to coalgebraic games, when a notion of parallel sum reminiscent of Conway's *selective sum* is considered on games, in place of *disjoint sum*. Here we recall the notion of coalgebraic game as it has been defined in [17]:

Definition 12 (Coalgebraic Games)

– *Let $\mathcal{A}$ be a set of atoms with functions:*
  (i) $\mu : \mathcal{A} \to \mathcal{N}$ yielding the name of the move (for a set $\mathcal{N}$ of names),
  (ii) $\lambda : \mathcal{A} \to \{L, R\}$ yielding the player who has moved.
  Assume $\mathcal{A}$ be closed under complementation.
– *Let $F_{\mathcal{A}} : Set^* \to Set^*$ be the functor defined by*

$$F_{\mathcal{A}}(A) = \mathcal{P}_{<\kappa}(\mathcal{A} \times A)$$

with the usual definition on morphisms, and let $(\mathbf{G}_{\mathcal{A}}, id)$ be the final $F_{\mathcal{A}}$-coalgebra.

A coalgebraic game *is an element x of the carrier $\mathbf{G}_{\mathcal{A}}$ of the final coalgebra.*

Plays and strategies are defined similarly as for multigames, see [17] for more details.

The above definition of coalgebraic games generalizes Conway games and games arising in Game Semantics, see [17] for more details. Following [11], coalgebraic games can be endowed with various notions of sum, the most studied being disjoint sum, which corresponds to tensor product in standard Game Semantics. Here we focus on a notion of parallel sum inspired by Conway's selective sum, where at each step the current player can perform a move either in the first or in the second component or in both components. This notion of sum admits a straightforward definition by coiteration.

Definition 13 (Selective Sum). *The* selective sum *of two games $\vee : \mathbf{G}_{\mathcal{A}} \times \mathbf{G}_{\mathcal{A}} \longrightarrow \mathbf{G}_{\mathcal{A}}$ is defined by:*

$$x \vee y = \{\langle a', x' \vee y \rangle \mid \langle a, x' \rangle \in x\} \cup \{\langle b', x \vee y' \rangle \mid \langle b, y' \rangle \in y\} \cup$$
$$\{\langle \langle a, b \rangle, x' \vee y' \rangle \mid \langle a, x' \rangle \in x \,\&\, \langle b, y' \rangle \in y\},$$

where a', b' are obtained from a, b by adding suitable tags, and $\langle a, b \rangle$ denotes the pairing of the moves a, b (we assume the set of moves to be closed under pairing).

As usual, one can define negation and then linear implication on games, *i.e.*:

$$\overline{x} = \{\langle \overline{a}, \overline{x'} \rangle \mid \langle a, x' \rangle \in x\} \qquad x \multimap y = \overline{x} \vee y \,.$$

Similarly as for multigames, when we restrict to polarized games, these can be endowed with a structure of symmetric monoidal category where tensor is selective sum. As in the case of multigames, also in the present case strategy composition is defined via the swivel-chair and a non-standard parallel application of strategies. We skip the details of this construction. Let call $\mathcal{Y}_{\mathbf{G}_A}$ be the category so obtained.

An analogous construction has been also carried out in [18] in the case of coalgebraic games and total strategies, and in [19] for a notion of game in the [1]-style.

5.1   Categorical Correspondence

The category $\mathcal{Y}_{\mathbf{G}_A}$ of games with selective sum turns out to be *equivalent* to the category $\mathcal{Y}_{\mathbf{M}_A}$ of multigames, *i.e.* there exist functors $S : \mathcal{Y}_{\mathbf{G}_A} \to \mathcal{Y}_{\mathbf{M}_A}$ and $T : \mathcal{Y}_{\mathbf{M}_A} \to \mathcal{Y}_{\mathbf{G}_A}$, and natural isomorphisms $\eta : G \circ F \xrightarrow{\cdot} Id_{\mathcal{Y}_{\mathbf{G}_A}}$ and $\eta' : Id_{\mathcal{Y}_{\mathbf{M}_A}} \xrightarrow{\cdot} F \circ G$.

Namely, given a multigame X in $\mathbf{M}_A$, this induces a game X_g, where the atomic moves are the sets of (multi)moves on the multigame X. Vice versa, given a game, x, one build a multigame x_m, where each atomic move a is replaced by the singleton multimove $\{a\}$. Clearly, for any multigame X, $(X_g)_m$ is isomorphic to X, and for any game x, $(x_m)_g$ is also isomorphic to x.

This allows us to define the object part of functors $S : \mathcal{Y}_{\mathbf{G}_A} \to \mathcal{Y}_{\mathbf{M}_A}$ and $T : \mathcal{Y}_{\mathbf{M}_A} \to \mathcal{Y}_{\mathbf{G}_A}$. Notice that S and T preserve tensor product on objects, up to isomorphism.

Functors S and T can be extended to strategies as follows.

For any strategy on multigames $\sigma : X \multimap Y$, we can associate a strategy $\sigma_g : X_g \multimap Y_g$, where the plays of σ_g are obtained from the plays of σ by splitting each multimove of σ containing atomic moves both in A and in B into a pair of moves on A and B, respectively. Vice versa, any strategy on games $\sigma : x \multimap y$ induces a strategy on multigames $\sigma_m : x_m \multimap y_m$, whose plays are obtained from the plays of σ by transforming each move instance of x or y into a singleton multimove, and each pair of moves $\langle a, b \rangle$ as the multimove $\{a, b\}$.

Summarizing, we can define functors $S : \mathcal{Y}_{\mathbf{G}_A} \to \mathcal{Y}_{\mathbf{M}_A}$ and $T : \mathcal{Y}_{\mathbf{M}_A} \to \mathcal{Y}_{\mathbf{G}_A}$ by:
for any game x, $Sx = x_m$, for any strategy $\sigma : X \multimap Y$, $S\sigma = \sigma_m$,
for any multigame X, $TX = X_g$, for any strategy $\sigma : X \multimap Y$, $T\sigma = \sigma_g$.

Then, we have:

Theorem 2. *The functors* $S : \mathcal{Y}_{\mathbf{G}_A} \to \mathcal{Y}_{\mathbf{M}_A}$ *and* $T : \mathcal{Y}_{\mathbf{M}_A} \to \mathcal{Y}_{\mathbf{G}_A}$ *are monoidal, and they give an equivalence between the categories* $\mathcal{Y}_{\mathbf{G}_A}$ *and* $\mathcal{Y}_{\mathbf{M}_A}$.

6   Final Remarks and Directions for Future Work

We have introduced coalgebraic multigames, where at each step the current player performs a multimove, *i.e.* a (finite) set of atomic moves. Coalgebraic multigames introduce a certain level of parallelism, and they are situated halfway between traditional sequential games and concurrent games. Multigame operations are smoothly defined in our coalgebraic framework as final morphisms via (generalized) coiteration schemata. A monoidal closed category of multigames and strategies is built, where tensor is sum. The relationship between coalgebraic multigames and games is expressed in categorical terms via an equivalence between the category of multigames and a monoidal closed category of coalgebraic games where tensor is selective sum.

Here is a list of comments and directions for future work.

- *Total strategies.* In this paper, coalgebraic (multi)games are endowed with a notion of partial strategy, whereby the given player can possibly refuse to provide an answer and give up the game. Alternatively, one could consider notions of *total strategies*, in the line of [15,17,18], where the player is forced to give an answer, if there exists any. On total strategies one can then define the notion of *winning/non-losing strategy* for a player, if it generates winning/non-losing plays against any possible counterstrategy. A finite play is taken to be winning for the player who performs the last move, while infinite plays are taken to be winning for L/R or draws. In order to formalize winning/non-losing strategies, one shall introduce a *payoff function* on plays, and enrich the notion of multigame with a payoff. We claim that the results of the present paper can be rephrased also in the context of total strategies.
- *Other notions of sum.* In [11], Chap. 14 *"How to Play Several Games at Once in a Dozen Different Ways"*, Conway introduces a number of different ways in which games can be played. Apart from disjunctive and selective sum, Conway defines the *conjunctive sum*, where at each step the current player makes a move in each (non-ended) component. A first attempt to extend Joyal's categorical construction to conjunctive sum fails, even in the case of polarized games, since trivially copy-cat strategies do not work. Alternative approaches are called for.
- *Semantics of concurrency.* In the literature, notions of concurrent games [2], asynchronous games [27], and distributed games [10,13,30] have been introduced as concurrent extensions of traditional games. Our categories of coalgebraic multigames and coalgebraic games with selective sum are more in the traditional line, but nonetheless, they reflect a form of parallelism. Namely, in [19] it has been shown that, in the context of functional languages, categories of multigames in the style of [1] accommodate *parallel or*. It would be interesting to explore to what extent multigames can be used for modeling concurrent and distributed languages, possibly featuring true concurrency. This would require an extension of our approach in order to account also for interference between moves/events.
- *Relating multigames to domains.* Since multigames introduce a level of parallelism, and as it is shown in [19], they accommodate for *e.g. parallel or* in functional programming, one may expect a tighter connection between multigames and traditional denotational semantics based on domains.

References

1. Abramsky, S., Jagadeesan, R., Malacaria, P.: Full abstraction for PCF. Inf. Comput. **163**, 404–470 (2000)
2. Abramsky, S., Mellies, P.A.: Concurrent games and full completeness. In: LICS'99, pp. 431–444. IEEE Computer Science Press (1999)
3. Abramsky, S., Winschel, V.: Coalgebraic analysis of subgame-perfect equilibria in infinite games without discounting. Mathematical Structures in Computer Science (2013) (to appear)
4. Aczel, P.: Non-wellfounded Sets. CSLI Lecture Notes, vol. 14. Stanford University, Stanford (1988)
5. Barwise, J., Moss, L.: Vicious Circles. CSLI Lecture Notes, vol. 60. Stanford University, Stanford (1996)
6. van Benthem, J.: Extensive games as process models. J. Logic Lang. Inform. **11**, 289–313 (2002)
7. Berlekamp, E., Conway, J., Guy, R.: Winning Ways. Academic Press, London (1982)
8. Bierman, G.: What is a categorical model of intuitionistic linear logic? In: Dezani-Ciancaglini, M., Plotkin, G. (eds.) TLCA 1995. LNCS, vol. 902, pp. 78–93. Springer, Heidelberg (1995)
9. Cancila, D., Honsell, F., Lenisa, M.: Generalized coiteration schemata. In: CMCS'03. ENTCS, vol. 82(1) (2003)
10. Clairambault, P., Gutierrez, J., Winskel, G.: The winning ways of concurrent games. In: LICS 2012. IEEE Computer Science Press (2012)
11. Conway, J.H.: On Numbers and Games. A K Peters Ltd, Natick (2001)
12. Forti, M., Honsell, F.: Set-theory with free construction principles. Ann. Scuola Norm. Sup. Pisa, Cl. Sci. **10**(4), 493–522 (1983)
13. Ghica, D.R., Menaa, M.N.: On the compositionality of round abstraction. In: Gastin, P., Laroussinie, F. (eds.) CONCUR 2010. LNCS, vol. 6269, pp. 417–431. Springer, Heidelberg (2010)
14. Honsell, F., Lenisa, M.: Conway games, coalgebraically. In: Kurz, A., Lenisa, M., Tarlecki, A. (eds.) CALCO 2009. LNCS, vol. 5728, pp. 300–316. Springer, Heidelberg (2009)
15. Honsell, F., Lenisa, M.: Conway games, algebraically and coalgebraically. Logical Meth. Comput. Sci. **7**(3) (2011)
16. Honsell, F., Lenisa, M., Redamalla, R.: Equivalences and congruences on infinite conway games. RAIRO Theor. Inf. Appl. **46**(2), 231–259 (2012)
17. Honsell, F., Lenisa, M., Redamalla, R.: Categories of coalgebraic games. In: Rovan, B., Sassone, V., Widmayer, P. (eds.) MFCS 2012. LNCS, vol. 7464, pp. 503–515. Springer, Heidelberg (2012)
18. Honsell, F., Lenisa, M., Pellarini, D.: Categories of coalgebraic games with selective sum. Fundamenta Informaticae (2013) (to appear)
19. Honsell, F., Lenisa, M.: Polarized multigames, draft paper (2013). http://sole.dimi.uniud.it/~marina.lenisa/Papers/Soft-copy-pdf/par.pdf
20. Hyland, M., Ong, L.: On full abstraction for PCF: I. II, and III. Inf. Comput. **163**, 285–408 (2000)
21. Jacobs, B., Rutten, J.: A tutorial on (co)algebras and (co)induction. Bull. EATCS **62**, 222–259 (1996)
22. Joyal, A.: Remarques sur la Theorie des Jeux a deux personnes. Gazette des Sciences Mathematiques du Quebec **1**(4), 46–52 (1977)

23. Lescanne, P.: Rationality and escalation in infinite extensive games (2011). Arxiv preprint arXiv:1112.1185
24. Lescanne, P., Perrinel, M.: Backward coinduction, Nash equilibrium and the rationality of escalation. Acta Informatica, 1–21 (2012)
25. Mellies, P.A.: Comparing hierarchies of types in models of linear logic. Inf. Comput. **189**(2), 202–234 (2004)
26. Mellies, P.A.: Asynchronous games 3: an innocent model of linear logic. In: CTCS2004. ENTCS, vol. 122 (2005)
27. Mellies, P.A.: Asynchronous games 2: the true concurrency of innocence. Theoret. Comput. Sci. **358**(2–3), 200–228 (2006)
28. Melliès, P.-A., Tabareau, N., Tasson, C.: An explicit formula for the free exponential modality of linear logic. In: Albers, S., Marchetti-Spaccamela, A., Matias, Y., Nikoletseas, S., Thomas, W. (eds.) ICALP 2009, Part II. LNCS, vol. 5556, pp. 247–260. Springer, Heidelberg (2009)
29. Pauly, M.: From programs to games: invariance and safety for bisimulation. In: Clote, P.G., Schwichtenberg, H. (eds.) CSL 2000. LNCS, vol. 1862, pp. 485–496. Springer, Heidelberg (2000)
30. Rideau, S., Winskel, G.: Concurrent strategies. In: LICS 2011. IEEE Computer Science Press (2011)

Regular Contributions

How to Kill Epsilons with a Dagger

A Coalgebraic Take on Systems with Algebraic Label Structure

Filippo Bonchi[1], Stefan Milius[2], Alexandra Silva[3,4,5], and Fabio Zanasi[1]([✉])

[1] ENS Lyon, U. de Lyon, CNRS, INRIA, UCBL, Lyon, France
fabio.zanasi@ens-lyon.fr
[2] Lehrstuhl Für Theoretische Informatik,
Friedrich-Alexander Universität Erlangen-Nürnberg, Erlangen, Germany
[3] Institute for Computing and Information Sciences,
Radboud University Nijmegen, Nijmegen, The Netherlands
[4] Centrum Wiskunde & Informatica, Amsterdam, The Netherlands
[5] HASLab/INESC TEC, Universidade Do Minho, Braga, Portugal

Abstract. We propose an abstract framework for modeling state-based systems with internal behavior as e.g. given by silent or ϵ-transitions. Our approach employs monads with a parametrized fixpoint operator $\dagger$ to give a semantics to those systems and implement a sound procedure of abstraction of the internal transitions, whose labels are seen as the unit of a free monoid. More broadly, our approach extends the standard coalgebraic framework for state-based systems by taking into account the algebraic structure of the labels of their transitions. This allows to consider a wide range of other examples, including Mazurkiewicz traces for concurrent systems.

1 Introduction

The theory of coalgebras provides an elegant mathematical framework to express the semantics of computing devices: the operational semantics, which is usually given as a state machine, is modeled as a coalgebra for a functor; the denotational semantics as the unique map into the final coalgebra of that functor. While the denotational semantics is often *compositional* (as, for instance, ensured by the bialgebraic approach of [24]), it is sometimes not *fully-abstract*, i.e., it discriminates systems that are equal from the point of view of an external observer. This is due to the presence of internal transitions (also called ϵ-transitions) that are not observable but that are not abstracted away by the usual coalgebraic semantics using the unique homomorphism into the final coalgebra.

In this paper, we focus on the problem of giving trace semantics to systems with internal transitions. Our approach stems from an elementary observation (pointed out in previous work, e.g. [23]): the labels of transitions form a monoid and the internal transitions are those labeled by the unit of the monoid. Thus, there is an *algebraic structure* on the labels that needs to be taken into

© IFIP International Federation for Information Processing 2014
M.M. Bonsangue (Ed.): CMCS 2014, LNCS 8446, pp. 53–74, 2014.
DOI: 10.1007/978-3-662-44124-4_4

account when modeling the denotational semantics of those systems. To illustrate this point, consider the following two non-deterministic automata (NDA).

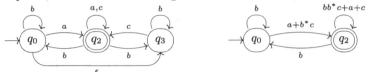

The one on the left (that we call $\mathbb{A}$) is an NDA with ϵ-transitions: its transitions are labeled either by the symbols of the alphabet $A = \{a, b, c\}$ or by the empty word $\epsilon \in A^*$. The one on the right (that we call $\mathbb{B}$) has transitions labeled by languages in $\mathcal{P}(A^*)$, here represented as regular expressions. The monoid structure on the labels is explicit on $\mathbb{B}$, while it is less evident in $\mathbb{A}$ since the set of labels $A \cup \{\epsilon\}$ does not form a monoid. However, this set can be trivially embedded into $\mathcal{P}(A^*)$ by looking at each symbols as the corresponding singleton language. For this reason each automaton with ϵ-transitions, like $\mathbb{A}$, can be regarded as an automaton with transitions labeled by languages, like $\mathbb{B}$. Furthermore, we can define the semantics of NDA with ϵ-transitions by defining the semantics of NDA with transitions labeled by languages: a word w is accepted by a state q if there is a path $q \xrightarrow{L_1} \cdots \xrightarrow{L_n} p$ where p is a final state, and there exist a decomposition $w = w_1 \cdots w_n$ such that $w_i \in L_i$. Observe that, with this definition, $\mathbb{A}$ and $\mathbb{B}$ accept the same language: all words over A that end with a or c. In fact, $\mathbb{B}$ was obtained from $\mathbb{A}$ in a well-known process to compute the regular expression denoting the language accepted by a given automaton [14].

We propose to define the semantics of systems with internal transitions following the same idea as in the above example. Given some transition type (i.e. an endofunctor) F, one first defines an embedding of F-systems with internal transitions into F^*-system, where F^* has been derived from F by making explicit the algebraic structure on the labels. Next one models the semantics of an F-system as the one of the corresponding F^*-system e. Naively, one could think of defining the semantics of e as the unique map $!_e$ into the final coalgebra for F^*. However, this approach turns out to be too fine grained, essentially because it ignores the underlying algebraic structure on the labels of e. The same problem can be observed in the example above: $\mathbb{B}$ and the representation of $\mathbb{A}$ as an automaton with languages as labels have different final semantics—they accept the same language only modulo the equations of monoids.

Thus we need to extend the standard coalgebraic framework by taking into account the algebraic structure on labels. To this end, we develop our theory for systems whose transition type F^* has a *canonical fixpoint*, i.e. its initial algebra and final coalgebra coincide. This is the case for many relevant examples, as observed in [12]. Our *canonical fixpoint semantics* will be given as the composite $¡ \circ !_e$, where $!_e$ is a coalgebra morphism given by finality and $¡$ is an algebra morphism given by initiality. The target of $¡$ will be an algebra for F^* encoding the equational theory associated with the labels of F^*-systems. Intuitively, $¡$ being an *algebra* morphism, will take the quotient of the semantics given by $!_e$

modulo those equations. Therefore the extension provided by ¡ is the technical feature allowing us to take into account the algebraic structure on labels.

To study the properties of our canonical fixpoint semantics, it will be convenient to formulate it as an operator $e \mapsto e^{\dagger}$ assigning to systems (seen as sets of equations) a certain *solution*. Within the same perspective we will implement a different kind of solution $e \mapsto e^{\ddagger}$ turning any system e with internal transitions into one $e^{\ddagger}$ where those have been abstracted away. By comparing the operators $e \mapsto e^{\dagger}$ and $e \mapsto e^{\ddagger}$, we will then be able to show that such a procedure (also called ϵ-*elimination*) is sound with respect to the canonical fixpoint semantics.

To conclude, we will explore further the flexibility of our framework. In particular, we will model the case in which the algebraic structure of the labels is quotiented under some equations, resulting in a coarser equivalence than the one given by the canonical fixpoint semantics. As a relevant example of this phenomenon, we give the first coalgebraic account of Mazurkiewicz traces.

Synopsis. After recalling the necessary background in Sect. 2, we discuss our motivating examples—automata with ϵ-transitions and automata on words—in Sect. 3. Section 4 is devoted to present the canonical fixpoint semantics and the sound procedure of ϵ-elimination. This framework is then instantiated to the examples of Sect. 3. Finally, in Sect. 5 we show how a quotient of the algebra on labels induces a coarser canonical fixpoint semantics. We propose Mazurkiewicz traces as a motivating example for such a construction. A full version of this paper with all proofs and extra material can be found in http://arxiv.org/abs/1402.4062.

2 Preliminaries

In this section we introduce the basic notions we need for our abstract framework. We assume some familiarity with category theory. We will use boldface capitals $\mathbf{C}$ to denote categories, $X, Y, \ldots$ for objects and $f, g, \ldots$ for morphisms. We use Greek letters and double arrows, e.g. $\eta \colon F \Rightarrow G$, for natural transformations, monad morphisms and any kind of 2-cells. If $\mathbf{C}$ has coproducts we will denote them by $X + Y$ and use inl, inr for the coproduct injections.

2.1 Monads

We recall the basics of the theory of monads, as needed here. For more information, see *e.g.* [18]. A monad is a functor $T \colon \mathbf{C} \to \mathbf{C}$ together with two natural transformations, a *unit* $\eta \colon \mathrm{id}_{\mathbf{C}} \Rightarrow T$ and a *multiplication* $\mu \colon T^2 \Rightarrow T$, which are required to satisfy the following equations, for every $X \in \mathbf{C}$: $\mu_X \circ \eta_{TX} = \mu_X \circ T\eta_X = \mathrm{id}$ and $T\mu_X \circ \mu_{TX} = \mu_X \circ \mu_X$.

A *morphism of monads* from (T, η^T, μ^T) to (S, η^S, μ^S) is a natural transformation $\gamma \colon T \Rightarrow S$ that preserves unit and multiplication: $\gamma_X \circ \eta_X^T = \eta_X^S$ and $\gamma_X \circ \mu_X^T = \mu_X^S \circ \gamma_{SX} \circ T\gamma_X$. A *quotient of monads* is a morphism of monads with epimorphic components.

Example 2.1. We briefly describe the examples of monads that we use in this paper.

1. Let $\mathbf{C} = \mathbf{Sets}$. The powerset monad $\mathcal{P}$ maps a set X to the set $\mathcal{P}X$ of subsets of X, and a function $f\colon X \to Y$ to $\mathcal{P}f\colon \mathcal{P}X \to \mathcal{P}Y$ given by direct image. The unit is given by the singleton set map $\eta_X(x) = \{x\}$ and multiplication by union $\mu_X(U) = \bigcup_{S \in U} S$.

2. Let $\mathbf{C}$ be a category with coproducts and E an object of $\mathbf{C}$. The exception monad $\mathcal{E}$ is defined on objects as $\mathcal{E}X = E + X$ and on arrows $f\colon X \to Y$ as $\mathcal{E}f = \mathrm{Id}_E + f$. Its unit and multiplication are given on $X \in \mathbf{C}$ respectively as $\mathrm{inr}_X\colon X \to E + X$ and $\nabla_E + \mathrm{Id}_X\colon E + E + X \to E + X$, where $\nabla_E = [\mathrm{id}_E, \mathrm{id}_E]$ is the codiagonal. When $\mathbf{C} = \mathbf{Sets}$, E can be thought as a set of *exceptions* and this monad is often used to encode computations that might fail throwing an exception chosen from the set E.

3. Let H be an endofunctor on a category $\mathbf{C}$ such that for every object X there exists a free H-algebra H^*X on X (equivalently, an initial $H + X$-algebra) with the structure $\tau_X \colon HH^*X \to H^*X$ and universal morphism $\eta_X \colon X \to H^*X$. Then as proved by Barr [5] (see also Kelly [16]) $H^*\colon \mathbf{C} \to \mathbf{C}$ is the functor part of a *free monad* on H with the unit given by the above η_X and the multiplication given by the freeness of H^*H^*X: μ_X is the unique H-algebra homomorphism from (H^*H^*X, τ_{H^*X}) to (H^*X, τ_X) such that $\mu_X \cdot \eta_{H^*X} = \eta_X$. Also notice that for a complete category every free monad arises in this way. Finally, for later use we fix the notation $\kappa = \tau \cdot H\eta \colon H \Rightarrow H^*$ for the universal natural transformation of the free monad.

Given a monad $M\colon \mathbf{C} \to \mathbf{C}$, its *Kleisli category* $\mathcal{K}\ell(M)$ has the same objects as $\mathbf{C}$, but morphisms $X \to Y$ in $\mathcal{K}\ell(M)$ are morphisms $X \to MY$ in $\mathbf{C}$. The identity map $X \to X$ in $\mathcal{K}\ell(M)$ is M's unit $\eta_X\colon X \to MX$; and composition $g \circ f$ in $\mathcal{K}\ell(M)$ uses M's multiplication: $g \circ f = \mu \circ Mg \circ f$. There is a forgetful functor $\mathcal{U}\colon \mathcal{K}\ell(T) \to \mathbf{C}$, sending X to TX and f to $\mu \circ Tf$. This functor has a left adjoint $\mathcal{J}$, given by $\mathcal{J}X = X$ and $\mathcal{J}f = \eta \circ f$. The Kleisli category $\mathcal{K}\ell(M)$ inherits coproducts from the underlying category $\mathbf{C}$. More precisely, for every objects X and Y their coproduct $X + Y$ in $\mathbf{C}$ is also a coproduct in $\mathcal{K}\ell(M)$ with the injections $\mathcal{J}\mathrm{inl}$ and $\mathcal{J}\mathrm{inr}$.

2.2 Distributive Laws and Liftings

The most interesting examples of the theory that we will present in Sect. 4 concern coalgebras for functors $\widehat{H}\colon \mathcal{K}\ell(M) \to \mathcal{K}\ell(M)$ that are obtained as liftings of endofunctors H on $\mathbf{Sets}$. Formally, given a monad $M\colon \mathbf{C} \to \mathbf{C}$, a *lifting* of $H\colon \mathbf{C} \to \mathbf{C}$ to $\mathcal{K}\ell(M)$ is an endofunctor $\widehat{H}\colon \mathcal{K}\ell(M) \to \mathcal{K}\ell(M)$ such that $\mathcal{J} \circ H = \widehat{H} \circ \mathcal{J}$. The lifting of a monad (T, η, μ) is a monad $(\widehat{T}, \widehat{\eta}, \widehat{\mu})$ such that $\widehat{T}$ is a lifting of T and $\widehat{\eta}, \widehat{\mu}$ are given on $X \in \mathcal{K}\ell(M)$ (i.e. $X \in \mathbf{Sets}$) respectively as $\mathcal{J}(\eta_X)$ and $\mathcal{J}(\mu_X)$.

A natural way of lifting functors and monads is by mean of distributive laws. A *distributive law* of a monad (T, η^T, μ^T) over a monad (M, η^M, μ^M) is a natural transformation $\lambda\colon TM \Rightarrow MT$, that commutes appropriately with the unit and multiplication of both monads; more precisely, the diagrams below commute:

$$
\begin{array}{ccc}
TX = \!\!= \!\!= TX & & TM^2X \xrightarrow{\lambda_{MX}} MTMX \xrightarrow{M\lambda_X} M^2TX \\
T\eta_X^M \downarrow \qquad \downarrow \eta_{TX}^M & & T\mu_X^M \downarrow \qquad\qquad\qquad \downarrow \mu_{TX}^M \\
TMX \xrightarrow{\lambda_X} MTX & & TMX \xrightarrow{\qquad \lambda_X \qquad} MTX \\
\eta_{MX}^T \uparrow \qquad \uparrow M\eta_X^T & & \mu_{MX}^T \uparrow \qquad\qquad\qquad \uparrow M\mu_X^T \\
MX = \!\!= \!\!= MX & & T^2MX \xrightarrow{T\lambda_{MX}} TMTX \xrightarrow{\lambda_{TX}} T^2MX
\end{array}
$$

A distributive law of a *functor* T over a *monad* (M, η^M, μ^M) is a natural transformation $\lambda\colon TM \Rightarrow MT$ such that only the two topmost squares above commute.

The following "folklore" result gives an alternative description of distributive laws in terms of liftings to Kleisli categories, see also [15,20] or [4].

Proposition 2.2 ([20]). *Let (M, η^M, μ^M) be a monad on a category $\mathbf{C}$. Then the following holds:*

1. *For every endofunctor T on $\mathbf{C}$, there is a bijective correspondence between liftings of T to $\mathcal{K}\ell(M)$ and distributive laws of T over M.*
2. *For every monad (T, η^T, μ^T) on $\mathbf{C}$, there is a bijective correspondence between liftings of (T, η^T, μ^T) to $\mathcal{K}\ell(M)$ and distributive laws of T over M.*

In what follows we shall simply write $\widehat{H}$ for the lifting of an endofunctor H.

Proposition 2.3 ([12]). *Let $M\colon \mathbf{C} \to \mathbf{C}$ be a monad and $H\colon \mathbf{C} \to \mathbf{C}$ be a functor with a lifting $\widehat{H}\colon \mathcal{K}\ell(M) \to \mathcal{K}\ell(M)$. If H has an initial algebra $\iota\colon HI \xrightarrow{\cong} I$ (in $\mathbf{C}$), then $\mathcal{J}\iota\colon \widehat{H}I \to I$ is an initial algebra for $\widehat{H}$ (in $\mathcal{K}\ell(M)$).*

In our examples, we will often consider the free monad (Example 2.1.3) $\widehat{H}^*$ generated by a lifted functor $\widehat{H}$. The following result will be pivotal.

Proposition 2.4. *Let $H\colon \mathbf{C} \to \mathbf{C}$ be a functor and $M\colon \mathbf{C} \to \mathbf{C}$ be a monad such that there is a lifting $\widehat{H}\colon \mathcal{K}\ell(M) \to \mathcal{K}\ell(M)$. Then the free monad $H^*\colon \mathbf{C} \to \mathbf{C}$ lifts to a monad $\widehat{H^*}\colon \mathcal{K}\ell(M) \to \mathcal{K}\ell(M)$. Moreover, $\widehat{H^*} = \widehat{H}^*$.*

Recall from [12] that for every polynomial endofunctor H on **Sets** there exists a canonical distributive law of H over any *commutative* monad M (equivalently, a canonical lifting of H to $\mathcal{K}\ell(M)$); this result was later extended to so-called analytic endofunctors of **Sets** (see [19]). This can be used in our applications since the power-set functor $\mathcal{P}$ is commutative, and so is the exception monad $\mathcal{E}$ iff $E = 1$.

2.3 Cppo-enriched Categories

For our general theory we are going to assume that we work in a category where the hom-sets carry a cpo structure. Recall that a *cpo* is a partially ordered set in which all ω-chains have a join. A cpo with bottom is a cpo with a least element $\bot$. A function between cpos is called *continuous* if it preserves joins of ω-chains. Cpos with bottom and continuous maps form a category that we denote by **Cppo**.

A **Cppo**-*enriched category* **C** is a category where (a) each hom-set $\mathbf{C}(X,Y)$ is a cpo with a bottom element $\bot_{X,Y} \colon X \to Y$ and (b) composition is continuous, that is:

$$g \circ \left(\bigsqcup_{n<\omega} f_n \right) = \bigsqcup_{n<\omega} (g \circ f_n) \qquad \text{and} \qquad \left(\bigsqcup_{n<\omega} f_n \right) \circ g = \bigsqcup_{n<\omega} (f_n \circ g).$$

The composition is called *left strict* if $\bot_{Y,Z} \circ f = \bot_{X,Z}$ for all arrows $f \colon X \to Y$.

In our applications, **C** will mostly be a Kleisli category for a monad on **Sets**. Throughout this subsection we assume that **C** is a **Cppo**-enriched category.

An endofunctor $H \colon \mathbf{C} \to \mathbf{C}$ is said to be *locally continuous* if for any ω-chain $f_n \colon X \to Y$, $n < \omega$ in $\mathbf{C}(X,Y)$ we have:

$$H \left(\bigsqcup_{n<\omega} f_n \right) = \bigsqcup_{n<\omega} H(f_n).$$

We are going to make use of the fact that a locally continuous endofunctor H on **C** has a *canonical fixpoint*, i.e. whenever its initial algebra exists it is also its final coalgebra:

Theorem 2.5 ([9]). *Let* $H \colon \mathbf{C} \to \mathbf{C}$ *be a locally continuous endofunctor on the* **Cppo**-*enriched category* **C** *whose composition is left-strict. If an initial* H-*algebra* $\iota \colon HI \xrightarrow{\cong} I$ *exists, then* $\iota^{-1} \colon I \xrightarrow{\cong} HI$ *is a final* H-*coalgebra.*

In the sequel, we will be interested in free algebras for a functor H on **C** and the free monad H^* (cf. Example 2.1.3). For this observe that coproducts in **C** are always **Cppo**-enriched, i.e. all copairing maps $[-, -] \colon \mathbf{C}(X,Y) \times \mathbf{C}(X',Y) \to \mathbf{C}(X + X', Y)$ are continuous; in fact, it is easy to show that this map is continuous in both of its arguments using that composition with the coproduct injections is continuous.

Proposition 2.6. *Let* **C** *be* **Cppo**-*enriched with composition left-strict. Furthermore, let* $H \colon \mathbf{C} \to \mathbf{C}$ *be locally continuous and assume that all free* H-*algebras exist. Then the free monad* H^* *is locally continuous.*

2.4   Final Coalgebras in Kleisli Categories

In our applications the **Cppo**-enriched category will be the Kleisli category $\mathbf{C} = \mathcal{K}\ell(M)$ of a monad on **Sets** and the endofunctors of interest are liftings $\widehat{H}$ of endofunctors H on **Sets**. It is known that in this setting a final coalgebra for the lifting $\widehat{H}$ can be obtained as a lifting of an initial H-algebra (see Hasuo et al. [12]). The following result is a variation of Theorem 3.3 in [12]:

Theorem 2.7. *Let* $M \colon \mathbf{Sets} \to \mathbf{Sets}$ *be a monad and* $H \colon \mathbf{Sets} \to \mathbf{Sets}$ *be a functor such that*

(a) $\mathcal{K}\ell(M)$ *is* **Cppo**-*enriched with composition left strict;*
(b) H *is accessible (i.e., H preserves λ-filtered colimits for some cardinal λ) and has a lifting $\widehat{H}\colon \mathcal{K}\ell(M) \to \mathcal{K}\ell(M)$ which is locally continuous.*

If $\iota\colon HI \overset{\cong}{\Rightarrow} I$ is the initial algebra for the functor H, then

1. *$\widehat{\jmath\iota}\colon \widehat{H}I \to I$ is the initial algebra for the functor $\widehat{H}$;*
2. *$\widehat{\jmath\iota}^{-1}\colon I \to \widehat{H}I$ is the final coalgebra for the functor $\widehat{H}$.*

The first item follows from Proposition 2.3 and the second one follows from Theorem 2.5. There are two differences with Theorem 3.3 in [12]:

(1) The functor $H\colon$ **Sets** $\to$ **Sets** is supposed to preserve ω-colimits rather that being accessible. We use the assumption of accessibility because it guarantees the existence of all free algebras for H and for $\widehat{H}$, which implies also that for all $Y \in \mathcal{K}\ell(M)$ an initial $\widehat{H}^*(\mathrm{Id}+Y)$-algebra exists. This property of $\widehat{H}^*$ will be needed for applying our framework of Sect. 4.
(2) We assume that the lifting $\widehat{H}\colon \mathcal{K}\ell(M) \to \mathcal{K}\ell(M)$ is locally continuous rather than locally monotone. We will need continuity to ensure the double dagger law in Remark 2.9. This assumption is not really restrictive since, as explained in Sect. 3.3.1 of [12], in all the meaningful examples where $\widehat{H}$ is locally monotone, it is also locally continuous.

Example 2.8 (NDA). Consider the powerset monad $\mathcal{P}$ (Example 2.1.1) and the functor $HX = A \times X + 1$ on **Sets** (with $1 = \{\checkmark\}$). The functor H lifts to $\widehat{H}$ on $\mathcal{K}\ell(\mathcal{P})$ as follows: for any $f\colon X \to Y$ in $\mathcal{K}\ell(\mathcal{P})$ (that is $f\colon X \to \mathcal{P}(Y)$ in **Sets**), $\widehat{H}f\colon A \times X + 1 \to A \times Y + 1$ is given by $\widehat{H}f(\checkmark) = \{\checkmark\}$ and $\widehat{H}f(\langle a, x\rangle) = \{\langle a, y\rangle \mid y \in f(x)\}$.

Non-deterministic automata (NDA) over the input alphabet A can be regarded as coalgebras for the functor $\widehat{H}\colon \mathcal{K}\ell(\mathcal{P}) \to \mathcal{K}\ell(\mathcal{P})$. Consider, on the left, a 3-state NDA, where the only final state is marked by a double circle.

$$X = \{1, 2, 3\} \quad A = \{a, b\}$$
$$e(1) = \{\langle a, 1\rangle, \langle b, 1\rangle, \langle b, 2\rangle\}$$
$$e(2) = \{\langle a, 2\rangle, \langle b, 3\rangle\} \quad e(3) = \{\checkmark, \langle a, 2\rangle, \langle b, 3\rangle\}$$

It can be represented as a coalgebra $e\colon X \to \widehat{H}X$, that is a function $e\colon X \to \mathcal{P}(A \times X + 1)$, given above on the right, which assigns to each state $x \in X$ a set which: contains $\checkmark$ if x is final; and $\langle a, y\rangle$ for all transitions $x \overset{a}{\to} y$.

It is easy to see that $M = \mathcal{P}$ and H above satisfy the conditions of Theorem 2.7 and therefore both the final $\widehat{H}$-coalgebra and the initial $\widehat{H}$-algebra are the lifting of the initial algebra for the functor $HX = A \times X + 1$, given by A^* with structure $\iota\colon A \times A^* + 1 \to A^*$ which maps $\langle a, w\rangle$ to aw and $\checkmark$ to ϵ.

For an NDA (X, e), the final coalgebra homomorphism $!_e\colon X \to A^*$ is the function $X \to \mathcal{P}A^*$ that maps every state in X to the language that it accepts. In $\mathcal{K}\ell(\mathcal{P})$:

$$X \dashrightarrow^{!_e} A^*$$

$$\epsilon \in !_e(x) \Leftrightarrow \checkmark \in e(x)$$
$$aw \in !_e(x) \Leftrightarrow \text{ for some } y \in X,\ (a, y) \in e(x) \text{ and } w \in !_e(y)$$

$$A \times X + 1 \xrightarrow{\quad A \times !_e + 1 \quad} A \times A^* + 1$$

with e on the left vertical, $\partial \iota^{-1}$ on the right vertical.

2.5   Monads with Fixpoint Operators

In order to develop our theory of systems with internal behavior, we will adopt an equational perspective on coalgebras. In the sequel we recall some preliminaries on this viewpoint.

Let $T : \mathbf{C} \to \mathbf{C}$ be a monad on any category $\mathbf{C}$. Any morphism $e : X \to T(X + Y)$ (i.e. a coalgebra for the functor $T(\mathrm{Id} + Y)$) may be understood as a system of mutually recursive equations. In our applications we are interested in the case where $\mathbf{C} = \mathcal{K}\ell(M)$ and $T = \widehat{H}^*$ is a (lifted) free monad. As in the example of NDA (Example 2.8) take $M = \mathcal{P}$ and $HX = 1 + A \times X$. Now, set $TX = A^* + A^* \times X$ and consider the following system of mutually recursive equations

$$x_0 \approx \{c, (ab, x_1)\}, \qquad x_1 \approx \{d, (a, x_0), (\epsilon, y)\},$$

where $x_0, x_1 \in X$ are *recursion variables*, $y \in Y$ is a *parameter* and $a, b, c, d \in A$. A *solution* assigns to each of the two variables x_0, x_1 an element of $\mathcal{P}(TY)$ such that the formal equations $\approx$ become actual identities in $\mathcal{K}\ell(\mathcal{P})$:

$$x_0 \mapsto \{(aba)^*c, (aba)^*abd, ((aba)^*ab, y)\}, \quad x_1 \mapsto \{(aab)^*d, (aab)^*ac, ((aab)^*, y)\}.$$

Observe that the above system of equations corresponds to an *equation morphism* $e : X \to T(X + Y)$ and the solution to a morphism $e^\dagger : X \to TY$, both in $\mathcal{K}\ell(M)$. The property that $e^\dagger$ is a solution for e is expressed by the following equation in $\mathcal{K}\ell(M)$:

$$e^\dagger = (X \xrightarrow{\ e\ } T(X + Y) \xrightarrow{T[e^\dagger, \eta_Y^T]} TTY \xrightarrow{\mu_Y^T} TY). \tag{1}$$

So $e \mapsto e^\dagger$ is a *parametrized fixpoint operator*, i.e. a family of fixpoint operators indexed by parameter sets Y.

Remark 2.9. In our applications we shall need a certain equational property of the operator $e \mapsto e^\dagger$: for all $Y \in \mathbf{C}$ and equation morphism $e : X \to T(X + X + Y)$, the following equation, called *double dagger law*, holds:

$$e^{\dagger\dagger} = (X \xrightarrow{\ e\ } T(X + X + Y) \xrightarrow{T(\nabla_X + Y)} T(X + Y))^\dagger.$$

This and other laws of parametrized fixpoint operators have been studied by Bloom and Ésik in the context of *iteration theories* [6]. A closely related notion is that of *Elgot monads* [1,2].

Example 2.10 (Least fixpoint solutions). Let $T : \mathbf{C} \to \mathbf{C}$ be a locally continuous monad on the **Cppo**-enriched category $\mathbf{C}$. Then T is equipped with a parametrized fixpoint operator obtained by taking least fixpoints: given a morphism $e : X \to T(X + Y)$ consider the function Φ_e on $\mathbf{C}(X, TY)$ given by $\Phi_e(s) = \mu_Y^T \circ T[s, \eta_Y^T] \circ e$. Then Φ_e is continuous and we take $e^\dagger$ to be the least fixpoint of Φ_e. Since $e^\dagger = \Phi_e(e^\dagger)$, Eq. (1) holds, and it follows from the argument in Theorem 8.2.15 and Exercise 8.2.17 in [6] that the operator $e \mapsto e^\dagger$ satisfies the axioms of iteration theories (or Elgot monads, respectively). In particular the double dagger law holds for the least fixpoint operator $e \mapsto e^\dagger$.

3   Motivating Examples

The work of [12] bridged a gap in the theory of coalgebras: for certain functors, taking the final coalgebra directly in **Sets** does not give the right notion of equivalence. For instance, for NDA, one would obtain bisimilarity instead of language equivalence. The change to Kleisli categories allowed the recovery of the usual language semantics for NDA and, more generally, led to the development of *coalgebraic trace semantics*.

In the Introduction we argued that there are relevant examples for which this approach still yields the unwanted notion of equivalence, the problem being that it does not consider the extra algebraic structure on the label set. In the sequel, we motivate the reader for the generic theory we will develop by detailing two case studies in which this phenomenon can be observed: NDA with ϵ-transitions and NDA with word transitions. Later on, in Example 5.7, we will also consider Mazurkiewicz traces [17].

NDA with ϵ-transition. In the world of automata, ϵ-transitions are considered in order to enable easy composition of automata and compact representations of languages. These transitions are to be interpreted as the empty word when computing the language accepted by a state. Consider, on the left, the following simple example of an NDA with ϵ-transitions, where states x and y just make ϵ transitions. The intended semantics in this example is that all states accept words in a^*.

$$
\begin{array}{ll}
e(x) = \{(\epsilon, y)\} & !_e(x) = \epsilon\epsilon a^* \\
e(y) = \{(\epsilon, z)\} & !_e(y) = \epsilon a^* \\
e(z) = \{(a, z), \checkmark\} & !_e(z) = a^*
\end{array}
$$

Note that, more explicitly, these are just NDA where the alphabet has a distinguished symbol ϵ. So, they are coalgebras for the functor $\widehat{H + \mathrm{Id}} \colon \mathcal{K}\ell(\mathcal{P}) \to \mathcal{K}\ell(\mathcal{P})$ (where H is the functor of Example 2.8), i.e. functions $e \colon X \to \mathcal{P}((A \times X + 1) + X) \cong \mathcal{P}((A + 1) \times X + 1)$, as made explicit for the above automaton in the middle.

The final coalgebra for $\widehat{H + \mathrm{Id}}$ is simply $(A + 1)^*$ and the final map $!_e \colon X \to (A + 1)^*$ assigns to each state the language in $(A + 1)^*$ that it accepts. However, the equivalence induced by $!_e$ is too fine grained: for the automata above, $!_e$ maps x, y and z to three different languages (on the right), where the number of ϵ plays an explicit role, but the intended semantics should disregard ϵ's.

NDA with word transitions. This is a variation on the motivating example of the introduction: instead of languages, transitions are labeled by words[1]. Formally, consider again the functor H from Example 2.8. Then NDA with word transitions are coalgebras for the functor $\widehat{H^*} \colon \mathcal{K}\ell(\mathcal{P}) \to \mathcal{K}\ell(\mathcal{P})$, that is, functions $e \colon X \to \mathcal{P}(A^* \times X + A^*) \cong \mathcal{P}(A^* \times (X + 1))$. We observe that they are like NDA but (1) transitions are labeled by words in A^*, rather than just symbols of the alphabet A, and (2) states have associated output languages, rather than just $\checkmark$. We will draw them as ordinary automata plus an arrow $\overset{L}{\Rightarrow}$ to denote the output language of a state (no $\Rightarrow$ stands for the empty language). For an example, consider the following word automaton and associated transition function e.

$$e(x) = \{(a,y)\} \quad e(y) = \{(b,z)\} \quad e(z) = \{c\}$$
$$e(u) = \{(\epsilon, v)\} \quad e(v) = \{(ab, z)\}$$

The semantics of NDA with word transitions is given by languages over A, obtained by concatenating the words in the transitions and ending with a word from the output language. For instance, x above accepts word abc but not ab.

However, if we consider the final coalgebra semantics we again have a mismatch. The initial H^*-algebra has carrier $(A^*)^* \times A^*$ that can be represented as the set of non-empty lists of words over A^*, where $(A^*)^*$ indicates possibly empty lists of words. Its structure $\iota \colon A^* \times ((A^*)^* \times A^*) + A^* \to (A^*)^* \times A^*$ maps w into $(\langle \rangle, w)$ and $(w', (l, w))$ into $(w' :: l, w)$. Here, we use $\langle \rangle$ to denote the empty list and $::$ is the append operation. By Theorem 2.7, the final $\widehat{H^*}$-coalgebra has the same carrier and structure $\mathcal{J}\iota^{-1}$. The final map, as a function $!_e \colon X \to \mathcal{P}((A^*)^* \times A^*)$, is then defined by commutativity of the following square (in $\mathcal{K}\ell(\mathcal{P})$):

$$
\begin{array}{ccc}
X & \dashrightarrow^{!_e} & (A^*)^* \times A^* \\
\downarrow{\scriptstyle e} & \quad (\langle \rangle, w) \in !_e(x) \Leftrightarrow w \in e(x) \quad & \downarrow{\scriptstyle \mathcal{J}\iota^{-1}} \\
& (w :: l, w') \in !_e(x) \Leftrightarrow \exists_y\ (w, y) \in e(x) \text{ and } (l, w') \in !_e(y). & \\
A^* \times X + A^* & \dashrightarrow[\text{id}_{A^*} \times !_e + \text{id}_{A^*}] & A^* \times ((A^*)^* \times A^*) + A^*
\end{array}
\tag{2}
$$

Once more, the semantics given by $!_e$ is too fine grained: in the above example, $!_e(x) = \{([a,b], c)\}$ and $!_e(u) = \{([\epsilon, ab], c)\}$ whereas the intended semantics would equate both x and u, since they both accept the language $\{abc\}$.

Note that any NDA can be regarded as word automaton. Recall the natural transformation $\kappa \colon \widehat{H} \Rightarrow \widehat{H^*}$ defined in Example 2.1.3: for the functor $\widehat{H}$ of NDA,

$$\kappa_X \colon A \times X + 1 \to A^* \times X + A^*$$

---

[1] More generally, one could consider labels from an arbitrary monoid.

maps any pair $(a, x) \in A \times X$ into $\{(a, x)\} \in \mathcal{P}(A^* \times X + A^*)$ and $\checkmark \in 1$ into $\{\epsilon\} \in \mathcal{P}(A^* \times X + A^*)$. Composing an NDA $e \colon X \to \widehat{H}X$ with $\kappa_X \colon \widehat{H}X \to \widehat{H^*}X$, one obtains the word automaton $\kappa_X \circ e$.

In the same way, every NDA with ϵ-transitions can also be seen as a word automaton by postcomposing with the natural transformation $[\kappa, \eta] \colon \widehat{H + \mathrm{Id}} \Rightarrow \widehat{H^*}$. Here, $\eta \colon \mathrm{Id} \Rightarrow \widehat{H^*}$ is the unit of the free monad $\widehat{H^*}$ defined on a given set X below (the multiplication $\mu \colon \widehat{H^*}\widehat{H^*} \Rightarrow \widehat{H^*}$ is shown on the right).

$$\eta_X \colon X \to A^* \times X + A^* \qquad \mu_X \colon A^* \times ((A^* \times X + A^*) + A^* \to A^* \times X + A^*$$
$$x \mapsto \{(\epsilon, x)\} \qquad (w, (w', x)) \mapsto \{(w \cdot w', x)\} \quad (w, w') \mapsto \{w \cdot w'\}$$
$$w \mapsto \{w\}$$

In the next section, we propose to define the semantics of $\widehat{H^*}$-coalgebras via a canonical fixpoint operator rather than with the final map which as we saw above might yield unwanted semantics. Then, using the observation above, the semantics of $\widehat{H}$-coalgebras and $\widehat{H + \mathrm{Id}}$-coalgebras will be defined by embedding them into $\widehat{H^*}$-coalgebras via the natural transformations κ and $[\kappa, \eta]$ described above.

4   Canonical Fixpoint Solutions

In this section we lay the foundations of our approach. A construction is introduced assigning canonical solutions to coalgebras seen as equation morphisms (*cf.* Sect. 2.5) in a **Cppo**-enriched setting. We will be working under the following assumptions.

Assumption 4.1. Let **C** be a **Cppo**-enriched category with coproducts and composition left-strict. Let T be a locally continuous monad on **C** such that, for all object Y, an initial algebra for $T(\mathrm{Id} + Y)$ exists.

As seen in Example 2.10, in this setting an equation morphism $e \colon X \to T(X + Y)$ may be given the least solution. Here, we take a different approach, exploiting the initial algebra-final coalgebra coincidence of Theorem 2.5.

For every parameter object $Y \in \mathbf{C}$, the endofunctor $T(\mathrm{Id} + Y)$ is a locally continuous monad because it is the composition of T with the (locally continuous) exception monad $\mathrm{Id} + Y$. Thus, by Theorem 2.5 applied to $T(\mathrm{Id} + Y)$, the initial $T(\mathrm{Id}+Y)$-algebra $\iota_Y \colon T(I_Y + Y) \xrightarrow{\cong} I_Y$ yields a final $T(\mathrm{Id}+Y)$-coalgebra $\iota_Y^{-1} \colon I_Y \xrightarrow{\cong} T(I_Y + Y)$. This allows us to associate with any equation morphism $e \colon X \to T(X + Y)$ a canonical morphism of type $X \to TY$ as in the following diagram.

$$
\begin{array}{c}
X \dashrightarrow^{!_e} I_Y \dashrightarrow^{i} TY \\
\end{array}
\tag{3}
$$

In (3), the map $!_e \colon X \to I_Y$ is the unique morphism of $T(\mathrm{Id} + Y)$-coalgebras given by finality of $\iota_Y^{-1} \colon I_Y \to T(I_Y + Y)$, whereas $\mathbf{i} \colon I_Y \to TY$ is the unique morphism of $T(\mathrm{Id} + Y)$-algebras given by initiality of $\iota_Y \colon T(I_Y + Y) \to I_Y$.

We call the composite $\mathbf{i} \circ !_e \colon X \to TY$ the *canonical fixpoint solution* of e. In the following we check that the canonical fixpoint solution is indeed a solution of e, in fact, it coincides with the least solution.

Proposition 4.2. *Given a morphism $e \colon X \to T(X + Y)$, then the least solution of e as in Example 2.10 is the canonical fixpoint solution: $e^\dagger = \mathbf{i} \circ !_e \colon X \to TY$ as in (3).*

As recalled in Example 2.10, the least fixpoint operator $e \mapsto e^\dagger$ satisfies the double dagger law. Thus Proposition 4.2 yields the following result[2].

Corollary 4.3. *Let $\mathbf{C}$ and $T \colon \mathbf{C} \to \mathbf{C}$ be as in Assumption 4.1. Then the canonical fixpoint operator $e \mapsto e^\dagger$ associated with T satisfies the double dagger law.*

We now introduce a factorisation result on the operator $e \mapsto e^\dagger$, which is useful for comparing solutions provided by different monads connected via a monad morphism.

Proposition 4.4 (Factorisation Lemma). *Suppose that T and T' are monads on $\mathbf{C}$ satisfying Assumption 4.1 and $\gamma \colon T \Rightarrow T'$ is a monad morphism. For any morphism $e \colon X \to T(X + Y)$:*

$$\gamma_Y \circ e^\dagger = (\gamma_{X+Y} \circ e)^\dagger \colon X \to T'Y,$$

where $e^\dagger$ is provided by the canonical fixpoint solution for T and $(\gamma_{X+Y} \circ e)^\dagger$ by the one for T'.

4.1   A Theory of Systems with Internal Behavior

We now use canonical fixpoint solutions to provide an abstract theory of systems with internal behavior, that we will later instantiate to the motivating examples of Sect. 3. Throughout this section, we will develop our framework for the following ingredients.

Assumption 4.5. Let $\mathbf{C}$ be a **Cppo**-enriched category with coproducts and composition left-strict and let $F \colon \mathbf{C} \to \mathbf{C}$ be a locally continuous functor for which all free F-algebras exist. Consider the following two monads derived from F:

– the free monad $F^* \colon \mathbf{C} \to \mathbf{C}$ (*cf.* Example 2.1.3), for which we suppose that an initial $F^*(\mathrm{Id} + Y)$-algebra exists for all $Y \in \mathbf{C}$;

---

[2] The equality of least and canonical fixpoint solutions can be used to state a stronger result, namely that canonical fixpoint solutions satisfy the axioms of iteration theories (*cf.* Example 2.10). However, the double dagger law is the only property that we need here, explaining the statement of Corollary 4.3.

- for a fixed $X \in \mathbf{C}$, the exception monad $FX + \mathrm{Id} \colon \mathbf{C} \to \mathbf{C}$ (*cf.* Example 2.1.2), for which we suppose that an initial $FX + \mathrm{Id} + Y$-algebra exists for all $Y \in \mathbf{C}$.

In the next proposition we verify that the construction introduced in the previous section applies to the two monads of Assumption 4.5.

Proposition 4.6. *Let* $\mathbf{C}$, F, F^* *and* $FX + \mathrm{Id}$ *be as in Assumption 4.5. Then* $\mathbf{C}$ *and the monads* $F^* \colon \mathbf{C} \to \mathbf{C}$ *and* $FX + \mathrm{Id} \colon \mathbf{C} \to \mathbf{C}$ *satisfy Assumption 4.1. Thus both* F^* *and* $FX + \mathrm{Id}$ *are monads with canonical fixpoint solution (which satisfy the double dagger law by Corollary 4.3).*

To avoid ambiguity, we denote with $e \mapsto e^{\dagger}$ the canonical fixpoint operator associated with F^* and with $e \mapsto e^{\ddagger}$ the one associated with $FX + \mathrm{Id}$.

We will employ the additional structure of those two monads for the analysis of *F-systems with internal transitions*. An *F-system* is simply an *F-coalgebra* $e \colon X \to FX$, where we take the operational point of view of seeing X as a space of states and F as the transition type of e. An *F-system with internal transitions* is an $(F + \mathrm{Id})$-coalgebra $e \colon X \to FX + X$, where the component X of the codomain is targeted by those transitions representing the internal (non-interacting) behavior of system e.

A key observation for our analysis is that *F-systems*—with or without internal transitions—enjoy a standard representation as *F*-systems*, that is, coalgebras of the form $e \colon X \to F^*X$.

Definition 4.7 (*F-systems as F*-systems*). *Let* $\kappa \colon F \to F^*$ *be as in Example 2.1.3. We introduce the following encoding* $e \mapsto \bar{e}$ *of F-systems and F-systems with internal transitions as F*-systems.*

- *Given an F-system* $e \colon X \to FX$, *define* $\bar{e} \colon X \to F^*X$ *as*

$$\bar{e} \colon X \xrightarrow{\ e\ } FX \xrightarrow{\ \kappa_X\ } F^*X.$$

- *Given an F-system with internal transitions* $e \colon X \to FX + X$, *define* $\bar{e} \colon X \to F^*X$ *as* $\bar{e} \colon X \xrightarrow{\ e\ } FX + X \xrightarrow{\ [\kappa_X, \eta_X^{F^*}]\ } F^*X$.

Thus *F-systems* (with or without internal transitions) may be seen as equation morphisms $X \to F^*(X + 0)$ for the monad F^* (with the initial object $Y = 0$ as parameter), with solutions by canonical fixpoint (*cf.* Sect. 2.5). This will allow us to achieve the following.

§1 We supply a uniform trace semantics for *F-systems*, possibly with internal transitions, and *F*-systems*, based on the canonical fixpoint solution operator of F^*.

§2 We use the canonical fixpoint operator of $FX + \mathrm{Id}$ to transform any *F-system* $e \colon X \to FX + X$ with internal transitions into an *F-system* $e \backslash \epsilon \colon X \to FX$ without internal transitions.

§3 We prove that the transformation of §2 is sound with respect to the semantics of §1.

§1: **Uniform trace semantics.** The canonical fixpoint semantics of F-systems, with or without internal transitions, and F^*-systems is defined as follows.

Definition 4.8 (Canonical Fixpoint Semantics).

- For an F^*-system $e\colon X \to F^*X$, its semantics $[\![e]\!]\colon X \to F^*0$ is defined as $e^\dagger$ (note that e can be seen as an equation morphism for F^* on parameter $Y = 0$).
- For an F-system $e\colon X \to FX$, its semantics $[\![e]\!]\colon X \to F0$ is defined as $\bar{e}^\dagger = (\kappa_X \circ e)^\dagger$.
- For an F-system with internal transitions $e\colon X \to FX + X$, its semantics $[\![e]\!]\colon X \to F0$ is defined as $\bar{e}^\dagger = ([\kappa_X, \eta_X^{F^*}] \circ e)^\dagger$.

The underlying intuition of Definition 4.8 is that canonical fixpoint solutions may be given an operational understanding. Given an F^*-system $e\colon X \to F^*X$, its solution $e^\dagger\colon X \to F^*0$ is formally defined as the composite $\mathrm{\textsf{¡}} \circ \,!_e$ (cf. (3)): we can see the coalgebra morphism $!_e$ as a map that gives the *behavior* of system e without taking into account the structure of labels and the algebra morphism $\mathrm{\textsf{¡}}$ as evaluating this structure, e.g. flattening words of words, using the initial algebra $\mu_0\colon F^*F^*0 \to F^*0$ for the monad F^*. In particular, the action of $\mathrm{\textsf{¡}}$ is what makes our semantics suitable for modeling "algebraic" operations on internal transitions such as ϵ-elimination, as we will see in concrete instances of our framework.

Remark 4.9. The canonical fixpoint semantics of Definition 4.8 encompasses the framework for traces in [12], where the semantics of an F-system $e\colon X \to FX$— without internal transitions—is defined as the unique morphism $!_e$ from X into the final F-coalgebra F^*0. Indeed, using finality of F^*0, it can be shown that $!_e = [\![e]\!]$. Theorem 4.10 below guarantees compatibility with Assumption 4.5.

The following result is instrumental in our examples and in comparing our theory with the one developed in [12] for trace semantics in Kleisli categories.

Theorem 4.10. *Let $M\colon \textbf{Sets} \to \textbf{Sets}$ be a monad and $H\colon \textbf{Sets} \to \textbf{Sets}$ be a functor satisfying the assumptions of Theorem 2.7, that is:*

(a) $\mathcal{K}\ell(M)$ is **Cppo**-*enriched and composition is left strict;*
(b) H is accessible and has a locally continuous lifting $\widehat{H}\colon \mathcal{K}\ell(M) \to \mathcal{K}\ell(M)$.

Then $\mathcal{K}\ell(M)$, $\widehat{H}$, $\widehat{H}^*$ and $\widehat{H}\mathcal{J}X + \mathrm{Id}$ (for a given set X) satisfy Assumption 4.5.

Example 4.11 (Semantics of NDA with word transitions). In Sect. 3, we have modeled NDA with word transitions as $\widehat{H}^*$-coalgebras on $\mathcal{K}\ell(M)$, where H and M are defined as for NDA (see Example 2.8). By Proposition 2.4, $\widehat{H^*} = \widehat{H}^*$ and thus, by virtue of Theorem 4.10, $\widehat{H}^*$ satisfies Assumption 4.5. Therefore we can define the semantics of NDA with word transitions $e\colon X \to \mathcal{P}(A^* \times X + A^*)$ via canonical fixpoint solutions as $[\![e]\!] = e^\dagger = \mathrm{\textsf{¡}} \circ \,!_e$:

$$X \dashrightarrow^{!_e} (A^*)^* \times A^* \dashrightarrow^{i} A^*$$

$$
\begin{array}{ccc}
i(\langle\rangle, w) & = & \{w\} \\
i(w :: l, w') & = & \{wu \mid u \in i(l, w')\}
\end{array}
$$

$$A^* \times X + A^* \xrightarrow[\mathrm{id}\times!_e + \mathrm{id}]{} A^* \times ((A^*)^* \times A^*) + A^* \xrightarrow[\mathrm{id}\times i + \mathrm{id}]{} A^* \times A^* + A^*$$

$$(4)$$

Observe that the above diagram is just (3) instantiated with $T = \widehat{H^*}$ and $Y = 0$. Moreover, this diagram is in $\mathcal{K\ell}(\mathcal{P})$ and hence the explicit definition of $e^\dagger$ as a function $X \to \mathcal{P}(A^*)$ is given by $e^\dagger(x) = \bigcup \mathcal{P}(i)(!_e(x))$.

Both $!_e$ and i can be defined uniquely by the commutativity of the above diagram. We have already defined $!_e$ in diagram (2) and the definition of i is given in the right-hand square of the above diagram. The isomorphism in the middle and μ_0 were defined in Sect. 3.

Using the above formula $e^\dagger(x) = \bigcup \mathcal{P}(i)(!_e(x))$ we now have the semantics of e:

$$w \in e^\dagger(x) \iff w \in e(x) \text{ or} \tag{5}$$

$$\exists_{y \in X, w_1, w_2 \in A^*} (w_1, y) \in e(x), w_2 \in e^\dagger(y) \text{ and } w = w_1 w_2.$$

This definition is precisely the language semantics: a word w is accepted by a state x if there exists a decomposition $w = w_1 \cdots w_n$ such that $x \xrightarrow{w_1} y_1 \xrightarrow{w_2} \cdots \xrightarrow{w_{n-1}} y_{n-1} \xrightarrow{w_n} \cdot$. Take again the automaton of the motivating example. We can calculate the semantics and observe that we now get exactly what was expected: $e^\dagger(u) = e^\dagger(v)$.

$$
\begin{aligned}
!_e(x) &= \{([a, b], c)\} & e^\dagger(x) &= \{abc\} \\
!_e(y) &= \{([b], c)\} & e^\dagger(y) &= \{bc\} \\
!_e(z) &= \{(\langle\rangle, c)\} & e^\dagger(z) &= \{c\} \\
!_e(u) &= \{([\epsilon, ab], c)\} & e^\dagger(u) &= \{abc\} \\
!_e(v) &= \{([ab], c)\} & e^\dagger(v) &= \{abc\}
\end{aligned}
$$

The key role played by the monad structure on A^* can be appreciated by comparing the graphs of $!_e$ and $e^\dagger = i \circ !_e$ as in the example above. The algebra morphism $i \colon (A^*)^* \times A^* \to A^*$ maps values from the initial algebra $(A^*)^* \times A^*$ for the *endofunctor* $\widehat{H^*}$ into the initial algebra A^* for the *monad* $\widehat{H^*}$: its action is precisely to take into account the additional equations encoded by the algebraic theory of the monad $\widehat{H^*}$. For instance, we can see the mapping of $!_e(u) = \{([\epsilon, ab], c)\}$ into the word abc as the result of concatenating the words ϵ, ab, c and then quotienting out of the equation $\epsilon abc = abc$ in the monoid A^*.

Remark 4.12 (Multiple Solutions). The canonical solution $e^\dagger$ is not the unique solution. Indeed, the uniqueness of $!_e$ in the left-hand square and of i in the right-hand square of the diagram above does not imply the uniqueness of $e^\dagger$. To see this, take for instance the automaton

Both $s(x) = \emptyset$ and $s'(x) = A^*$ are solutions. The canonical one is the least one, i.e., $e^\dagger(x) = s(x) = \emptyset$.

Example 4.13 (Semantics of NDA with ϵ-transitions). NDA with ϵ-transitions are modeled as $\widehat{H + \mathrm{Id}}$-coalgebras on $\mathcal{K}\ell(M)$, where H and M are defined as for NDA (see Example 2.8). We can define the semantics of NDA with ϵ-transitions via canonical fixpoint solutions as $[\![e]\!] = \bar{e}^\dagger$, where $\bar{e}$ is the automaton with word transitions corresponding to e (see Definition 4.7). The first example in Sect. 3 would be represented as follows,

$$\bar{e}(x) = [\kappa_X, \eta_X] \circ e(x) \;=\; \{(\epsilon, y)\}$$
$$\bar{e}(y) = [\kappa_X, \eta_X] \circ e(y) \;=\; \{(\epsilon, z)\}$$
$$\bar{e}(z) = [\kappa_X, \eta_X] \circ e(z) \;=\; \{(a, z), \epsilon\}$$

where η and κ are defined as at the end of Sect. 3. By using (5), it can be easily checked that the semantics $[\![e]\!] = \bar{e}^\dagger \colon X \to \mathcal{P}A^*$ maps x, y and z into a^*.

§2: Elimination of internal transitions. We view an F-system $e \colon X \to FX + X$ with internal transitions as an equation morphism for the monad $FX + \mathrm{Id}$, with parameter $Y = 0$. Thus we can use the canonical fixpoint solution of $FX + \mathrm{Id}$ to obtain an F-system $e^\dagger \colon X \to FX + 0 = FX$, which we denote by $e \backslash \epsilon$. The construction is depicted below.

(6)

Example 4.14 (ϵ-elimination). Using the automaton of Example 4.13, we can perform ϵ-elimination, as defined in (6), using the canonical solution for the monad $\widehat{H_{\mathcal{J}}X} + \mathrm{Id}$:

We obtain the following NDA $e\backslash\epsilon \overset{\text{def}}{=} \text{¡} \circ \text{!}_e : X \to A \times X + 1$.

$$
\begin{aligned}
\text{!}_e(x) &= \{(2, a, z), (2, \checkmark)\} & e\backslash\epsilon(x) &= \{(a, z), \checkmark\} \\
\text{!}_e(y) &= \{(1, a, z), (1, \checkmark)\} & e\backslash\epsilon(y) &= \{(a, z), \checkmark\} \\
\text{!}_e(z) &= \{(0, a, z), (0, \checkmark)\} & e\backslash\epsilon(x) &= \{(a, z), \checkmark\}
\end{aligned}
$$

The semantics $\llbracket e\backslash\epsilon \rrbracket$ is defined as $\overline{e\backslash\epsilon}^\dagger$, where $\overline{e\backslash\epsilon} = \kappa_X \circ e\backslash\epsilon$ is the representation of the NDA $e\backslash\epsilon$ as an automaton with word transitions (Definition 4.7). It is immediate to see, in this case, that $\llbracket e\backslash\epsilon \rrbracket = \llbracket e \rrbracket$. This fact is an instance of Theorem 4.17 below.

Remark 4.15. Note that ϵ-elimination was recently defined using a trace operator on a Kleisli category [3, 11, 22]. These works are based on the trace semantics of Hasuo et al. [12] and tailored for ϵ-elimination. They do not take into account any algebraic structure of the labels and are hence not applicable to the other examples we consider in this paper.

§3: Soundness of ϵ-elimination. We now formally prove that the canonical fixpoint semantics of e and $e\backslash\epsilon$ coincide. To this end, first we show how the construction $e \mapsto e\backslash\epsilon$ can be expressed in terms of the canonical fixpoint solution of F^*. This turns out to be an application of the factorisation lemma (Proposition 4.4), for which we introduce the natural transformation $\pi \colon FX + \mathrm{Id} \Rightarrow F^*(X + \mathrm{Id})$ defined at $Y \in \mathbf{C}$ by

$$
\pi_Y \colon FX + Y \xrightarrow{[\kappa_X,\, \eta_Y^{F^*}]} F^*X + F^*Y \xrightarrow{[F^*\mathsf{inl}, F^*\mathsf{inr}]} F^*(X + Y) \, .
$$

Since F^* is a monad with canonical fixpoint solutions, it can be verified that so is $F^*(X + \mathrm{Id})$. Moreover, π is a monad morphism between $FX + \mathrm{Id}$ and $F^*(X + \mathrm{Id})$. These observations allow us to prove the following.

Proposition 4.16 (Factorisation property of $e \mapsto e\backslash\epsilon$). *For any F-system $e : X \to FX + X$ with internal transitions, consider the equation morphism $\pi_X \circ e : X \to F^*(X + X)$. Then:*

$$
\pi_0 \circ e\backslash\epsilon = (\pi_X \circ e)^\dagger : X \to F^*X.
$$

Proof. This follows simply by an application of Proposition 4.4 to $e\backslash\epsilon = e^\ddagger$ and $\gamma = \pi$ with $Y = 0$. $\qquad\square$

We are now in position to show point §3: soundness of ϵ-elimination.

Theorem 4.17 (Eliminating internal transitions is sound). *For any F-system $e : X \to FX + X$ with internal transitions,*

$$
\llbracket e\backslash\epsilon \rrbracket = \llbracket e \rrbracket.
$$

Proof. The statement is shown by the following derivation.

$$
\begin{aligned}
[\![e \backslash \epsilon]\!] &= [\![e^\ddagger]\!] && \text{Definition of } e \backslash \epsilon \\
&= (\kappa_X \circ e^\ddagger)^\dagger && \text{Definition of } [\![-]\!] \text{ (Definition 4.8)} \\
&= (\pi_0 \circ e^\ddagger)^\dagger && \text{Definition of } \pi_0 \\
&= (\pi_X \circ e)^{\dagger\dagger} && \text{Proposition 4.16} \\
&= (F^*(\nabla_X) \circ (\pi_X \circ e))^\dagger && \text{double dagger law} \\
&= \bar{e}^\dagger && \text{Definition of } \bar{e} \text{ (Definition 4.7) and } \pi_X \\
&= [\![e]\!] && \text{Definition of } [\![-]\!].
\end{aligned}
$$

$\square$

5   Quotient Semantics

When considering behavior of systems it is common to encounter spectrums of successively coarser equivalences. For instance, in basic process algebra trace equivalence can be obtained by quotienting bisimilarity with an axiom stating the distributivity of action prefixing by non-determinism [21]. There are many more examples of this phenomenon, including Mazurkiewicz traces, which we will describe below.

In this section we develop a variant of the canonical fixpoint semantics, where we can encompass in a uniform manner behaviors which are quotients of the canonical behaviors of the previous section (that is, the object F^*0).

Assumption 5.1. Let $\mathbf{C}$, F, F^* and $FX + \mathrm{Id}$ be as in Assumption 4.5 and $\gamma \colon F^* \Rightarrow Q$ a monad quotient for some monad Q. Moreover, suppose that for all $Y \in \mathbf{C}$ an initial $Q(\mathrm{Id} + Y)$-algebra exists.

Observe that, as Assumption 5.1 subsumes Assumption 4.5, we are within the framework of previous section, with the canonical fixpoint solution of F^* providing semantics for F^*- and F-systems. For our extension, one is interested in $Q0$ as a semantic domain coarser than F^*0 and we aim at defining an interpretation for F-systems in $Q0$. To this aim, we first check that Q has canonical fixpoint solutions.

Proposition 5.2. *Let $\mathbf{C}$, F, Q and $\gamma \colon F^* \Rightarrow Q$ be as in Assumption 5.1. Then Assumption 4.1 holds for $\mathbf{C}$ and Q, meaning that Q is a monad with canonical fixpoint solutions (which satisfy the double dagger law by Corollary 4.3).*

We use the notation $e \mapsto e^\sim$ for the canonical fixpoint operator of Q. This allows us to define the semantics of Q-systems, analogously to what we did for F^*-systems in Definition 4.8. Moreover, the connecting monad morphism $\gamma \colon F^* \Rightarrow Q$ yields an extension of this semantics to include also systems of transition type F^* and F.

Definition 5.3 (Quotient Semantics). *The quotient semantics of F-systems, with or without internal transitions, F^*-systems and Q-systems is defined as follows.*

- *For a Q-system $e\colon X \to QX$, its semantics $[\![e]\!]_\sim\colon X \to Q0$ is defined as $e^\sim$ (note that e can be regarded as an equation morphism for Q with $Y = 0$).*
- *For an F^*-system $e\colon X \to F^*X$, its semantics $[\![e]\!]_\sim\colon X \to Q0$ is defined as $(\gamma_X \circ e)^\sim$.*
- *For an F-system e—with or without internal transitions—its semantics $[\![e]\!]_\sim\colon X \to Q0$ is defined as $(\gamma_X \circ \bar{e})^\sim$, where $\bar{e}$ is as in Definition 4.7.*

The Factorisation Lemma (Proposition 4.4) allows us to establish a link between the canonical fixpoint semantics $[\![-]\!]$ and the quotient semantics $[\![-]\!]_\sim$.

Proposition 5.4 (Factorisation for the quotient semantics). *Let e be either an F^*-system or an F-system (with or without internal transitions). Then:*

$$[\![e]\!]_\sim = \gamma_0 \circ [\![e]\!]. \tag{7}$$

As a corollary we obtain that eliminating internal transitions is sound also for quotient semantics.

Corollary 5.5. *For any F-system $e : X \to FX + X$ with internal transitions,*

$$[\![e]\!]_\sim = [\![e \backslash \epsilon]\!]_\sim.$$

The quotient semantics can be formulated in a Kleisli category $\mathcal{K\ell}(M)$ by further assuming (c) below. This is needed to lift a quotient of monads from **Sets** to $\mathcal{K\ell}(M)$.

Theorem 5.6. *Let $M\colon$ **Sets** $\to$ **Sets** be a monad and $H\colon$ **Sets** $\to$ **Sets** be an accessible functor satisfying the assumptions of Theorem 2.7. By Proposition 2.4 the free monad H^* on H lifts to a monad $\widehat{H}^*\colon \mathcal{K\ell}(M) \to \mathcal{K\ell}(M)$ via a distributive law $\lambda\colon H^*M \Rightarrow MH^*$ with $\widehat{H}^* = \widehat{H}^*$. Let $R\colon$ **Sets** $\to$ **Sets** be a monad and $\xi\colon H^* \Rightarrow R$ a monad quotient such that*

(c) for each set X, there is a map $\lambda'_X\colon RMX \to MRX$ making the following commute.

$$
\begin{array}{ccc}
H^*MX & \xrightarrow{\ \lambda_X\ } & MH^*X \\
{\scriptstyle \xi_{MX}}\downarrow & & \downarrow{\scriptstyle M\xi_X} \\
RMX & \xrightarrow[\ \lambda'_X\ ]{} & MRX
\end{array}
$$

Then the following hold:

1. *there is a monad $\widehat{R}\colon \mathcal{K\ell}(M) \to \mathcal{K\ell}(M)$ lifting R and a monad morphism $\widehat{\xi}\colon \widehat{H}^* \Rightarrow \widehat{R}$ defined as $\widehat{\xi}_X = \mathcal{J}(\xi_X)$;*

2. $\mathcal{K}\ell(M)$, $\widehat{H}$, $\widehat{H}^*$, $\widehat{H}\partial X + \mathrm{Id}$ *(for a given set X)*, $\widehat{R}$ *and* $\widehat{\xi}\colon \widehat{H}^* \Rightarrow \widehat{R}$ *satisfy Assumption 5.1.*

Notice that condition (c) and the first part of statement 1 are related to [7, Theorem 1]; however, that paper treats distributive laws of monads over endofunctors.

Example 5.7 (Mazurkiewicz traces). This example, using a known equivalence in concurrency theory, illustrates the use of the quotient semantics developed in Sect. 5.

The trace semantics proposed by Mazurkiewicz [17] accounts for concurrent actions. Intuitively, let A be the action alphabet and $a, b \in A$. We will call a and b concurrent, and write $a \equiv b$, if the order in which these actions occur is not relevant. This means that we equate words that only differ in the order of these two actions, e.g. $uabv$ and $ubav$ denote the same Mazurkiewicz trace.

To obtain the intended semantics of Mazurkiewicz traces we use the quotient semantics defined above[3]. In particular, for Mazurkiewisz traces one considers a symmetric and irreflexive "independence" relation I on the label set A. Let $\equiv$ be the least congruence relation on the free monoid A^* such that

$$(a, b) \in I \Rightarrow ab \equiv ba.$$

We now have two monads on **Sets**, namely $H^*X = A^* \times X + A^*$ and $RX = A^*/_\equiv \times X + A^*/_\equiv$. There is the canonical quotient of monads $\xi\colon H^* \Rightarrow R$ given by identifying words of the same $\equiv$-equivalence class. It can be checked that those data satisfy the assumptions of Theorem 5.6 and thus we are allowed to apply the quotient semantics $[\![-]\!]_\sim$. This will be given on an NDA $e\colon X \to \widehat{H}X$ by first embedding it into $\widehat{H}^*$ as $\bar{e} = \kappa_X \circ e\colon X \to \widehat{H}^*X$ and then into $\widehat{R}$ as $\widehat{\xi}_X \circ \bar{e}\colon X \to \widehat{R}X$. To this morphism we apply the canonical fixpoint operator of $\widehat{R}$ to obtain $(\widehat{\xi}_X \circ \bar{e})^{\tilde{\ }}$, that is, the semantics $[\![e]\!]_\sim\colon X \to R0 = A^*/_\equiv$. It is easy to see that this definition captures the intended semantics: for all states $x \in X$

$$[\![e]\!]_\sim(x) = \{[w]_\equiv \mid w \in [\![e]\!](x)\}.$$

Indeed, by Proposition 5.4, $[\![e]\!]_\sim = \widehat{\xi}_0 \circ [\![e]\!]$ and $\widehat{\xi}_0\colon \widehat{H}^*0 \to \widehat{R}0$ is just $\partial\xi_0$ where $\xi_0\colon A^* \to A^*/_\equiv$ maps every word w into its equivalence class $[w]_\equiv$.

6   Discussion

The framework introduced in this paper provides a uniform way to express the semantics of systems with internal behaviour via canonical fixpoint solutions. Moreover, these solutions are exploited to eliminate internal transitions in a sound way, i.e., preserving the semantics. We have shown our approach at work

---

[3] Mazurkiewicz traces were defined over labelled transition systems which are similar to NDA but where every state is final. For simplicity, we consider LTS here immediately as NDA.

on NDA with ϵ-transitions but, by virtue of Theorem 4.10, it also covers all the examples in [12] (like probabilistic systems) and more (like the weighted automata on positive reals of [22]).

It is worth noticing that, in principle, our framework is applicable also to examples that do not arise from Kleisli categories. Indeed the theory of Sect. 4 is formulated for a general category **C**: Assumption 4.5 only requires **C** to be **Cppo**-enriched and the monad T to be locally continuous. The role of these assumptions is two-fold: (a) ensuring the initial algebra-final coalgebra coincidence and (b) guaranteeing that the canonical fixpoint operator $e \mapsto e^\dagger$ satisfies the *double dagger law*. If (a) implies (b), we could have formulated our theory just assuming the coincidence of initial algebra and final coalgebra and without any **Cppo**-enrichment. Condition (a) holds for some interesting examples not based on Kleisli categories, e.g. for examples in the category of join semi-lattices. Therefore it is of relevance to investigate the following question: given a monad T with initial algebra-final coalgebra coincidence, under which conditions does the canonical fixpoint solution provided by T satisfy the double dagger law?

As a concluding remark, let us recall that our original question concerned the problem of modeling the semantics of systems where labels carry an algebraic structure. In this paper we have mostly been focusing on automata theory, but there are many other examples in which the information carried by the labels has relevance for the semantics of the systems under consideration: in logic programming labels are substitutions of terms; in (concurrent) constraint programming they are elements of a lattice; in process calculi they are actions representing syntactical contexts and in tile systems [10] they are morphisms in a category. We believe that our approach provides various insights towards a coalgebraic semantics for these computational models.

Acknowledgments. We are grateful to the anonymous referees for valuable comments. The work of Alexandra Silva is partially funded by the ERDF through the Programme COMPETE and by the Portuguese Foundation for Science and Technology, project ref. FCOMP-01-0124-FEDER-020537 and SFRH/BPD/71956/2010. The first and the fourth author acknowledge support by project ANR 12ISO 2001 PACE.

References

1. Adámek, J., Milius, S., Velebil, J.: Equational properties of iterative monads. Inform. Comput. **208**, 1306–1348 (2010). doi:10.1016/j.ic.2009.10.006
2. Adámek, J., Milius, S., Velebil, J.: Elgot theories: a new perspective of the equational properties of iteration. Math. Structures Comput. Sci. **21**(2), 417–480 (2011)
3. Asada, K., Hidaka, S., Kato, H., Hu, Z., Nakano, K.: A parameterized graph transformation calculus for finite graphs with monadic branches. In: Peña, R., Schrijvers, T. (eds.) PPDP, pp. 73–84. ACM (2013)
4. Balan, A., Kurz, A.: On coalgebras over algebras. Theoret. Comput. Sci. **412**(38), 4989–5005 (2011)
5. Barr, M.: Coequalizers and free triples. Math. Z. **116**, 307–322 (1970)
6. Bloom, S.L., Ésik, Z.: Iteration Theories: The Equational Logic of Iterative. EATCS Monographs on Theoretical Computer Science. Springer, Heidelberg (1993)

7. Bonsangue, M.M., Hansen, H.H., Kurz, A., Rot, J.: Presenting distributive laws. In: Heckel and Milius [15], pp. 95–109

8. Main, M.G., Melton, A.C., Mislove, M.W., Schmidt, D., Brookes, S.D. (eds.): MFPS 1993. LNCS, vol. 802. Springer, Heidelberg (1994)

9. Freyd, P.J.: Remarks on Algebraically Compact Categories. London Mathematical Society Lecture Notes Series, vol. 177. Cambridge University Press, London (1992)

10. Gadducci, F., Montanari, U.: The tile model. In: Plotkin, G.D., Stirling, C., Tofte, M. (eds.) Proof, Language, and Interaction, pp. 133–166. MIT Press, Boston (2000)

11. Hasuo, I., Jacobs, B., Sokolova, A.: Generic forward and backward simulations. In: (Partly in Japanese) Proceedings of JSSST Annual Meeting (2006)

12. Hasuo, I., Jacobs, B., Sokolova, A.: Generic trace semantics via coinduction. Log. Methods Comput. Sci. **3**(4:11), 1–36 (2007)

13. Heckel, R., Milius, S. (eds.): Algebra and Coalgebra in Computer Science. Lecture Notes in Computer Science, vol. 8089. Springer, Heidelberg (2013)

14. Hopcroft, J., Motwani, R., Ullman, J.: Introduction to Automata Theory, Languages, and Computation, 3rd edn. Wesley, Lebanon (2006)

15. Johnstone, P.: Adjoint lifting theorems for categories of algebras. Bull. London Math. Soc. **7**, 294–297 (1975)

16. Kelly, G.M.: A unified treatment of transfinite constructions for free algebras, free monoids, colimits, associated sheaves, and so on. Bull. Austral. Math. Soc. **22**, 1–83 (1980)

17. Mazurkiewicz, A.: Concurrent program schemes and their interpretation. DAIMI PB-78, Computer Science Department, Aarhus University (1977)

18. Mac Lane, S.: Categories for the Working Mathematician. Springer, Berlin (1971)

19. Milius, S., Palm, T., Schwencke, D.: Complete iterativity for algebras with effects. In: Kurz, A., Lenisa, M., Tarlecki, A. (eds.) CALCO 2009. LNCS, vol. 5728, pp. 34–48. Springer, Heidelberg (2009)

20. Mulry, P.S.: Lifting theorems for Kleisli categories. In: Brookes et al. [10], pp. 304–319

21. Rabinovich, A.M.: A complete axiomatisation for trace congruence of finite state behaviors. In: Brookes et al. [10], pp. 530–543

22. Silva, A., Westerbaan, B.: A coalgebraic view of ε-transitions. In: Heckel and Milius [15], pp. 267–281

23. Sobociński, P.: Relational presheaves as labelled transition systems. In: Pattinson, D., Schröder, L. (eds.) CMCS 2012. LNCS, vol. 7399, pp. 40–50. Springer, Heidelberg (2012)

24. Turi, D., Plotkin, G.: Towards a mathematical operational semantics. In: Proceedings of Logic in Computer Science (LICS'97). IEEE Computer Society (1997)

On Coalgebras with Internal Moves

Tomasz Brengos[(✉)]

Faculty of Mathematics and Information Science,
Warsaw University of Technology, Koszykowa 75, 00-662 Warszawa, Poland
t.brengos@mini.pw.edu.pl

Abstract. In the first part of the paper we recall the coalgebraic approach to handling the so-called invisible transitions that appear in different state-based systems semantics. We claim that these transitions
are always part of the unit of a certain monad. Hence, coalgebras with
internal moves are exactly coalgebras over a monadic type. The rest of
the paper is devoted to supporting our claim by studying two important
behavioural equivalences for state-based systems with internal moves,
namely: weak bisimulation and trace semantics. We continue our research
on weak bisimulations for coalgebras over order enriched monads. The
key notions used in this paper and proposed by us in our previous work
are the notions of an order saturation monad and a saturator. A saturator operator can be intuitively understood as a reflexive, transitive
closure operator. There are two approaches towards defining saturators
for coalgebras with internal moves. Here, we give necessary conditions
for them to yield the same notion of weak bisimulation. Finally, we propose a definition of trace semantics for coalgebras with silent moves via a
uniform fixed point operator. We compare strong and weak bisimilation
together with trace semantics for coalgebras with internal steps.

Keywords: Bisimulation · Coalgebra · Conway operator · Epsilon transition · Fixed point operator · Internal transition · Logic · Monad · Saturation · Trace · Trace semantics · Traced monoidal category · Uniform
fixed point operator · Weak bisimulation · Weak trace semantics · van
Glabbeek spectrum

1 Introduction

In recent years we have witnessed a rapid development of the theory of coalgebras as a unifying theory for state-based systems [11,14,19,31]. Coalgebras to
some extent are one-step entities in their nature. They can be thought of and
understood as a representation of a single step of visible computation of a given
process. Yet, for many state-based systems it is useful to consider a part of computation branch that is allowed to take several steps and in some sense remains

This work has been supported by the grant of Warsaw University of Technology no.
504M for young researchers.

© IFIP International Federation for Information Processing 2014
M.M. Bonsangue (Ed.): CMCS 2014, LNCS 8446, pp. 75–97, 2014.
DOI: 10.1007/978-3-662-44124-4_5

neutral (invisible) to the structure of the process. For instance, the so-called τ-transitions also called *invisible transitions* for labelled transition systems [25, 26] or ε-transitions for non-deterministic automata [15]. As will be witnessed here, these special branches of computation are the same in their nature, yet they are used in order to develop different notions of equivalence of processes, e.g. weak bisimulation for LTS [25] or trace semantics for non-deterministic automata with ε-moves, we call ε-NA in short [15]. These are not the only state-based systems considered in the literature with a special invisible computational branch. Fully probabilistic systems [3] or Segala systems [33, 34] are among those, to name a few. All these systems are instances of a general notion of a coalgebra. If so, then how should we consider the invisible part of computation coalgebraically? As we will see further on, the invisible part of the computation can be and should be, in our opinion, considered as part of the unit of a monad. Before we state basic results let us summarize known literature on the topic of invisible transitions from perspective of weak bisimulation, trace semantics and coalgebra.

Weak Bisimulation. The notion of a strong bisimulation for different transition systems plays an important role in theoretical computer science. A weak bisimulation is a relaxation of this notion by allowing silent, unobservable transitions. Here, we focus on the weak bisimulation and weak bisimilarity proposed by R. Milner [25, 26] (see also [32]). Analogues of Milner's weak bisimulation are established for different deterministic and probabilistic transition systems (e.g. [3, 32–34]). It is well known that one can introduce Milner's weak bisimulation for LTS in several different but equivalent ways.

The notion of a strong bisimulation, unlike the weak bisimulation, has been well captured coalgebraically (see e.g. [11, 31, 39]). Different approaches to defining weak bisimulations for coalgebras have been presented in the literature. The earliest paper is [30], where the author studies weak bisimulations for while programs. In [28] the author introduces a definition of weak bisimulation for coalgebras by translating a coalgebraic structure into an LTS. This construction works for coalgebras over a large class of functors but does not cover the distribution functor, hence it is not applicable to different types of probabilistic systems. In [29] weak bisimulations are introduced via weak homomorphisms. As noted in [38] this construction does not lead to intuitive results for probabilistic systems. In [38] the authors present a definition of weak bisimulation for classes of coalgebras over functors obtained from bifunctors. Here, weak bisimulation of a system is defined as a strong bisimulation of a transformed system. In [5] we proposed a new approach to defining weak bisimulation in two different ways. Two definitions of weak bisimulation described by us in [5] were proposed in the setting of coalgebras over ordered functors. The key ingredient of the definitions is the notion of a saturator. As noted in [5] the saturator is sometimes too general to model only weak bisimulation and may be used to define other known equivalences, e.g. delay bisimulation [32]. Moreover, the saturators from [5] do not arise in any natural way. To deal with this problem we have presented a canonical way to consider weak bisimulation saturation in our previous paper [6]. Part of the results from [6] are recalled in this paper.

We should also mention [10,24] which appeared almost at the same time as our previous paper [6]. The former is a talk on the on-going research by S. Goncharov and D. Pattinson related to weak bisimulation for coalgebras. Their approach is similar to ours as it uses fixed points. It is worth noting that the authors cover some examples that do not fit our framework (e.g. fully probabilistic systems). However, they do not hide the invisible steps inside a monadic structure. The latter is a paper in which the authors study weak bisimulation for labelled transition systems weighted over semirings. They propose a coalgebraic approach towards defining weak bisimulation which relies on ε-elimination procedure presented in [35].

Weak Trace Semantics. Trace semantics is a standard behavioural equivalence for many state-based systems. Generic trace semantics for coalgebras has been proposed in [14,19]. If T is a monad on a category C and $F : \mathsf{C} \to \mathsf{C}$ is an endofunctor then the trace semantics of TF-coalgebras is final semantics for coalgebras considered in a different category, namely the Kleisli category for the monad T [14,17]. It is worth noting that trace semantics can also be defined for GT-coalgebras for an endofunctor $G : \mathsf{C} \to \mathsf{C}$ [19,36] via the so-called $\mathcal{EM}$-extension semantics. In our paper however, we focus only on TF-coalgebras and do not consider GT-coalgebras. Trace semantics can also be defined for different state-based systems with internal, invisible moves. In order to distinguish trace semantics for systems with and without silent steps we will sometimes call the former "weak trace semantics". One coalgebraic approach towards defining trace semantics for systems with ε-moves (invisible moves) is based on a very simple idea, has been presented in [13,35] and can be summarized as follows. In the first step we consider invisible moves as visible. Then we find the trace semantics for an "all-visible-steps" coalgebra and finally, we remove all occurrences of the invisible label and get the desired weak trace semantics. We discuss this approach in our paper and call it the "top-down" approach. The term "top-down" refers to the fact that we somewhat artificially treat the invisible moves as if they were visible and then we remove their occurrences from the trace. Such an approach does not use any structural properties of silent moves. A dual approach, a "bottom-up" method, should make use of their structural properties. Here, we present a "bottom-up" method for coalgebras with internal steps that treats silent moves as part of the unit of a certain monad.

Content and Organization of the Paper. The paper is organized as follows. Section 2 recalls basic notions in category theory, algebra and coalgebra. Section 3 describes two very general methods for dealing with silent steps via a monadic structure that have been proposed in our previous work [6]. We will see that these two methods appear in classical definitions of a weak bisimulation for LTS's. In Sect. 4 we recall the definition of an order saturation monad that comes from [6] and claim that this object is suitable for defining weak bisimulations for coalgebras. An order saturation monad is an order enriched monad equipped with an extra operator, a saturator $(-)^*$, that assigns to any coalgebra $\alpha : X \to TX$ a coalgebra $\alpha^* : X \to TX$ and can be thought of as a reflexive, transitive closure operator. It turns out that in the classical literature on labelled transition systems and weak bisimulation one can find two different saturators yelding the same notion of equivalence. These two saturators are natural consequences of the two stategies towards handling invisible steps via monadic structure. What is new in this section is the following:

- Weak bisimulation is defined as a kernel bisimulation [39] on a saturated structure and not via lax- and oplax-homomorphisms in Aczel-Mendler style as it was done in [6].
- We present both saturators in a general setting and ask when they yield the same notion of weak bisimulation. We give sufficient conditions functors should satisfy so that weak bisimulation coincides for both approaches.

In Sect. 5 we discuss a novel approach towards defining trace semantics for coalgebras with internal moves. Here, weak trace semantics morphism is obtained axiomatically by the so-called coalgebraic trace operator, i.e. a uniform fixed point operator. For **Cppo**-enriched monads, a coalgebraic trace operator is given by the least fixed point operator $\mu x.(x \cdot \alpha)$. Moreover, we show that the coalgebraic trace operator for ε-NA's arises from properties of the so-called free LTS monad. To be more precise, Kleisli category for the free LTS monad is traced monoidal category in the sense of Joyal et al. [16]. In Sect. 6, in a fairly general setting, we formulate how strong bisimulation, weak bisimulation and weak trace semantics are related. Hence, according to our knowledge we present the first paper that considers a comparison of three different behaviour equivalences in van Glabbeek's spectrum for systems with internal moves [8] from coalgebraic perspective.

2   Basic Notions and Properties

Algebras and Coalgebras. Let C be a category and let $F \colon \mathsf{C} \to \mathsf{C}$ be a functor. An F-algebra is a morphism $a : FA \to A$ in C. A *homomorphism* between algebras $a : FA \to A$ and $b : FB \to B$ is a morphism $f : A \to B$ in C such that $b \circ F(f) = f \circ a$. Dually, an F-*coalgebra* is a morphism $\alpha : X \to FX$ in C. The domain X of α is called *carrier* and the morphism α is sometimes also called *structure*. A *homomorphism* from an F-coalgebra $\alpha : X \to FX$ to an F-coalgebra $\beta : Y \to FY$ is a morphism $f \colon X \to Y$ in C such that $F(f) \circ \alpha = \beta \circ f$. The category of all F-coalgebras (F-algebras) and homomorphisms between them is denoted by C_F (resp. C^F). Many transition systems can be captured by the notion of coalgebra. In this paper we mainly focus on labelled transition systems with a silent label and non-deterministic automata with ε-moves. These two structures have been defined and thoroughly studied in the computer science literature (see e.g. [15,25,26,32]). Let Σ be a fixed set of alphabet letters. A *labelled transition system* over the alphabet $\Sigma_\tau = \Sigma + \{\tau\}$ (or an *LTS* in short) is a triple $\langle X, \Sigma_\tau, \to \rangle$, where X is called a *set of states* and $\to \subseteq X \times \Sigma_\tau \times X$ is a *transition*. The label τ is considered a special label sometimes called silent or invisible label. For an LTS $\langle X, \Sigma_\tau, \rightarrow \rangle$ instead of writing $(x, \sigma, x') \in \rightarrow$ we write $x \xrightarrow{\sigma} x'$. Labelled transition systems can be viewed as coalgebras over the type $\mathcal{P}(\Sigma_\tau \times \mathcal{I}d)$ [31]. From coalgebraic perspective, a *non-deterministic automaton with ε- transitions*, or ε-NA in short, over alphabet Σ is a coalgebra of the type $\mathcal{P}(\Sigma_\varepsilon \times \mathcal{I}d + 1)$, where $1 = \{\checkmark\}$ is fixed one element set and $\Sigma_\varepsilon = \Sigma + \{\varepsilon\}$. Note that LTS's differ from ε-NA's in the presence of 1 in the type. It is responsible for specifying which states are *final* and which are not. To be more precise for ε-NA $\alpha : X \rightarrow \mathcal{P}(\Sigma_\varepsilon \times X + 1)$ we call a state $x \in X$ *final* if $\checkmark \in \alpha(x)$. For more information on automata the reader is referred to e.g. [15].

Strong Bisimulation for Coalgebras. Notions of strong bisimulation have been well captured coalgebraically [2,11,31,39]. Let F be a Set-endofunctor and consider an F-coalgebra $\alpha : X \rightarrow FX$. In Aczel-Mendler style [2,39], a *(strong) bisimulation* is a relation $R \subseteq X \times X$ for which there is a structure $\gamma : R \rightarrow TR$ making $\pi_1 : R \rightarrow X$ and $\pi_2 : R \rightarrow X$ homomorphisms between γ and α. In this paper however we consider defining bisimulation as the so-called kernel bisimulation [39]. Let $F : \mathsf{C} \rightarrow \mathsf{C}$ be an endofunctor on an arbitrary category. Let $\alpha : X \rightarrow FX$ and $\beta : Y \rightarrow FY$ be F-coalgebras. A relation R on X and Y (i.e. a jointly-monic span $X \xleftarrow{\pi_1} R \xrightarrow{\pi_2} Y$ in C) is *kernel bisimulation* or *bisimulation* in short if there is a coalgebra $\gamma : Z \rightarrow FZ$ and homomorphisms f from α to γ and g from β to γ such that R with π_1, π_2 is the pullback of $X \xrightarrow{f} Z \xleftarrow{g} Y$. For a thorough study of the relation between Aczel-Mendler style of defining bisimulation and kernel bisimulation the reader is referred to [39] for details.

Monads. A *monad* on C is a triple (T, μ, η), where $T : \mathsf{C} \rightarrow \mathsf{C}$ is an endofunctor and $\mu : T^2 \implies T, \eta : \mathcal{I}d \implies T$ are two natural transformations for which the following two diagrams commute: The transformation μ is called *multiplication* and η *unit*. Each monad gives rise to a canonical category - Kleisli category for T. If (T, μ, η) is a monad on category C then *Klesli category* $\mathcal{K}l(T)$ for T has the class of objects equal to the class of objects of C and for two objects X, Y in $\mathcal{K}l(T)$ we have $\mathrm{Hom}_{\mathcal{K}l(T)}(X, Y) = \mathrm{Hom}_\mathsf{C}(X, TY)$ with the composition $\cdot$ in $\mathcal{K}l(T)$ defined between two morphisms $f : X \rightarrow TY$ and $g : Y \rightarrow TZ$ by $g \cdot f := \mu_Z \circ T(g) \circ f$ (here, $\circ$ denotes the composition in C).

$$\begin{array}{ccc} T^3 & \xrightarrow{\mu T} & T^2 \\ {\scriptstyle T\mu} \downarrow & & \downarrow {\scriptstyle \mu} \\ T^2 & \xrightarrow{\mu} & T \end{array}$$

$$\begin{array}{ccc} T & \xrightarrow{T\eta} & T^2 \\ {\scriptstyle \eta T} \downarrow & \searrow & \downarrow {\scriptstyle \mu} \\ T^2 & \xrightarrow{\mu} & T \end{array}$$

Example 1. The powerset endofunctor $\mathcal{P} : \mathsf{Set} \rightarrow \mathsf{Set}$ is a monad with the multiplication $\mu : \mathcal{P}^2 \implies \mathcal{P}$ and the unit $\eta : \mathcal{I}d \implies \mathcal{P}$ given on their X-components by $\mu_X : \mathcal{PP}X \rightarrow \mathcal{P}X; S \mapsto \bigcup S$ and $\eta_X : X \rightarrow \mathcal{P}X; x \mapsto \{x\}$. For any category C with binary coproducts and an object $A \in \mathsf{C}$ define $\mathcal{M}_A : \mathsf{C} \rightarrow \mathsf{C}$ as $\mathcal{M}_A = \mathcal{I}d + A$. The functor carries a monadic structure $(\mathcal{M}_A, \mu, \eta)$, where the X-components of the multiplication and the unit are the following: $\mu_X : (X + A) + A \rightarrow X + A; \mu_X = [id_{X+A}, \iota^2]$ and $\eta_X : X \rightarrow X + A; \eta_X = \iota^1$. Here, ι^1 and ι^2 denote the coprojections into the first and the second component of $X + A$ respectively. The monad $\mathcal{M}_A$ is sometimes called *exception monad*.

Since in many cases we will work with two categories at once: C and $\mathcal{K}l(T)$, morphisms in C will be denoted using standard arrow $\to$, whereas for morphisms in $\mathcal{K}l(T)$ we will use the symbol $\multimap$. For any object X in C (or equivalently in $\mathcal{K}l(T)$) the identity map from X to itself in C will be denoted by id_X and in $\mathcal{K}l(T)$ by 1_X or simply 1 if the domain can be deduced from the context.

The category C is a subcategory of $\mathcal{K}l(T)$ where the inclusion functor $^\sharp$ sends each object $X \in C$ to itself and each morphism $f : X \to Y$ in C to the morphism $f^\sharp : X \multimap Y$ given by $f^\sharp : X \to TY; f^\sharp = \eta_Y \circ f$. Each monad (T, μ, η) on a category C arises as the composition of left and right adjoint:

Here, $U_T : \mathcal{K}l(T) \to C$ is a functor defined as follows. For any object $X \in \mathcal{K}l(T)$ (i.e. $X \in C$) the object $U_T X$ is given by $U_T X := TX$ and for any morphism $f : X \multimap Y$ in $\mathcal{K}l(T)$ (i.e. $f : X \to TY$ in C) the morphism $U_T f : TX \to TY$ is given by $U_T f = \mu_Y \circ Tf$.

We say that a functor $F : C \to C$ *lifts to* an endofunctor $\overline{F} : \mathcal{K}l(T) \to \mathcal{K}l(T)$ provided that the following diagram commutes [14,19]:

There is a one-to-one correspondence between liftings $\overline{F}$ and *distributive laws* $\lambda : FT \implies TF$ [19,23]. Given a distributive law $\lambda : FT \implies TF$ a lifting $\overline{F} : \mathcal{K}l(T) \to \mathcal{K}l(T)$ is defined by:

$$\overline{F}X := FX \text{ for any object } X \in \mathcal{K}l(T),$$

$$\overline{F}f : FX \to TFY; \overline{F}f = \lambda_Y \circ Ff \text{ for any morphism } f : X \to TY.$$

Conversely, a lifting $\overline{F} : \mathcal{K}l(T) \to \mathcal{K}l(T)$ of F gives rise to a distributive law $\lambda : FT \implies TF$ defined by $\lambda_X : FTX \to TFX; \lambda_X = \overline{F}(id_{TX})$. A monad T on a cartesian closed category C is called *strong* if there is a transformation $st_{X,Y} : X \times TY \to T(X \times Y)$ called *tensorial strength* satisfying the strength laws listed in e.g. [20]. Existence of strength guarantees that for any object Σ the functor $\Sigma \times \mathcal{I}d : C \to C$ admits a lifting $\overline{\Sigma} : \mathcal{K}l(T) \to \mathcal{K}l(T)$. To be more precise we define a functor $\overline{\Sigma} : \mathcal{K}l(T) \to \mathcal{K}l(T)$ as follows. For any object $X \in \mathcal{K}l(T)$ (i.e. $X \in C$) we put $\overline{\Sigma}X := \Sigma \times X$, and for any morphism $f : X \multimap Y$ (i.e. $f : X \to TY$ in C) we define $\overline{\Sigma}f : \Sigma \times X \to T(\Sigma \times Y)$ by $\overline{\Sigma}f := st_{\Sigma,Y} \circ (id_\Sigma \times f)$. Existence of the transformation $st_{X,Y}$ is not a strong assumption. For instance all monads on Set are strong.

A category is *order enriched* if each hom-set is a poset with order preserved by composition. An endofunctor on an order enriched category is *locally monotonic* if it preserves order. A category C is ***Cppo**-enriched* if for any objects X, Y:

- the hom-set $Hom_C(X, Y)$ is a poset with a least element $\bot$,
- for any ascending ω-chain $f_0 \leqslant f_1 \leqslant \ldots$ in $Hom_C(X, Y)$ the supremum $\bigvee_{i \in \mathbb{N}} f_i$ exists,
- $g \circ \bigvee_{i \in \mathbb{N}} f_i = \bigvee_{i \in \mathbb{N}} g \circ f_i$ and $(\bigvee_{i \in \mathbb{N}} f_i) \circ h = \bigvee_{i \in \mathbb{N}} f_i \circ h$ for any ascending ω-chain $f_0 \leqslant f_1 \leqslant \ldots$ and g, h with suitable domain and codomain.

Note that it is *not* necessarily the case that $f \circ \bot = \bot$ or $\bot \circ f = \bot$ for any morphism f. An endofunctor on a **Cppo**-enriched category is called *locally* *continuous* if it preserves suprema of ascending ω-chains. For more details on **Cppo**-enriched categories the reader is referred to e.g. [1,14].

Example 2. The Kleisli category for the powerset monad $\mathcal{P}$ is **Cppo**-enriched [14]. The order on the hom-sets is imposed by the natural point-wise order. The strength map for $\mathcal{P}$ is given by

$$\mathrm{st}_{X,Y} : X \times \mathcal{P}Y \to \mathcal{P}(X \times Y); (x, S) \mapsto \{(x, y) \mid y \in S\}.$$

The lifting $\overline{\Sigma} : \mathcal{K}l(\mathcal{P}) \to \mathcal{K}l(\mathcal{P})$ of $\Sigma \times \mathcal{I}d : \mathsf{Set} \to \mathsf{Set}$ is a locally continuous functor [14]. The Kleisli category for the monad $\mathcal{M}_1$ on Set is also **Cppo**-enriched [14]. Order on hom-sets is imposed by the point-wise order and for any X the set $\mathcal{M}_1 X = X + 1 = X + \{\bot\}$ is a poset whose partial order $\leqslant$ is given by $x \leqslant y$ iff $x = \bot$ or $x = y$.

Monads on Kleisli Categories. In this paper we will often work with monads on Kleisli categories. Here we list basic properties of such monads. Everything presented below with the exception of the last theorem follows easily by classical results in category theory (see e.g. [21]). Assume that (T, μ, η) is a monad on C and $S : \mathsf{C} \to \mathsf{C}$ is a functor that lifts to $\overline{S} : \mathcal{K}l(T) \to \mathcal{K}l(T)$ with the associated distributive law $\lambda : ST \implies TS$. Moreover, let $(\overline{S}, m, e)$ be a monad on $\mathcal{K}l(T)$. We have the following two adjoint situations whose composition is an adjoint situation [21].

$$\mathsf{C} \;\; {}^{\sharp}\!\!\overset{\frown}{\underset{U_T}{\rightleftarrows}} \bot \mathcal{K}l(T) \;\; \bot \;\; {}^{\sharp}\!\!\overset{\frown}{\underset{U_{\overline{S}}}{\rightleftarrows}} \mathcal{K}l(\overline{S})$$

This yields a monadic structure on the functor $TS : \mathsf{C} \to \mathsf{C}$. The X-components of the multiplication $\mathfrak{m}$ and the unit $\mathfrak{e}$ of the monad TS are given by:

$$\mathfrak{m}_X = \mu_{SX} \circ T\mu_{SX} \circ TTm_X \circ T\lambda_{SX} \quad \text{and} \quad \mathfrak{e}_X = e_X.$$

The composition $\cdot$ in $\mathcal{K}l(TS) = \mathcal{K}l(\overline{S})$ is given in terms of the composition in C as follows. For $f : X \to TSY$ and $g : Y \to TSZ$ we have:

$$
\begin{array}{ccccccc}
X & \xrightarrow{f} & TSY & \xrightarrow{TSg} & TSTSZ & \xrightarrow{T\lambda_{SZ}} & T^2 S^2 Z \\
g\cdot f \downarrow & & & & & & \downarrow T^2(m_Z) \\
TSZ & \xleftarrow{\mu_{SZ}} & & T^2 SZ & \xleftarrow{T\mu_{SZ}} & T^3 SZ &
\end{array}
$$

The following result can be proved by straightforward verification.

Theorem 1. *Assume that $\mathcal{K}l(T)$ is **Cppo**-enriched and $\overline{S}$ is locally continuous. Then $\mathcal{K}l(TS) = \mathcal{K}l(\overline{S})$ is **Cppo**-enriched.*

3   Hiding Internal Moves Inside a Monadic Structure

Throughout this paper we assume that (T, μ, η) is a monad on a category C with binary coproducts. Let $+$ denote the binary coproduct operator in C. Assume that $F : \mathsf{C} \to \mathsf{C}$ is a functor and let $F_\tau = F + \mathcal{I}d$. In this paper we deal with functors of the form $TF_\tau = T(F + \mathcal{I}d)$. Labelled transition system and ε-NA functor are of this form since

$$\mathcal{P}(\Sigma_\tau \times \mathcal{I}d) \cong \mathcal{P}(\Sigma \times \mathcal{I}d + \mathcal{I}d) = \mathcal{P}(F + \mathcal{I}d) \text{ for } F = \Sigma \times \mathcal{I}d \text{ and}$$
$$\mathcal{P}(\Sigma_\varepsilon \times \mathcal{I}d + 1) \cong \mathcal{P}(\Sigma \times \mathcal{I}d + 1 + \mathcal{I}d) = \mathcal{P}(F + \mathcal{I}d) \text{ for } F = \Sigma \times \mathcal{I}d + 1.$$

The functor F represents the visible part of the structure, whereas the functor $\mathcal{I}d$ represents silent moves. Functors of this type were used to consider ε-elimination from coalgebraic perspective in [13,35]. In [6] we noticed that given some mild assumptions on the monad T, the functor TF_τ can itself be turned into a monad or embedded into one. The aim of this section is to recall these results here. Before we do it, we will list basic definitions and properties concerning categories and monads used in the construction.

Basic Definitions and Properties. For a family of objects $\{X_k\}_{k \in I}$ if the coproduct $\coprod_i X_i$ exists then by $\iota^k : X_k \to \coprod_k X_k$ we denote the coprojection into k-th component of $\coprod_k X_k$.

We say that a category is *a category with zero morphisms* if for any two objects X, Y there is a morphism $0_{X,Y}$ which is an annihilator w.r.t. composition. To be more precise $f \circ 0 = 0 = 0 \circ g$ for any morphisms f, g with suitable domain and codomain.

Example 3. For the monad $T \in \{\mathcal{P}, \mathcal{M}_1\}$ on Set the category $\mathcal{K}l(T)$ is a category with zero morphisms given by $\bot : X \to \mathcal{P}Y; x \mapsto \varnothing$ for $\mathcal{P}$ and $\bot : X \to \mathcal{M}_1 Y; x \mapsto \bot$ for the monad $\mathcal{M}_1$.

Given two monads (S, μ^S, η^S) and $(S', \mu^{S'}, \eta^{S'})$ a *monad morphism* h is a natural transformation $h : S \implies S'$ which preserves unit and multiplication of monads, i.e. $h \circ \eta^S = \eta^{S'}$ and $h \circ \mu^S = \mu^{S'} \circ hh$. A *free monad* over a functor $F : \mathsf{C} \to \mathsf{C}$ [7,22] is a monad (F^*, m, e) together with a natural transformation $\nu : F \implies F^*$ such that for any monad (S, m^S, e^S) on C and a natural transformation $s : F \implies S$ there is a unique monad morphism $h : (F^*, m, e) \to (S, m^S, e^S)$ such that the following diagram commutes:

Theorem 2. *[7] Assume that for an endofunctor $F : \mathsf{C} \to \mathsf{C}$ and any object X the free F-algebra over X (=initial $F(-) + X$-algebra) i_X exists in C^F. For an object X and a morphism $f : X \to Y$ in C let F^*X denote the carrier of i_X and $F^*f : F^*X \to F^*Y$ denote the unique morphism for which the following diagram commutes:*

$$
\begin{array}{ccc}
FF^*X + X & \xrightarrow{\quad i_X \quad} & F^*X \\
{\scriptstyle F(F^*f)+id_X}\Big\downarrow & & \Big\downarrow{\scriptstyle F^*f} \\
FF^*Y + X \xrightarrow[id+f]{} FF^*Y + Y & \xrightarrow[i_Y]{} & F^*Y
\end{array}
$$

The assignment F^* is functorial and can be naturally equipped with a monadic structure (F^*, m, e) which is a consequence of the universal properties of i_X. Moreover, this monad is the free monad over F.

In the sequel we assume the following:

- The functor $F : \mathsf{C} \to \mathsf{C}$ lifts to $\overline{F} : \mathcal{K}l(T) \to \mathcal{K}l(T)$. As a direct consequence we get that $F_\tau = F + \mathcal{I}d$ lifts to a functor $\overline{F_\tau} = \overline{F} + \mathcal{I}d$ on $\mathcal{K}l(T)$. This follows by the fact that coproducts in $\mathcal{K}l(T)$ come from coproducts in the base category (see also e.g. [14] for a discussion on liftings of coproducts of functors).
- The functor F admits the free F-algebra i_X in C^F for any object X. By theorem above this yields the free monad (F^*, m, e) over F in C.

Monadic Structure on TF_τ. The aim of this subsection is to present the first strategy towards handling the invisible part of computation by a monadic structure. Note that in the following result all morphisms, in particular all coprojections and mediating morphisms, live in $\mathcal{K}l(T)$.

Theorem 3. *[6] If $\mathcal{K}l(T)$ is a category with zero morphisms then the triple $(\overline{F_\tau}, m', e')$, where $e' : \mathcal{I}d \multimap \overline{F} + \mathcal{I}d; e'_X = \iota^2$ and*

$$
m' : \overline{F}(\overline{F} + \mathcal{I}d) + (\overline{F} + \mathcal{I}d) \xrightarrow{\overline{F}([0,id])+id} \overline{F} + (\overline{F} + \mathcal{I}d) \xrightarrow{[\iota^1, id]} \overline{F} + \mathcal{I}d
$$

is a monad on $\mathcal{K}l(T)$. Two adjoint situations $\mathsf{C} \rightleftarrows \mathcal{K}l(T) \rightleftarrows \mathcal{K}l(\overline{F_\tau})$ yield a monadic structure on TF_τ.

The composition $\cdot$ in $\mathcal{K}l(TF_\tau) = \mathcal{K}l(\overline{F_\tau})$ is given as follows. Let $\lambda : F_\tau T \Longrightarrow TF_\tau$ denote the distributive law associated with the lifting $\overline{F_\tau}$ of F_τ. For any $f : X \to TF_\tau Y$, $g : Y \to TF_\tau Z$ we have:

$$
g \cdot f = \mu_{F_\tau Z} \circ T\mu_{F_\tau Z} \circ TTm'_Z \circ T\lambda_{F_\tau Z} \circ TF_\tau g \circ f.
$$

We illustrate the above construction in the following example, where $T = \mathcal{P}$ and $F_\tau = \Sigma_\tau \times \mathcal{I}d$.

Example 4. As mentioned before, the monad $\mathcal{P}$ (as any other monad on Set) comes with strength st which lifts the functor $\Sigma_\tau \times \mathcal{I}d : \mathsf{Set} \to \mathsf{Set}$ to the functor $\overline{\Sigma_\tau} : \mathcal{K}l(\mathcal{P}) \to \mathcal{K}l(\mathcal{P})$. For the functor $\overline{\Sigma_\tau} \cong \overline{\Sigma} + \mathcal{I}d$ we define the multiplication m' and the unit e' as in Theorem 3. For any set $X \in \mathcal{K}l(\mathcal{P})$ we put $m'_X : \overline{\Sigma_\tau}\,\overline{\Sigma_\tau}X \multimap \overline{\Sigma_\tau}X$ and $e'_X : X \multimap \overline{\Sigma_\tau}X$ to be:

$$
m'_X(\sigma_1, \sigma_2, x) = \begin{cases} \{(\sigma_1, x)\} & \text{if } \sigma_2 = \tau, \\ \{(\sigma_2, x)\} & \text{if } \sigma_1 = \tau, \\ \varnothing & \text{otherwise} \end{cases} \qquad e'_X(x) = \{(\tau, x)\}.
$$

By Theorem 3 the triple $(\overline{\Sigma_\tau}, m', e')$ is a monad on $\mathcal{K}l(\mathcal{P})$. By composing the two adjoint situations we get a monadic structure on the LTS functor. The composition in $\mathcal{K}l(\mathcal{P}(\Sigma_\tau \times \mathcal{I}d))$ is given as follows. For $f : X \to \mathcal{P}(\Sigma_\tau \times Y)$ and $g : Y \to \mathcal{P}(\Sigma_\tau \times Z)$ we have $g \cdot f : X \to \mathcal{P}(\Sigma_\tau \times Z)$:

$$g \cdot f(x) = \{(\sigma, z) \mid x \xrightarrow{\sigma}_f y \xrightarrow{\tau}_g z \text{ or } x \xrightarrow{\tau}_f y \xrightarrow{\sigma}_g z \text{ for some } y \in Y\}.$$

The construction provided by Theorem 3 can be applied only when $\mathcal{K}l(T)$ is a category with zero morphisms. Some monads fail to have this property. For example, if instead of considering the monad $\mathcal{P}$ we consider the non-empty powerset monad $\mathcal{P}_{\neq \varnothing}$. In what follows we focus on the second strategy for handling internal transitions by a monadic structure on the functor which does not require from $\mathcal{K}l(T)$ to be a category with zero morphisms.

Monadic Structure on TF^*. Here, we present an approach towards dealing with silent moves which uses free monads. At the beginning of this section we stated clearly that the coalgebras we are dealing with are of the type TF_τ. Any TF_τ-coalgebra $\alpha : X \to TF_\tau X$ can be turned into a TF^*-coalgebra $\underline{\alpha} : X \to TF^* X$ by putting

$$\underline{\alpha} = T([\nu_X, e_X]) \circ \alpha,$$

where the mono-transformation $[\nu, e] : F_\tau \implies F^*$ comes from the definition of a free monad.

Example 5. Consider the LTS functor $\mathcal{P}(\Sigma_\tau \times \mathcal{I}d) \cong \mathcal{P}(\Sigma \times \mathcal{I}d + \mathcal{I}d)$ and let $F = \Sigma \times \mathcal{I}d$. The free monad over F in Set is given by $(\Sigma^* \times \mathcal{I}d, m, e)$, where Σ^* is the set of finite words over Σ together with the empty string $\varepsilon \in \Sigma^*$ and m and e are given for any set X as follows:

$$m_X : \Sigma^* \times \Sigma^* \times X \to \Sigma^* \times X; (s, s', x) \mapsto (ss', x) \text{ and}$$
$$e_X : X \to \Sigma^* \times X; x \mapsto (\varepsilon, x).$$

For any $\alpha : X \to \mathcal{P}(\Sigma_\tau \times X)$ we define $\underline{\alpha} : X \to \mathcal{P}(\Sigma^* \times X)$ by

$$\underline{\alpha}(x) = \{(a, y) \mid (a, y) \in \alpha(x) \text{ and } a \in \Sigma\} \cup \{(\varepsilon, y) \mid (\tau, y) \in \alpha(x)\}.$$

Example 6. The ε-NA's are coalgebras of the type TF_τ for the monad $T = \mathcal{P}$ and $F = \Sigma \times \mathcal{I}d + 1$. The functor $F = \Sigma \times \mathcal{I}d + 1$ lifts to $\mathcal{K}l(\mathcal{P})$ [14] and admits all free F-algebras. Let F^* denote the free monad over F. The functor $F^* : \mathsf{Set} \to \mathsf{Set}$ is defined on objects and morphisms by

$$F^* X = \Sigma^* \times X + \Sigma^*,$$
$$F^* f : \Sigma^* \times X + \Sigma^* \to \Sigma^* \times Y + \Sigma^*; F^* f = (id_{\Sigma^*} \times f) + id_{\Sigma^*} \text{ for } f : X \to Y.$$

The monadic structure (F^*, m, e) is given by:

$$m_X : \Sigma^* \times (\Sigma^* \times X + \Sigma^*) + \Sigma^* \to \Sigma^* \times X + \Sigma^*;$$
$$m_X(s_1, s_2, x) = (s_1 s_2, x) \quad m_X(s_1, s_2) = s_1 s_2 \quad m_X(s_1) = s_1,$$
$$e_X : X \to \Sigma^* \times X + \Sigma^*; x \mapsto (\varepsilon, x).$$

For any ε-NA coalgebra $\alpha : X \to \mathcal{P}(\Sigma_\varepsilon \times X + 1)$ we define

$$\underline{\alpha} : X \to \mathcal{P}(\Sigma^* \times X + \Sigma^*); x \mapsto \{(a, y) \in \Sigma^* \times X \mid (a, y) \in \alpha(x)\} \cup A_x,$$

where $A_x = \mathbf{if}\ \checkmark \in \alpha(x)\ \mathbf{then}\ \{\varepsilon\}\ \mathbf{else}\ \varnothing$.

In order to proceed with the construction we need one additional lemma.

Lemma 1. *[6] The algebra* $i^\sharp_X = \eta_{F^*X} \circ i_X : FF^*X + X \to TF^*X$ *is the free* $\overline{F}$*-algebra over* X *in* $\mathcal{K}l(T)^{\overline{F}}$.

Let $\overline{F}^* : \mathcal{K}l(T) \to \mathcal{K}l(T)$ be the functor obtained by following the guidelines of Theorem 2 using the family $\{i^\sharp_X\}_{X \in \mathcal{K}l(T)}$ of free algebras in $\mathcal{K}l(T)^{\overline{F}}$.

Theorem 4. *[6] We have the following:*

1. $F^* : \mathsf{C} \to \mathsf{C}$ *lifts to* $\overline{F}^* : \mathcal{K}l(T) \to \mathcal{K}l(T)$,
2. $(\overline{F}^*, m^\sharp, e^\sharp)$ *is the free monad over* $\overline{F}$ *in* $\mathcal{K}l(T)$.

Two adjoint situations $\mathsf{C} \rightleftarrows \mathcal{K}l(T) \rightleftarrows \mathcal{K}l(\overline{F}^*)$ *yield a monadic structure on* TF^*.

The composition $\cdot$ in $\mathcal{K}l(TF^*) = \mathcal{K}l(\overline{F}^*)$ is given as follows. Let $\lambda : F^*T \Longrightarrow TF^*$ denote the distributive law associated with the lifting $\overline{F}^*$ of F^*. The composition of $f : X \to TF^*Y$, $g : Y \to TF^*Z$ in $\mathcal{K}l(TF^*)$ is given by:

$$g \cdot f = \mu_{F^*Z} \circ T\mu_{F^*Z} \circ TTm^\sharp_Z \circ T\lambda_{F^*Z} \circ TF^*g \circ f =$$
$$\mu_{F^*Z} \circ T\mu_{F^*Z} \circ TT(\eta_Z \circ m_Z) \circ T\lambda_{F^*Z} \circ TF^*g \circ f =$$
$$\mu_{F^*Z} \circ TTm_Z \circ T\lambda_{F^*Z} \circ TF^*g \circ f.$$

Example 7. The composition $\cdot$ in $\mathcal{K}l(\mathcal{P}(\Sigma^* \times \mathcal{I}d))$ is given by the following formula. For $f : X \to \mathcal{P}(\Sigma^* \times Y)$ and $g : Y \to \mathcal{P}(\Sigma^* \times Z)$ we have $g \cdot f : X \to \mathcal{P}(\Sigma^* \times Z)$:

$$g \cdot f(x) = \{(s_1 s_2, z) \mid x \xrightarrow{s_1}_f y \xrightarrow{s_2}_g z \text{ for some } y \in Y \text{ and } s_1, s_2 \in \Sigma^*\}.$$

We call the monad $\mathcal{P}(\Sigma^* \times \mathcal{I}d)$ *free LTS monad*.

Example 8. The composition $\cdot$ in $\mathcal{K}l(\mathcal{P}(\Sigma^* \times \mathcal{I}d + \Sigma^*))$ is given by the following formula. For $f : X \to \mathcal{P}(\Sigma^* \times Y + \Sigma^*)$ and $g : Y \to \mathcal{P}(\Sigma^* \times Z + \Sigma^*)$ we have $g \cdot f : X \to \mathcal{P}(\Sigma^* \times Z + \Sigma^*)$:

$$g \cdot f(x) = \{(s_1 s_2, z) \mid x \xrightarrow{s_1}_f y \xrightarrow{s_2}_g z \text{ for some } y \in Y \text{ and } s_1, s_2 \in \Sigma^*\} \cup$$
$$\{s_1 s_2 \mid x \xrightarrow{s_1}_f y \text{ and } s_2 \in g(y) \cap \Sigma^*, \text{ for some } y \in Y\} \cup$$
$$\{s_1 \in \Sigma^* \mid s_1 \in f(x)\}.$$

We call $\mathcal{P}(\Sigma^* \times \mathcal{I}d + \Sigma^*)$ monad *free ε-NA monad* or *ε-NA monad* in short.

We see that if we deal with functors of the form $T(F + \mathcal{I}d)$, where T is a monad, given some mild assumptions on T and F we may deal with the silent and observable part of computation inside a monadic structure on the functor TF_τ itself or by embedding the functor TF_τ into the monad TF^* by the natural transformation $F_\tau \implies F^*$. Therefore, from now on the term "coalgebras with internal moves" becomes synonymous to "coalgebras over a monadic type". Weak bisimulation and, as we will also see, trace equivalence are defined for coalgebras over monadic types, without the need for specifying visible and silent part of the structure.

4   Weak Bisimulation

In this section we recall classical definition(s) of weak bisimulation for labelled transition systems and coalgebraic constructions from [6]. Weak bisimulation for labelled transition systems can be defined as a strong bisimulation on a saturated structure. Process of saturation can be described as taking the reflexive and transitive closure of a given structure w.r.t. the suitable composition and order. First of all we present a paragraph devoted to classical definitions of weak bisimulation for LTS. Then we show how Kleisli compositions from Examples 4 and 7 play role in the LTS saturation. These examples motivate the definition of an order saturation monad and weak bisimulation [6]. What is essentially new in this section is the following. First of all we present a definition of weak bisimulation in terms of a kernel bisimulation on the saturated structure and not via lax- and oplax-homomorphisms in Aczel-Mendler style as it was done in [6]. Second of all, the last paragraph compares the two generalizations of the strategies towards saturation from the point of view of weak bisimulation which was not done in [6].

Weak Bisimulation for LTS. Let $\alpha : X \to \mathcal{P}(\Sigma_\tau \times X)$ be a labelled transition system coalgebra. For $\sigma \in \Sigma_\tau$ and $s \in \Sigma^*$ define the relations $\overset{\sigma}{\implies}, \overset{s}{\to}, \overset{s}{\implies} \subseteq X \times X$ by

$$\overset{\sigma}{\implies} = \begin{cases} (\overset{\tau}{\to})^* & \text{if } \sigma = \tau \\ (\overset{\tau}{\to})^* \circ \overset{\sigma}{\to} \circ (\overset{\tau}{\to})^* & \text{otherwise,} \end{cases} \qquad \overset{s}{\to} = \begin{cases} \overset{\tau}{\to} & \text{if } s = \varepsilon \\ \overset{\sigma_1}{\to} \circ \ldots \circ \overset{\sigma_n}{\to} & \text{for } s = \sigma_1 \ldots \sigma_n, \end{cases}$$

$$\overset{s}{\implies} = \begin{cases} (\overset{\tau}{\to})^* & \text{if } s \text{ is the empty word} \\ (\overset{\tau}{\to})^* \circ \overset{\sigma_1}{\to} \circ (\overset{\tau}{\to})^* \circ \ldots \circ (\overset{\tau}{\to})^* \circ \overset{\sigma_n}{\to} \circ (\overset{\tau}{\to})^* & \text{for } s = \sigma_1 \ldots \sigma_n \end{cases}$$

where, given any relation $R \subseteq X \times X$, the symbol R^* denotes the reflexive and transitive closure of R. We now present four different but equivalent definitions of weak bisimulation for LTS's. Due to limited space we do so in one definition block.

Definition 1. *[25, 26, 32] A relation $R \subseteq X \times X$ is called* weak bisimulation *on α if the following condition holds. If $(x, y) \in R$ then*

(i) for any $\sigma \in \Sigma_\tau$ the condition $x \xrightarrow{\sigma} x'$ implies $y \xRightarrow{\sigma} y'$
(ii) for any $\sigma \in \Sigma_\tau$ the condition $x \xRightarrow{\sigma} x'$ implies $y \xRightarrow{\sigma} y'$
(iii) for any $s \in \Sigma^*$ the condition $x \xrightarrow{s} x'$ implies $y \xRightarrow{s} y'$
(iv) for any $s \in \Sigma^*$ the condition $x \xRightarrow{s} x'$ implies $y \xRightarrow{s} y'$

and $y' \in X$ such that $(x', y') \in R$ and a symmetric statement holds.

In this paper we will focus on Definitions 1.ii and 1.iv and their generalization. They both suggest that weak bisimulation can be defined as a strong bisimulation on a *saturated model*. It is worth noting that in our previous paper we focused on analogues of Definitions 1.i and 1.iii and comparison with Definitions 1.ii and 1.iv respectively (see [6] for details).

Saturation for LTS Coalgebraically. Let us assume that $\cdot$ is a composition in $\mathcal{K}l(\mathcal{P}(\Sigma_\tau \times \mathcal{I}d))$ as in Example 4. Given an LTS coalgebra $\alpha : X \to \mathcal{P}(\Sigma_\tau \times X)$ the saturated LTS $\alpha^* : X \to \mathcal{P}(\Sigma_\tau \times X)$ is obtained as follows: $\alpha^* = 1_X \vee \alpha \vee \alpha \cdot \alpha \vee \ldots = \bigvee_{n=0,1,2\ldots} \alpha^n$, where $\bigvee$ denotes supremum in the complete lattice $(\mathcal{P}(\Sigma_\tau \times X)^X, \leqslant)$, where the relation $\leqslant$ is given by $\alpha \leqslant \beta \iff \alpha(x) \subseteq \beta(x)$ for any $x \in X$. We see that for $(\sigma, y) \in \Sigma_\tau \times X$: $(\sigma, y) \in \alpha^*(x)$ if and only if $x \xRightarrow{\sigma}_\alpha y$. Weak bisimulation on α according to Definition 1.ii is a strong bisimulation on α^*.

If we now consider $\cdot$ to be composition in $\mathcal{K}l(\mathcal{P}(\Sigma^* \times \mathcal{I}d))$ as in Example 7 for an LTS considered as a $\mathcal{P}(\Sigma^* \times \mathcal{I}d)$-coalgebra $\alpha : X \to \mathcal{P}(\Sigma^* \times X)$ define $\alpha^* : X \to \mathcal{P}(\Sigma^* \times X)$ to be $\alpha^* = 1_X \vee \alpha \vee \alpha \cdot \alpha \vee \ldots = \bigvee_{n=0,1,2\ldots} \alpha^n$. Then $(s, y) \in \alpha^*(x)$ if and only if $x \xRightarrow{s}_\alpha y$ for any $s \in \Sigma^*$. Weak bisimulation from Definition 1.iv is a strong bisimulation on α^*.

Saturation for T-coalgebras. A monad T whose Kleisli category is order-enriched is called *ordered $*$-monad* or *ordered saturation monad* [6] provided that in $\mathcal{K}l(T)$ for any morphism $\alpha : X \multimap X$ there is a morphism $\alpha^* : X \multimap X$ satisfying the following conditions:

(a) $1 \leqslant \alpha^*$,
(b) $\alpha \leqslant \alpha^*$,
(c) $\alpha^* \cdot \alpha^* \leqslant \alpha^*$,
(d) if $\beta : X \multimap X$ satisfies $1 \leqslant \beta$, $\alpha \leqslant \beta$ and $\beta \cdot \beta \leqslant \beta$ then $\alpha^* \leqslant \beta$,
(e) for any $f : X \to Y$ in C and any $\beta : Y \multimap Y$ in $\mathcal{K}l(T)$ we have:

$$f^\sharp \cdot \alpha \square \beta \cdot f^\sharp \implies f^\sharp \cdot \alpha^* \square \beta^* \cdot f^\sharp \text{ for } \square \in \{\leqslant, \geqslant\}.$$

For the rest of the section we assume that T is an order saturation monad with the saturator operator $(-)^*$.

Remark 1. We could try and define α^* as the least fix point $\mu x.(1 \vee x \cdot \alpha)$. Indeed, if T is e.g. complete join-semilatice enriched monad then the saturated structure is defined this way. We believe that our definition is slightly more general as it does not require for the mapping $x \mapsto 1 \vee x \cdot \alpha$ to be well defined. Intuitively however, α^* should and will be associated with $\mu x.(1 \vee x \cdot \alpha)$.

Example 9. The powerset monad $\mathcal{P}$ and the non-empty powerset monad $\mathcal{P}_{\neq\varnothing}$ are examples of order saturation monads [6]. The monads from Examples 4 and 7 are order saturation monads [6]. Also the $\mathcal{CM}$ monad of convex distributions described in [17] is an order saturation monad [6]. Although we will not focus on $\mathcal{CM}$ in this paper it is a very important monad that is used to model Segala systems, their trace semantics and probabilistic weak bisimulations [6,17,33,34]. Any Kleene monad [9] is also an order saturation monad [6].

Since $\mathcal{P}$, $\mathcal{P}(\Sigma_\tau \times \mathcal{I}d)$ and $\mathcal{P}(\Sigma^* \times \mathcal{I}d)$ are order saturation monads, the following question arises: is the saturation operator for LTS monads related to saturation in $\mathcal{Kl}(\mathcal{P})$? The following theorem answers that question in general and shows the relation between a saturation operator in $\mathcal{Kl}(T)$ and $\mathcal{Kl}(TS)$ for a monad $\overline{S}$ on $\mathcal{Kl}(T)$.

Theorem 5. *[6] Assume $S : \mathsf{C} \to \mathsf{C}$ lifts to $\overline{S} : \mathcal{Kl}(T) \to \mathcal{Kl}(T)$ and $(\overline{S}, m, e)$ is a monad on $\mathcal{Kl}(T)$. If $\overline{S}$ is locally monotonic and satisfies the equation*

$$m_X \cdot \overline{S}[(m_X \cdot \overline{S}\alpha)^* \cdot e_X] = (m_X \cdot \overline{S}\alpha)^*$$

for any $\alpha : X \multimap \overline{S}X$, then the monad TS is an order saturation monad with the saturation operator $(-)^\star$ given by $\alpha^\star = (m_X \cdot \overline{S}\alpha)^* \cdot e_X$.

If $T = \mathcal{P}$ and S is taken either to be $\Sigma_\tau \times \mathcal{I}d$ or $\Sigma^* \times \mathcal{I}d$, then the lifting $\overline{S}$ exists and is equipped with a monadic structure as in Sect. 3. Moreover, $\overline{S}$ satisfies the assumptions of Theorem 5 [6]. In other words, the LTS saturations for $\mathcal{P}(\Sigma_\tau \times \mathcal{I}d)$ and $\mathcal{P}(\Sigma^* \times \mathcal{I}d)$ are obtained respectively by

$$(m'_X \cdot \overline{\Sigma_\tau}\alpha)^* \cdot e'_X \text{ and } (m^\sharp_X \cdot \overline{\Sigma}^*\alpha)^* \cdot e^\sharp_X.$$

In sections to come we will deal with generalizations of these two saturations and check under which conditions they yield the same notion of weak bisimulation.

Weak Bisimulation for T-coalgebras. The following slogan should be in our opinion considered the starting point to the theory of weak bisimulation for T-coalgebras: *weak bisimulation on $\alpha : X \to TX$ = bisimulation on $\alpha^* : X \to TX$.*

Definition 2. *Let $\alpha : X \to TX$ be a T-coalgebra. A relation $X \xleftarrow{\pi_1} R \xrightarrow{\pi_2} X$ is weak bisimulation on α if it is a bisimulation on α^*.*

We see that the above definition coincides with the standard definition of weak bisimulation for LTS considered as $\mathcal{P}(\Sigma_\tau \times \mathcal{I}d)$- and $\mathcal{P}(\Sigma^* \times \mathcal{I}d)$-coalgebras.

Weak Bisimulation for TF_τ-and TF^*-coalgebras. This subsection will be devoted to comparing both approaches towards defining weak bisimulation for TF_τ-coalgebras that generalize Definitions 1.ii and 1.iv for LTS. Here, we additionally assume that $\mathcal{Kl}(T)$ is a category with zero morphisms. Then we may either define a monadic structure on TF_τ or embed the functor into the monad TF^*. These two approaches applied for LTS give two different saturations, yet the weak bisimulations coincide. It is natural to suspect that given some mild assumptions it will also be the case in a more general setting. We will now list all the necessary ingredients.

We assume $(\overline{F}_\tau, m', e')$ and $(\overline{F}^*, m^\sharp, e^\sharp)$ are monads as in Sect. 3 and that both satisfy the assumptions of Theorem 5 for the monad $\overline{S}$. For sake of simplicity and clarity of notation we will drop $\sharp$ and write $(\overline{F}^*, m, e)$ instead of $(\overline{F}^*, m^\sharp, e^\sharp)$. The consequences of these assumptions are the following:

- A natural transformation $\nu : \overline{F} \implies \overline{F}^*$ which arises by the definition of a free monad.
- A natural transformation $\iota^1 : \overline{F} \implies \overline{F}_\tau = \overline{F + Id} = \overline{F} + \mathcal{I}d$. This transformation is given regardless of the assumptions.
- Unique monad morphism $h : (\overline{F}^*, m, e) \multimap (\overline{F}_\tau, m', e')$ in $Kl(T)$ making the first three diagrams commute:

$$
\begin{array}{ccc}
\overline{F} \xrightarrow{\nu} \overline{F}^* & \mathcal{I}d \xrightarrow{e} \overline{F}^* & \overline{F}^*\overline{F}^* \xrightarrow{m} \overline{F}^* \\
\searrow{\scriptstyle\iota^1}\ \ \downarrow{\scriptstyle h} & \searrow{\scriptstyle e'=\iota^2}\ \ \downarrow{\scriptstyle h} & {\scriptstyle hh}\downarrow\ \ \ \ \ \downarrow{\scriptstyle h} \\
\overline{F}_\tau & \overline{F}_\tau & \overline{F}_\tau\overline{F}_\tau \xrightarrow{m'} \overline{F}_\tau
\end{array}
\qquad
\begin{array}{c}
\overline{F}_\tau \xrightarrow{[\nu, e]} \overline{F}^* \\
{\scriptstyle[\iota^1, e']=id}\searrow\ \ \downarrow{\scriptstyle h} \\
\overline{F}_\tau
\end{array}
$$

Commutativity of the first two diagrams implies commutativity of the forth. Existence and uniqueness of h follows by the fact that $\overline{F}^*$ is a free monad over $\overline{F}$ and $\iota^1 : \overline{F} \implies \overline{F}_\tau$ is a natural transformation.

- The monads TF_τ and TF^* are order saturation monads. The saturation operators $(-)^\star$ and $(-)^*$ for TF_τ- and TF^*-coalgebras resp. are given as follows. Let $\alpha : X \multimap \overline{F}_\tau X$ (i.e. $\alpha : X \to TF_\tau X$) and $\beta : Y \multimap \overline{F}^* Y$ (i.e. $\beta : Y \to TF^*Y$). We have:

$$
\alpha^\star = (m_X' \cdot \overline{F}_\tau \alpha)^* \cdot e_X' \qquad \text{and} \qquad \beta^* = (m_Y \cdot \overline{F}^* \beta)^* \cdot e_Y.
$$

Example 10. Let $T = \mathcal{P}$ and $F_\tau = \Sigma_\tau \times \mathcal{I}d$, $F^* = \Sigma^* \times \mathcal{I}d$. The morphism $h_X : \overline{\Sigma^*} X \multimap \overline{\Sigma_\tau} X$ is given by:

$$
h_X : \Sigma^* \times X \to \mathcal{P}(\Sigma_\tau \times X); (s, x) \mapsto
\begin{cases}
\{(\tau, x)\} & \text{if } |s| = 0, \\
\{(s, x)\} & \text{if } |s| = 1, \\
\varnothing & \text{otherwise.}
\end{cases}
$$

Consider any coalgebra $\alpha : X \multimap \overline{F}_\tau X$ and let $\underline{\alpha} : X \multimap \overline{F}^* X$ be given by $\underline{\alpha} = [\nu_X, e_X] \cdot \alpha$. Note that this is the same coalgebra as in the paragraph on monadic structure on TF^* in Sect. 3. Here, however, it is defined in terms of the composition in $Kl(T)$ and not C, and all superscripts $\sharp$ are dropped to simplify the notation. By commutativity of the last diagram above we have:

$$
h_X \cdot \underline{\alpha} = h_X \cdot [\nu_X, e_X] \cdot \alpha = \alpha.
$$

We will now try to compare bisimulations for $\alpha^\star$ and $\underline{\alpha}^*$. In case of labelled transition systems a relation is a bisimulation on $\alpha^\star$ if and only if it is a bisimulation on $\underline{\alpha}^*$. Below we verify how general is this statement and what conditions are required to be satisfied for it to remain true.

Lemma 2. *Assume that for any* $\phi : \overline{F}^*X \multimap \overline{F}^*X$ *and* $\psi : \overline{F_\tau}X \multimap \overline{F_\tau}X$ *if* $\psi \cdot h_X = h_X \cdot \phi$ *then* $\psi^* \cdot h_X = h_X \cdot \phi^*$. *In this case we have* $h_X \cdot \underline{\alpha}^* = \alpha^\star$.

Remark 2. Note that the assumption in Lemma 2 about the natural transformation h is crucial even though T is assumed to be an order saturation monad. Assumption (e) in the definition of order saturatiom monad does not guarantee that h satisfies the desired property since it is not in general of the form $h'^\sharp$ for some $h' : F^*X \to F_\tau X$ in C. However, if T is a Kleene monad [6,9] then this assumption is always satisfied. The powerset monad $\mathcal{P}$ is an example of a Kleene monad.

The following theorem follows directly from the above lemma.

Theorem 6. *Assume that for any* $\phi : \overline{F}^*X \multimap \overline{F}^*X$ *and* $\psi : \overline{F_\tau}X \multimap \overline{F_\tau}X$ *if* $\psi \cdot h_X = h_X \cdot \phi$ *then* $\psi^* \cdot h_X = h_X \cdot \phi^*$. *Any bisimulation on* $\underline{\alpha}^*$ *is a bisimulation on* $\alpha^\star$.

Our aim now will be to prove the converse.

Lemma 3. *We have* $\underline{\alpha}^* \leqslant (\underline{\alpha}^\star)^*$.

Remark 3. Before we state the next result we have to make one essential remark. Note that the technical condition concerning the transformation $[\nu, e]$ in the lemma below would follow from $[\nu, e]$ being a monad morphism. However, $[\nu, e] : \overline{F_\tau} \implies \overline{F}^*$ is *not* a monad morphism. It does not satisfy the 2nd axiom of a monad morphism. To see this consider $T = \mathcal{P}$, $(\overline{\Sigma_\tau}, m', e')$, $(\overline{\Sigma}^*, m, e)$ as in Examples 4 and 7 and a visible label $a \in \Sigma$. We have

$$[\nu_X, e_X] \cdot m'_X(a, a, x) = \varnothing \text{ and}$$
$$m_X \cdot [\nu_{\overline{\Sigma}^*X}, e_{\overline{\Sigma}^*X}] \cdot \overline{\Sigma_\tau}[\nu_X, e_X](a, a, x) = \{(aa, x)\}.$$

Lemma 4. *Assume* $[\nu_X, e_X] \cdot m'_X \cdot \overline{F_\tau}\alpha \leqslant m_X \cdot \overline{F}^*\underline{\alpha} \cdot [\nu_X, e_X]$. *Then* $\underline{\alpha}^\star \leqslant \underline{\alpha}^*$.

Theorem 7. *Let* α *satisfy the inequality from the assumptions of the previous statement. Any bisimulation on* $\alpha^\star$ *is a bisimulation on* $\underline{\alpha}^*$.

Proof. We have $\alpha \leqslant \alpha^\star$ and hence $\underline{\alpha} \leqslant \underline{\alpha}^\star$. This, together with Lemma 4, implies that $\underline{\alpha}^* \leqslant (\underline{\alpha}^\star)^* \leqslant (\underline{\alpha}^*)^* \overset{\diamond}{=} \underline{\alpha}^*$ (see [6] for a proof of the equality marked with (◇)). Assume $X \overset{\pi_1}{\leftarrow} R \overset{\pi_2}{\rightarrow} X$ is a bisimulation on $\alpha^\star$. It is also a bisimulation on $\underline{\alpha}^\star$. Finally, since $\underline{\alpha}^* = (\underline{\alpha}^\star)^*$ the relation R is a bisimulation on $\underline{\alpha}^*$.

Theorem 8. *Assume that cotupling* $[-, -]$ *in* $\mathcal{K}l(T)$ *is monotonic w.r.t. both arguments and the zero morphisms* $0_{X,Y} : X \multimap Y$ *are the least elements of the posets* $Hom_{\mathcal{K}l(T)}(X, Y)$. *Then any bisimulation on* $\alpha^\star$ *is a bisimulation on* $\underline{\alpha}^*$.

Remark 4. The powerset monad $\mathcal{P}$ satisfies assumptions of the above theorem. It is worth mentioning that the $\mathcal{CM}$ monad used to model Segala systems does not satisfy them as the zero morphisms in $\mathcal{K}l(\mathcal{CM})$ are not least elements of the partially ordered hom-sets [17]. The monad $\mathcal{CM}$ deserves a separate treatment and we leave this for future research.

5   Trace Semantics for Coalgebras with Internal Moves

The aim of this section is to present some ideas on how to approach the notion
of trace semantics for structures with invisible moves. As mentioned before in
order to distinguish the trace semantics for coalgebras with and without silent
steps we will often use the term *weak trace semantics* or *trace semantics for
structures with internal moves* to refer to the former.

Before we go into details we start this section by recalling a basic example
of trace semantics for ε-NA's [15].

Definition 3. *Given a non-deterministic automaton with ε-transitions $\alpha : X \rightarrow
\mathcal{P}(\Sigma_\varepsilon \times X + 1)$ its* trace semantics *is a morphism $tr_\alpha : X \rightarrow \mathcal{P}(\Sigma^*)$ which maps
any state $x \in X$ to the set of words over Σ it accepts. To be more precise, for a
word $w \in \Sigma^*$ we have $w \in tr_\alpha(x)$ provided that either $w = \varepsilon$ and $\checkmark \in \alpha(x)$ or
$w = a_1 \ldots a_n$ for $a_i \in \Sigma$ and there is $x' \in X$ such that*

$$x(\xrightarrow{\varepsilon})^* \circ \xrightarrow{a_1} \circ (\xrightarrow{\varepsilon})^* \ldots (\xrightarrow{\varepsilon})^* \circ \xrightarrow{a_n} \circ (\xrightarrow{\varepsilon})^* x'$$

with $\checkmark \in \alpha(x')$.

The above definition is an instance of what we call a "bottom-up" approach
towards trace semantics for non-deterministic automata with internal moves.
This approach considers ε steps as invisible steps that can wander around a
structure freely. In other words, from our perspective ε-steps that are used in this
definition are what they should be, i.e. are part of the unit of the ε-NA monad.
There is a second obvious approach towards defining trace semantics for ε-NA's.
We call this approach "top-down", since at first we treat ε steps artificially as
if they were standard visible steps. Given an ε-NA $\alpha : X \rightarrow \mathcal{P}(\Sigma_\varepsilon \times X + 1)$ we
find its trace $tr'_\alpha : X \rightarrow \mathcal{P}((\Sigma \cup \{\varepsilon\})^*)$ and then map all words from $(\Sigma \cup \{\varepsilon\})^*$
to words in Σ^* by removing all occurrences of the ε label. As a result we obtain
the same trace as in Definition 3. Since in many cases we know how to find
finite trace semantics for coalgebras with only visible steps [14] it is easy to
generalize the "top-down" approach to coalgebras with internal activities. This
is exactly how authors of [13,35] do it in their papers. We, however, will present a
bottom-up approach towards weak trace semantics that works for a large family
of coalgebras whose type is a monad.

Coalgebraic View on Weak Trace Semantics for ε-NA. In this subsection
we focus on coalgebras for the monad $\mathcal{P}(\Sigma^* \times X + \Sigma^*)$. Recall that by Example 6
any ε-NA coalgebra $\alpha : X \rightarrow \mathcal{P}(\Sigma_\varepsilon \times X + 1)$ can be considered a $\mathcal{P}(\Sigma^* \times X + \Sigma^*)$-
coalgebra. For simplicity and clarity of notation put $F = \Sigma \times \mathcal{I}d + 1$ and $F^* =
\Sigma^* \times \mathcal{I}d + \Sigma^*$. Let us list two basic facts concerning ε-NA monad:

 – The lifting $\overline{F}^* : \mathcal{K}l(\mathcal{P}) \rightarrow \mathcal{K}l(\mathcal{P})$ is locally continuous [14].
 – The ε-NA monad $\mathcal{P}F^*$ is **Cppo**-enriched. This follows by Theorem 1.

For any $\alpha : X \multimap X$ in $\mathcal{K}l(\mathcal{P}F^*)$ (i.e. $\alpha : X \rightarrow \mathcal{P}F^*X$) define the following
mapping $tr_\alpha : X \multimap \varnothing$ (i.e. $tr_\alpha : X \rightarrow \mathcal{P}(\Sigma^*)$):

$$\mathrm{tr}_\alpha = \bigvee_{n\in\mathbb{N}} \bot \cdot \alpha^n,$$

where $\bot \colon X \multimap \varnothing$ is given by $\bot \colon X \to \mathcal{P}(\Sigma^*); x \mapsto \varnothing$ and $\cdot$ denotes the composition in $\mathcal{Kl}(\mathcal{P}F^*)$ as in Example 8. It is simple to see that tr_α is the least morphism in $Hom_{\mathcal{Kl}(\mathcal{P}F^*)}(X,\varnothing) = Hom_{\mathsf{Set}}(X,\mathcal{P}(\Sigma^*))$ satisfying $\mathrm{tr}_\alpha = \mathrm{tr}_\alpha \cdot \alpha$. In other words,

$$\mathrm{tr}_\alpha = \mu x.x \cdot \alpha.$$

Recursively, if we put $\mathrm{tr}_0 = \bot$ and $\mathrm{tr}_n = \mathrm{tr}_{n-1} \cdot \alpha$ then $\mathrm{tr}_\alpha = \bigvee_n \mathrm{tr}_n$.

Example 11. Let $\Sigma = \{a,b\}$ and let $\alpha : X \to \mathcal{P}(\Sigma_\varepsilon \times X + 1)$ be given by the following diagram (ε-labels are omitted). We have $\mathrm{tr}_0 : X \to \mathcal{P}(\Sigma^*), x \mapsto \varnothing$ and

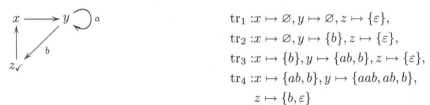

$$\mathrm{tr}_1 : x \mapsto \varnothing, y \mapsto \varnothing, z \mapsto \{\varepsilon\},$$
$$\mathrm{tr}_2 : x \mapsto \varnothing, y \mapsto \{b\}, z \mapsto \{\varepsilon\},$$
$$\mathrm{tr}_3 : x \mapsto \{b\}, y \mapsto \{ab,b\}, z \mapsto \{\varepsilon\},$$
$$\mathrm{tr}_4 : x \mapsto \{ab,b\}, y \mapsto \{aab,ab,b\},$$
$$z \mapsto \{b,\varepsilon\}$$

The following result can be shown by straightforward verification.

Theorem 9. *For any ε-NA coalgebra considered as $\mathcal{P}(\Sigma^* \times \mathcal{Id} + \Sigma^*)$-coalgebra the trace semantics morphism from Definition 3 and the morphism tr_α above coincide.*

Weak Coalgebraic Trace Semantics via Fixed Point Operator. We see that for ε-NA's their weak trace semantics is obtained as the least fixed point of the assignment $x \mapsto x \cdot \alpha$ in $\mathcal{Kl}(\mathcal{P}(\Sigma^* \times \mathcal{Id} + \Sigma^*))$. Interestingly, such a fixed point is not unique.

Example 12. Let $\Sigma = \{a\}, X = \{x\}$ and let ε-NA $\alpha : X \to \mathcal{P}(\Sigma_\varepsilon \times X + 1)$ be defined by the following diagram: $x \circlearrowright \varepsilon$. It is easy to check that the morphism $g : X \to \mathcal{P}(\Sigma^*); x \mapsto \{a\}$ satisfies $g = g \cdot \alpha$ and it is not the least fixed point since the least fixed point is given by $\mathrm{tr}_\alpha(x) = \varnothing$.

Here we generalize the ideas presented in the previous subsection to T-coalgebras. It should be noted at the very beginning that this section should serve as merely a starting point for future research.

Let us first focus on a known approach for defining trace semantics via coinduction in Kleisli category [14] and translating these results to our setting. In [14] the authors present trace semantics definition via coinduction for TF-coalgebras, where T is a monad and F satisfies some reasonable assumptions. In our setting however, we do not consider a special functor F or in other words $F = \mathcal{Id}$ and our coalgebras are T-coalgebras. Consider the category $\mathcal{Kl}(T)_{\mathcal{Id}}$ of $\mathcal{Id}$-coalgebras in $\mathcal{Kl}(T)$. Note that any T-coalgebra $\alpha : X \to TX$ is $\alpha : X \multimap X$ and is a member of $\mathcal{Kl}(T)_{\mathcal{Id}}$. Based on the approach from [14] trace semantics of α

should be obtained via coinduction in $\mathcal{K}l(T)_{\overline{F}}$. In our setting however, i.e. when $F = \mathcal{I}d$, the category $\mathcal{K}l(T)_{\mathcal{I}d}$ rarely admits the terminal object. For instance if we consider our ε-NA monad $\mathcal{P}(\Sigma^* \times \mathcal{I}d + \Sigma^*)$, the category of $\mathcal{I}d$-coalgebras $\mathcal{K}l(\mathcal{P}(\Sigma^* \times \mathcal{I}d + \Sigma^*))_{\mathcal{I}d}$ has no terminal object. However, it still makes sense to talk about trace for coalgebras for the monad $\mathcal{P}(\Sigma^* \times \mathcal{I}d + \Sigma^*)$. We did it via the least fixed point of the assignment $x \mapsto x \cdot \alpha$. In the general case we do it via uniform fixed point operator [37].

Assume that C is a category with the initial object 0 (this object is also initial in $\mathcal{K}l(T)$). A *fixed point operator* f on $\mathcal{K}l(T)$ is a family of morphisms:

$$\mathrm{f} : Hom_{\mathcal{K}l(T)}(X, X) \rightarrow Hom_{\mathcal{K}l(T)}(X, 0)$$

satisfying $\mathrm{f}(\alpha) \cdot \alpha = \mathrm{f}(\alpha)$ for any $\alpha : X \multimap X$. A fixed point operator f on $\mathcal{K}l(T)$ is *uniform* w.r.t. $(-)^\sharp : \mathsf{C} \rightarrow \mathcal{K}l(T)$ [37] if

$$h^\sharp \cdot \alpha = \beta \cdot h^\sharp \implies \mathrm{f}(\beta) \cdot h^\sharp = \mathrm{f}(\alpha)$$

for any $\alpha : X \multimap X$, $\beta : Y \multimap Y$ in $\mathcal{K}l(T)$ and $h : X \rightarrow Y$ in C. Coalgebraically speaking, the premise of the above implication says that the morphism h is a homomorphism between coalgebras $\alpha : X \rightarrow TX$ and $\beta : Y \rightarrow TY$ in C_T. We call a uniform fixed point operator on $\mathcal{K}l(T)$ a *coalgebraic trace operator* and we denote it by $\mathrm{tr}_{(-)}$.

Theorem 10. *Assume that* $\mathcal{K}l(T)$ *is a* ***Cppo****-enriched category and assume that for any* $f : X \rightarrow Y$ *in* C *we have* $\perp \cdot f^\sharp = \perp$. *For* $\alpha : X \multimap X$ *define* $\mathrm{tr}_\alpha : X \multimap 0$ *by* $\mathrm{tr}_\alpha = \mu x.(x \cdot \alpha) = \bigvee_{n \in \mathbb{N}} \perp \cdot \alpha^n$. *Then* $\mathrm{tr}_{(-)}$ *is a coalgebraic trace operator on* $\mathcal{K}l(T)$.

It may not be instantly clear for the reader why we choose uniformity as a property of a coalgebraic trace operator. Uniformity is a powerful notion which, in some forms, determines the least fixed point to be the unique uniform fixed point operator [37]. For the ε-NA monad $\mathcal{P}(\Sigma^* \times \mathcal{I}d + \Sigma^*)$ the least fixed point operator is a uniform fixed point operator w.r.t.

$$\sharp : \mathsf{Set} \rightarrow \mathcal{K}l(\mathcal{P}(\Sigma^* \times \mathcal{I}d + \Sigma^*)).$$

However, as we will see further on (Theorem 11 and Example 13), it is uniform also with respect to a richer category than Set, namely, it is uniform w.r.t.:

$$\sharp : \mathcal{K}l(\mathcal{P}(\Sigma^* \times \mathcal{I}d)) \rightarrow \mathcal{K}l(\mathcal{M}_1) \cong \mathcal{K}l(\mathcal{P}(\Sigma^* \times \mathcal{I}d + \Sigma^*)).$$

Uniqueness of a uniform fixed point operator on $\mathcal{K}l(T)$ can be imposed by inital algebra = final coalgebra coincidence in the base category C [37]. This coincidence is the core of generic coalgebraic trace semantics theory [14]. This is why we believe that the uniform fixed point operators can and will serve as an extension of the generic coalgebraic trace semantics to weak trace semantics.

We end this section with a result that links weak trace semantics for ε-NA's to uniform traced monoidal categories in the sense of Joyal et al. [16].

However, instead of a uniform categorical trace operator on a monoidal category with binary coproducts and initial object we will equivalently work with a uniform Conway operator [12,16]. The following theorem (modulo the uniformity) can be found in [4].

Theorem 11. *Assume* C *is equipped with a uniform Conway operator*

$$(-)^\dagger_{X,A} : Hom(X, X + A) \to Hom(X, A).$$

Let A be an object in C *and $\mathcal{M}_A = \mathcal{I}d + A$ the exception monad on* C*. Then the operator $tr_{(-)} : Hom_{\mathcal{K}l(\mathcal{M}_A)}(X, X) \to Hom_{\mathcal{K}l(\mathcal{M}_A)}(X, 0)$ defined by $tr_\alpha = \alpha^\dagger$ for $\alpha : X \to X + A$ in* C *(or equivalently $\alpha : X \multimap X$ in $\mathcal{K}l(\mathcal{M}_A)$) is a coalgebraic trace operator on the category $\mathcal{K}l(\mathcal{M}_A)$ which is uniform w.r.t. $^\sharp :$ C $\to \mathcal{K}l(\mathcal{M}_A)$.*

Example 13. The ε-NA's and their trace semantics fits into the above setting since the ε-NA monad satisfies:

$$\mathcal{P}(\Sigma^* \times \mathcal{I}d + \Sigma^*) \cong \mathcal{P}(\Sigma^* \times (\mathcal{I}d + 1)).$$

Hence, if we put $T = \mathcal{P}(\Sigma^* \times \mathcal{I}d)$ to be the free LTS monad then the ε-NA monad is given by $T(\mathcal{I}d + 1) = T\mathcal{M}_1$. Since the free LTS monad $\mathcal{P}(\Sigma^* \times \mathcal{I}d) \cong \mathcal{P}(\Sigma^*)^{\mathcal{I}d}$ is an example of a quantale monad [18] on Set its Kleisli category $\mathcal{K}l(\mathcal{P}(\Sigma^* \times \mathcal{I}d))$ with binary coproducts and initial object is equipped with a uniform Conway operator (or equivalently a uniform categorical trace operator) [12,18]. Therefore, if we put C $= \mathcal{K}l(\mathcal{P}(\Sigma^* \times \mathcal{I}d))$ then the Kleisli category for the exception monad $\mathcal{M}_1 = \mathcal{I}d + 1$ defined on C is isomorphic to the Kleisli category for ε-NA monad, i.e. $\mathcal{K}l(\mathcal{M}_1) \cong \mathcal{K}l(\mathcal{P}(\Sigma^* \times \mathcal{I}d + \Sigma^*))$. The analysis of the Conway operator for the Kleisli category for the monad $\mathcal{P}(\Sigma^* \times \mathcal{I}d)$ [18] leads to a conclusion that tr_α obtained for ε-NA's via Theorem 11 is exactly the least fixed point operator we introduced in the previous subsection.

To conclude, when allowing invisible steps into our setting, i.e. considering coalgebras over monadic types, weak trace semantics becomes a categorical fixed point operator. Moreover, as the above example states, there is a strong connection between coalgebraic trace operator for ε-NA coalgebras and traced monoidal categories. Although traced categories have been studied from coalgebraic perspective in [18] they were considered a special instance of the generic coalgebraic trace theory. With Example 13 at hand we believe that it should be the other way around in many cases, i.e. coalgebraic trace semantics for coalgebras with internal moves is a direct consequence of the fact that certain Kleisli categories are traced monoidal categories.

6   Weak Bisimulation and Weak Trace Semantics

We have shown that two behavioural relations, namely, weak bisimulation and weak trace equivalence can be defined using fixed points of certain maps. In case of trace equivalence this map is given by $x \mapsto x \cdot \alpha$, in case of weak bisimulation it is $x \mapsto 1 \vee x \cdot \alpha$. We see that both equivalences should be considered individually, as they require different assumptions. Yet, in a restrictive enough setting we should be able to compare these notions at once. Indeed, in the setting of monads whose Kleisli category has hom-sets being complete join semilattices and whose composition preserves all non-empty joins, it is possible for us to talk about three behavioural equivalences at once, namely, weak trace semantics, weak bisimilarity and bisimilarity. In this case we can prove the following.

Theorem 12. *Let T be a monad as above and let $\bot = \bot \cdot f^\sharp$ for any $f : X \to Y$ in C. A strong bisimulation on $\alpha : X \to TX$ is also a weak bisimulation on α. Moreover, if we define the trace map to be $tr_\alpha = \mu x.x \cdot \alpha$ then $tr_\alpha = tr_{\alpha^*}$. In other words, weak bisimilarity implies weak trace equivalence.*

7 Summary and Future Work

This paper shows that coalgebras with internal moves can be understood as coalgebras over a type which is a monad. We believe that such a treatment makes formulation of many different properties and behavioural equivalences simpler. It is natural to suspect that many other types of different behavioural equivalences can be translated into the coalgebraic setting this way. One of these is dynamic bisimulation [27] which should be obtained as a strong bisimulation on $\mu x.(\alpha \vee x \cdot \alpha)$ (i.e. a transitive closure of α). We believe that this paper may serve as a starting point for a larger project to translate some of the equivalences from van Glabbeek's spectrum of different equivalences for state-based systems with silent labels [8, 32] into the setting of coalgebras with internal activities.

Finally, as mentioned in Sect. 5 we should aim at extending the coalgebraic trace semantics theory for systems without internal transitions [14] to systems with silent moves. Uniform fixed point operator could serve as such an extension. Moreover, we should build a more traced monoidal category oriented theory of coalgebraic traces and refer it to known results for generic coalgebraic trace.

Acknowledgements. I would like to thank Alexandra Silva for inspiring me with the literature on categorical fixed points. I am also very grateful to anonymous referees for various comments and remarks that hopefully made this work more interesting and easier to follow.

References

1. Abramsky, S., Jung, A.: Domain theory. In: Abramsky, S., Gabbay, D.M., Maibaum, T.S.E. (eds.) Handbook of Logic in Computer Science, pp. 1–168. Oxford Univ. Press, Oxford (1994)
2. Aczel, P., Mendler, N.: A final coalgebra theorem. In: Pitt, D.H., Rydeheard, D.E., Dybjer, P., Pitts, A.M., Poigné, A.M. (eds.) CTCS 1989. LNCS, vol. 389, pp. 357–365. Springer, Heidelberg (1989)

3. Baier, C., Hermanns, H.: Weak bisimulation for fully probabilistic processes. In: Grumberg, O. (ed.) CAV 1997. LNCS, vol. 1254, pp. 119–130. Springer, Heidelberg (1997)
4. Benton, N., Hyland, M.: Traced premonoidal categories. Theor. Inf. Appl. **37**(4), 273–299 (2003)
5. Brengos, T.: Weak bisimulations for coalgebras over ordered functors. In: Baeten, J.C.M., Ball, T., de Boer, F.S. (eds.) TCS 2012. LNCS, vol. 7604, pp. 87–103. Springer, Heidelberg (2012)
6. Brengos, T.: Weak bisimulations for coalgebras over ordered monads. CoRR abs/1310.3656 (2013) (submitted)
7. Barr, M.: Coequalizers and free triples. Math. Z. **116**, 307–322 (1970)
8. van Glabbeek, R.J.: The linear time-branching time spectrum II - the semantics of sequential systems with silent moves. In: Best, E. (ed.) CONCUR 1993. LNCS, vol. 715, pp. 66–81. Springer, Heidelberg (1993)
9. Goncharov, S.: Kleene monads. Ph.D. thesis (2010)
10. Goncharov, S., Pattinson, D.: Weak Bisimulation for Monad-Type Coalgebras (2013) Slides. http://www8.cs.fau.de/sergey/talks/weak-talk.pdf
11. Gumm, H.P.: Elements of the general theory of coalgebras. In: LUATCS 99. Rand Afrikaans University, Johannesburg (1999)
12. Hasegawa, M.: Models of Sharing Graphs. A Categorical Semantics of let and letrec. Ph.D. thesis. University of Edinburgh (1997)
13. Hasuo, I., Jacobs, B., Sokolova, A.: Generic forward and backward simulations. In: (Partly in Japanese) Proceedings of the JSSST Annual Meeting (2006)
14. Hasuo, I., Jacobs, B., Sokolova, A.: Generic trace semantics via coinduction. Logical Meth. Comput. Sci. **3**(4:11), 1–37 (2007)
15. Hopcroft, J.E., Motwani, R., Ullman, J.D.: Introduction to Automata Theory, Languages, and Computation, 3rd edn. Prentice Hall, Englewood Cliffs (2006)
16. Joyal, A., Street, R., Verity, D.: Traced monoidal categories. Math. Proc. Cambridge Philos. Soc. **3**, 447–468 (1996)
17. Jacobs, B.: Coalgebraic trace semantics for combined possibilitistic and probabilistic systems. Electron. Notes Theore. Comput. Sci. **203**(5), 131–152 (2008)
18. Jacobs, B.: From coalgebraic to monoidal traces. Electron. Notes Theor. Comput. Sci. **264**(2), 125–140 (2010)
19. Jacobs, B., Silva, A., Sokolova, A.: Trace semantics via determinization. In: Pattinson, D., Schröder, L. (eds.) CMCS 2012. LNCS, vol. 7399, pp. 109–129. Springer, Heidelberg (2012)
20. Kock, A.: Strong functors and monoidal monads. Archiv der Mathematik **23**(1), 113–120 (1972)
21. Mac Lane, S.: Categories for working mathematician, 2nd edn. Springer, New York (1998)
22. Manes, E.: Algebraic Theories, 1st edn. Springer, New York (1976)
23. Manes, E., Mulry, P.: Monad compositions I: general constructions and recursive distributive laws. Theory Appl. Categories **18**(7), 172–208 (2007)
24. Miculan, M., Peressotti, M.: Weak bisimulations for labelled transition systems weighted over semirings. CoRR abs/1310.4106 (2013)
25. Milner, R. (ed.): A Calculus of Communicating Systems. LNCS, vol. 92. Springer, Heidelberg (1980)
26. Milner, R.: Communication and Concurrency. Prentice Hall, New York (1989)
27. Montanari, U., Sassone, V.: Dynamic congruence vs. progressing bisimulation for CCS^*. Fundamenta Informaticae **16**(2), 171–199 (1992)

28. Rothe, J.: A syntactical approach to weak (bi)-simulation for coalgebras. In: Proceedings of the CMCS02. Electronic Notes in Theoretical Computer Science, vol. 65, pp. 270–285 (2002)
29. Rothe, J., Masulović, D.: Towards weak bisimulation for coalgebras. In: Proceedings of the Categorical Methods for Concurrency, Interaction and Mobility. Electronic Notes in Theoretical Computer Science, vol. 68(1), pp. 32–46 (2002)
30. Rutten, J.: A note on coinduction and weak bisimilarity for while programs. RAIRO - Theore. Inf. Appl. **33**(4–5), 393–400 (1999)
31. Rutten, J.: Universal coalgebra: a theory of systems. Theoret. Comput. Sci. **249**, 3–80 (2000)
32. Sangiorgi, D.: Introduction to Bisimulation and Coinduction. Cambridge University Press, Cambridge (2011)
33. Segala, R., Lynch, N.: Probabilistic simulations for probabilistic processes. In: Jonsson, B., Parrow, J. (eds.) CONCUR 1994. LNCS, vol. 836, pp. 481–496. Springer, Heidelberg (1994)
34. Segala, R.: Modeling and verification of randomized distributed real-time systems. Ph.D. thesis, MIT (1995)
35. Silva, A., Westerbaan, B.: A coalgebraic view of ε-transitions. In: Heckel, R. (ed.) CALCO 2013. LNCS, vol. 8089, pp. 267–281. Springer, Heidelberg (2013)
36. Silva, A., Bonchi, F., Bonsangue, M., Rutten, J. :Generalizing the powerset construction, coalgebraically. In: Proceedings of the FSTTCS 2010, LIPIcs 8, pp. 272–283 (2010)
37. Simpson, A., Plotkin, G.: Complete axioms for categorical fixed-point operators. In: Proceedings of the 15th Annual Symposium on Logic in Computer Science, pp. 30–41 (2000)
38. Sokolova, A., de Vink, E., Woracek, H.: Coalgebraic weak bisimulation for action-type systems. Sci. Ann. Comp. Sci. **19**, 93–144 (2009)
39. Staton, S.: Relating coalgebraic notions of bisimulation. Logical Meth. Comput. Sci. **7**(1), 1–21 (2011)

A Coalgebraic View of Characteristic Formulas in Equational Modal Fixed Point Logics

Sebastian Enqvist[1,2(✉)] and Joshua Sack[1]

[1] Institute of Logic, Language, and Computation, Universiteit van Amsterdam,
P.O. Box 94242, 1090 GE Amsterdam, The Netherlands
[2] Department of Philosophy, Lund University,
Kungshuset, Lundagard, 222 22 Lund, Sweden
Sebastian.Enqvist@fil.lu.se

Abstract. The literature on process theory and structural operational semantics abounds with various notions of behavioural equivalence and, more generally, simulation preorders. An important problem in this area from the point of view of logic is to find formulas that characterize states in finite transition systems with respect to these various relations. Recent work by Aceto et al. shows how such characterizing formulas in equational modal fixed point logics can be obtained for a wide variety of behavioural preorders using a single method. In this paper, we apply this basic insight from the work by Aceto et al. to Baltag's "logics for coalgebraic simulation" to obtain a general result that yields characteristic formulas for a wide range of relations, including strong bisimilarity, simulation, as well as bisimulation and simulation on Markov chains and more. Hence this paper both generalizes the work of Aceto et al. and makes explicit the coalgebraic aspects of their work.

1 Introduction

The literature on process theory and structural operational semantics contains a multitude of various notions of behavioural equivalence and, more generally, simulation preorders. The most prominent example, perhaps, is the notion of *strong bisimulation*: given labelled transition systems S and T, a relation Z between states of S and states of T is said to be a strong bisimulation if the following conditions hold:

Forth: If uZv and $u \xrightarrow{a} u'$ for some action a, then there is v' with $v \xrightarrow{a} v'$ and $u'Zv'$.
Back: If uZv and $v \xrightarrow{a} v'$, then there is u' with $u \xrightarrow{a} u'$ and $u'Zv'$.

The weaker notion of *simulation* is like bisimulation except that the "Back" condition is dropped. Another way to weaken the notion of strong bisimulation is to "truncate" the silent τ-transitions, according to the intuition that bisimulation

Supported by The Netherlands Organisation for Scientific Research VIDI project 639.072.904.

M.M. Bonsangue (Ed.): CMCS 2014, LNCS 8446, pp. 98–117, 2014.
DOI: 10.1007/978-3-662-44124-4_6

should capture equivalence of *observable* behaviour. The resulting concept of behavioural equivalence is called *weak* bisimulation.

An important problem in this area from the point of view of logic is to find formulas that characterize states in finite transition systems with respect to these various relations. For example, in the case of strong bisimilarity, we want to find a formula φ that characterizes a given state u in a finite labelled transition system S "up to bisimilarity", in the sense that a state v in a transition system T satisfies φ if and only if (T, v) is bisimilar with (S, u). Recent work by Aceto et al. shows how such characterizing formulas in equational modal fixed point logics can be obtained for a wide variety of behavioural preorders using a single method. In such equational fixed point logics, the semantics of formulas is parametric in a system of equations, which are to be read (in this context) as greatest fixed point definitions of variables. For example, in Hennessy-Milner logic, the equation

$$p := \varphi \wedge [a]p$$

assigns to the variable p the meaning: "the formula φ is true throughout every a-path starting from the current state". Generally, a fixed point language allows us to characterize infinite or looping behaviour of a model using finitary formulas.

In this paper, we apply the basic insight from the work by Aceto et al. to Baltag's "logics for coalgebraic simulation", which generalize the original coalgebraic languages introduced by Moss in the seminal paper [13], to obtain a general result that yields characteristic formulas for a wide range of relations. These include strong bisimilarity, simulation, as well as bisimulation and simulation on Markov chains and more. The key observations that will drive the result are:

1. The semantics for the modal operators and the various notions of simulation both arise from the same concept of relation lifting via a lax extension.
2. A finite coalgebra can itself be viewed as a system of equations.

These features of the logics make the construction of the characteristic formulas particularly direct and natural. However, the syntax of these languages directly involve the functor T, and can be somewhat difficult to grasp intuitively. Therefore, we also provide conditions that allow us to automatically derive characteristic formulas in the language of *predicate liftings* for a given (finitary) functor. These latter languages have become increasingly popular in the coalgebraic logic community, and they have the advantage of staying closer to the more conventional syntax of languages like Hennessy-Milner logic.

2 Basics

2.1 Set Coalgebras and Lax Extensions

In this section we introduce some basic concepts from coalgebra theory that will be used later on. We assume familiarity with basic category theoretic concepts. We fix a functor $T : \mathbf{Set} \rightarrow \mathbf{Set}$, where $\mathbf{Set}$ is the category of sets and mappings

For simplicity, we assume that T preserves set inclusions, so that a set inclusion $\iota : X \to Y$ is mapped to a set inclusion $T\iota : TX \to TY$. This assumption is actually more innocent than it may seem at first, since every set functor is naturally isomorphic "up-to-$\emptyset$" to one that preserves set inclusions. More precisely, for every set functor T there is a functor T' such that the restrictions of these two functors to the full subcategory of non-empty sets are naturally isomorphic, see [1] for details.

We will make use of an approach to coalgebraic logic developed by Alexandru Baltag, based on certain methods of extending the signature functor T to relations [3]. This approach is a generalization of the original formulation of coalgebraic logic due to Moss [13]. While Baltag uses "weak T-relators" (for more on relators and simulations [7,9,17]), we shall here use the slightly more general notion of "lax extension" [11], which works just as well. Besides that, our approach is the same as Baltag's.

Definition 1. *Given a function $f : X \to Y$, let $\widehat{f} = \{(x, f(x)) \mid x \in X\}$ be the graph of f. Let $\Delta_X = \widehat{Id_X}$ be the graph of the identity map on X.*

The concept of a lax extension is defined as follows:

Definition 2. *A* lax extension *of a set functor T is a relation lifting (i.e. a mapping that sends every relation $R \subseteq X \times Y$ to a relation $LR \subseteq TX \times TY$) subject to the constraints:*

L1: $R \subseteq S$ *implies* $LR \subseteq LS$,
L2: $LR; LS \subseteq L(R; S)$,
L3: $\widehat{Tf} \subseteq L\widehat{f}$ *for any mapping f.*

Thus lax extensions are lax endofunctors on the 2-category of sets and relations with inclusions between relations as the 2-cells. Note that condition **L3** implies that

$$\Delta_{TX} \subseteq L\Delta_X \tag{1}$$

for all sets X, since $\Delta_{TX} = \widehat{Id_{TX}} = \widehat{T(Id_X)}$. A good example of a lax extension (that also happens to be a weak relator), which we will come back to several times, is the following:

Example 1. The finitary covariant powerset functor $(T = \mathcal{P}_\omega)$ has a lax extension given by

$$L_{sim}R := \{(A, B) \in \mathcal{P}_\omega X \times \mathcal{P}_\omega Y \mid \forall a \in A \; \exists b \in B : aRb\}.$$

In other words, $L_{sim}R$ consists of all pairs (A, B), such that there is a function from A to B whose graph is a subset of R. It is easy to see that **L1** and **L2** hold. For **L3**, let $f : X \to Y$, and let $A\widehat{\mathcal{P}_\omega f}B$. This means that $B = f[A]$. As f maps A to B and its graph is $\widehat{f}$, we have that $AL_{sim}\widehat{f}B$.

Definition 3 (L-simulation). *An L-simulation* from a T-coalgebra (X, α) to (Y, β) *is a binary relation* $Z \subseteq X \times Y$ *such that uZv implies $\alpha(u)(LZ)\beta(v)$.* *Given pointed T-coalgebras $(\mathfrak{A}, u)$ and $(\mathfrak{B}, v)$, we write $(\mathfrak{A}, u) \preceq_L (\mathfrak{B}, v)$ to say that there is an L-simulation Z from $\mathfrak{A}$ to $\mathfrak{B}$ with uZv.*

Note that L_{sim}-simulation (with L_{sim} from Example 1) is simulation on Kripke frames.

2.2 Symmetric Lax Extensions and Bisimulation

We write the converse of a relation R as R°. Given a relation lifting L, let $L^\circ : R \mapsto (L(R^\circ))^\circ$. Call a relation lifting L *symmetric* if $L = L^\circ$.

Example 2 (Barr extension). Given sets X, Y and a binary relation $R \subseteq X \times Y$, the relation $\overline{T}R \subseteq TX \times TY$ is defined by

$$a\overline{T}Rb \Leftrightarrow \exists c \in T(R) : T\pi_X(c) = a \ \& \ T\pi_Y(c) = b.$$

Here, π_X and π_Y are the projections from the product $X \times Y$. Then $\overline{T}$ is a symmetric relation lifting, and in the case where T preserves weak pullbacks, $\overline{T}$ is a symmetric lax extension of T called the *Barr extension* of T [4].

Definition 4. *Given a set functor T, a T-bisimulation is an L-simulation, where $L = \overline{T}$.*

We could also call a T-bisimulation a $\overline{T}$-bisimulation, and we will more generally define what is meant by L-bisimulation for any lax extension L (not necessarily symmetric) that *extends* $\overline{T}$ in that $\overline{T}R \subseteq LR$ for each relation R. We first observe the following.

Observation 1. *If L is a lax extension that extends $\overline{T}$, then L° is a lax extension.*

Proof. We prove each case in turn.

– By **L1** in Definition 2, we have

$$R \subseteq S \Rightarrow R^\circ \subseteq S^\circ \Rightarrow L(R^\circ) \subseteq L(S^\circ) \Rightarrow L^\circ(R) \subseteq L^\circ(S).$$

– We reason as follows:

$$L^\circ(R); L^\circ(S) = (L(R^\circ))^\circ; (L(S^\circ))^\circ = (L(S^\circ); L(R^\circ))^\circ$$
$$\subseteq (L(S^\circ; R^\circ))^\circ = (L((R; S)^\circ))^\circ = L^\circ(R; S).$$

– Let $f : X \to Y$. For $a \in TX$, we have

$$a(\overline{T}\widehat{f})(Tf(a)) \Rightarrow (Tf(a))(\overline{T}(\widehat{f^\circ}))a \qquad (\overline{T} \text{ is a symmetric relation lifting})$$
$$\Rightarrow (Tf(a))(L(\widehat{f^\circ}))a \qquad (\overline{T}R \subseteq LR \text{ for all relations } R)$$
$$\Rightarrow a(L\widehat{f})(Tf(a)).$$

Thus $\widehat{Tf} \subseteq L^\circ(\widehat{f})$.   □

Definition 5. *Given a lax extension L, let $B(L)$ be defined[1] by*

$$B(L) = L \cap L^\circ.$$

We call $B(L)$ the bisimulator *of L.*

The reader can easily check that the following holds:

Observation 2. *If L_1 and L_2 are lax extensions, then $L_1 \cap L_2$ is a lax extension.*

Hence we get:

Observation 3. *If L is a lax extension, such that $\overline{T}R \subseteq LR$ for each relation R, then its bisimulator $B(L)$ is a symmetric lax extension.*

Proof. First note that as L extends $\overline{T}$, we have by Observation 1 that L° is a lax extension. Then by Observation 2, $B(L)$ is a lax extension. By definition $B(L)$ is symmetric.   □

Remark 1. If L_{sim} is the lax extension for the finitary power set functor $\mathcal{P}_\omega$ that was given in Example 1, then $B(L_{sim}) = \overline{\mathcal{P}_\omega}$.

Definition 6. *If L is a lax extension, such that $\overline{T}R \subseteq LR$ for every relation R, then an L-bisimulation is a $B(L)$-simulation.*

By the previous remark, Z is an L_{sim}-bisimulation if and only if Z is a $\overline{\mathcal{P}_\omega}$-bisimulation.

3   Coalgebraic Logic with Fixed Point Equations

3.1   Basic Coalgebraic Modal Logic

Definition 7. *Given a (finite) set of variables V, the syntax of the* basic coalgebraic logic *over V is defined as the smallest set $\mathcal{L}$ such that*

- *$p \in \mathcal{L}$ for all $p \in V$,*
- *$\varphi, \psi \in \mathcal{L}$ implies $\varphi \wedge \psi \in \mathcal{L}$ and $\varphi \vee \psi \in \mathcal{L}$,*
- *if Φ is a finite subset of $\mathcal{L}$ then $\Box a \in \mathcal{L}$ and $\Diamond a \in \mathcal{L}$ for each $a \in T\Phi$.*

The following observation is made in [18]:

Proposition 1. *Let T be a set functor that preserves inclusions. Then for any $a \in TX$ where X is a finite set, there is a unique smallest set $Y \subseteq X$ with $a \in TY$.*

In particular, this guarantees that for any formula $\Box a$ or $\Diamond a$, there is a unique smallest set of formulas Φ with $a \in T\Phi$. We denote this set by $SPT(a)$, for "support of a". When X is understood by context, we write SPT for SPT_X.

Given a fixed *valuation*, a function $v : V \to \mathcal{P}(X)$ (equivalently $v \in \mathcal{P}(X)^V$), we define the satisfaction relation $\vDash_v$ between pointed coalgebras $(\mathfrak{A}, u)$ (where $\mathfrak{A} = (X, \alpha)$) and formulas in $\mathcal{L}(V)$, relative to the valuation v, with the following inductive clauses:

---

[1] Here, for lax extensions L_1 and L_2 we define $L_1 \cap L_2$ by $R \mapsto L_1 R \cap L_2 R$.

- $(\mathfrak{A}, u) \vDash_v p$ iff $u \in v(p)$, for a propositional variable p,
- $(\mathfrak{A}, u) \vDash_v \varphi \wedge \psi$ iff $(\mathfrak{A}, u) \vDash_v \varphi$ and $(\mathfrak{A}, u) \vDash_v \psi$,
- $(\mathfrak{A}, u) \vDash_v \varphi \vee \psi$ iff $(\mathfrak{A}, u) \vDash_v \varphi$ or $(\mathfrak{A}, u) \vDash_v \psi$,
- $(\mathfrak{A}, u) \vDash_v \Box a$ iff $\alpha(u)(L \vDash_v)a$,
- $(\mathfrak{A}, u) \vDash_v \Diamond a$ iff $\alpha(u)(L^\circ(\vDash_v))a$.

Sometimes the definition of the semantics of $\Box a$ and $\Diamond a$ are given with $\vDash_v$ replaced with $\vDash_v \restriction_{X \times SPT(a)} = \vDash_v \cap X \times SPT(a)$. We show that both definitions are equivalent.

Observation 4. *If L is a lax extension, then $\alpha(u)(L \vDash_v \restriction_{X \times SPT(a)})a$ if and only if $\alpha(u)(L \vDash_v)a$.*

Proof. First note that $\vDash_v \restriction_{X \times SPT(a)} \subseteq \vDash_v$ and hence by **L1**, $\alpha(u)(L \vDash_v \restriction_{X \times SPT(a)})a$ implies $\alpha(u)(L \vDash_v)a$. For the other direction, suppose that $\alpha(u)(L \vDash_v)a$. By definition of SPT, $a \in TSPT(a)$, and hence $(a, a) \in \Delta_{TSPT(a)}$. By (1), $(a, a) \in L\Delta_{SPT(a)}$. Then $\alpha(u)(L \vDash_v); L(\Delta_{SPT(a)})a$. The desired result follows from this, **L2**, and the fact that $\vDash_v \restriction_{X \times SPT(a)} = (\vDash_v); (\Delta_{SPT(a)})$. $\qquad\square$

Remark 2. If L is a symmetric lax extension, then the formulas $\Box a$ and $\Diamond a$ are equivalent. In this case, we might write ∇a instead of $\Box a$ to emphasize that $\Box$ and $\Diamond$ are the same. If $L = \overline{T}$, then these modalities are the same as the ∇-modality from the (finitary version of) Moss' presentation of coalgebraic logic [13].

3.2   Fixed Point Semantics

In this section we introduce the (greatest) fixed point semantics for the logic $\mathcal{L}(V)$, relative to a system of equations. First, we have to say more precisely what a system of equations is:

Definition 8 (System of equations). *Given a set of variables V, a system of fixed-point equations is defined to be a mapping $s : V \to \mathcal{L}(V)$.*

We shall construct a fixed point semantics using a system of equations as a parameter. First note that the set $\mathcal{P}(X)^V$ of V-indexed tuples of subsets of a set X, or "valuations in X", forms a complete lattice under the relation $\sqsubseteq$ of point-wise set inclusion. That is, given $v, v' : V \to \mathcal{P}(X)$ we set

$$v \sqsubseteq v' \text{ iff } \forall x \in V : v(x) \subseteq v'(x).$$

We denote the arbitrary (potentially infinite) join operation in this lattice by $\bigvee$. The reader can now easily check that a system of fixed point equations s defines a monotone operation $\mathcal{O}_s$ on the lattice of valuations $\mathcal{P}(X)^V$, by letting (for $v : V \to \mathcal{P}(X)$ and $x \in V$):

$$\mathcal{O}_s(v)(x) = \{w \in X \mid (\mathfrak{A}, w) \vDash_v s(x)\}.$$

By the Knaster-Tarski fixed point theorem, $\mathcal{O}_s$ is guaranteed a greatest fixed point, which we denote $GFP(s)$, so that

$$GFP(s) := GFP(\mathcal{O}_s) = \bigvee\{\sigma : V \to \mathcal{P}(X) \mid \sigma \sqsubseteq \mathcal{O}_s(\sigma)\}.$$

For a pointed coalgebra $(\mathfrak{A}, u)$ we write $(\mathfrak{A}, u) \vDash_s \varphi$ as an abbreviation for $(\mathfrak{A}, u) \vDash_{GFP(s)} \varphi$, and say in this case that $(\mathfrak{A}, u)$ satisfies the formula φ relative to the system of equations s. We could, of course, introduce a least fixed point semantics relative to s in the same way, but since we will not have any use for that here we refrain from doing so.

The following observation will be used for proving the correctness of the characteristic formula for mutual simulation in Example 7. It plays a somewhat similar role as [2, Lemma 4.6] toward this goal.

Observation 5. *Let s and t be systems of equations, such that $s = t \restriction V_0$ for some subset V_0 of the variables used in t. Given a T-coalgebra $(\mathfrak{A}, u)$, and a variable $x \in V_0$,*

$$(\mathfrak{A}, u) \vDash_s x \text{ iff } (\mathfrak{A}, u) \vDash_t x.$$

Proof. The proof of this is a straightforward induction. The key observation is that as $s = t \restriction V_0$, for each $p \in V_0$, $t(p)$ is a formula over the variables in V_0, and hence all variables not in V_0 are "unreachable" in t from variables in V_0.   □

4   Characteristic Formulas

Definition 9. *Given a lax extension L and coalgebras $\mathfrak{A} = (X, \alpha)$ and $\mathfrak{B} = (Y, \beta)$, let $\mathcal{F}_L$ be the endofunction on $\mathcal{P}(X \times Y)$ defined by*

$$(x, y) \in \mathcal{F}_L(R) \Leftrightarrow \alpha(x)(LR)\beta(y).$$

Note that $\mathcal{F}_L$ is a monotone increasing function on the complete lattice of relations in $\mathcal{P}(X \times Y)$, and hence by the Knaster-Tarski fixed point theorem, $\mathcal{F}_L$ has a greatest fixed point. It is clear that a relation $R \subseteq X \times Y$ is a post-fixed point of $\mathcal{F}_L$ iff it is an L-simulation, and so the greatest fixed point of $\mathcal{F}_L$ is the relation $\preceq_L$, i.e. we have $(u, v) \in GFP(\mathcal{F}_L)$ iff there is an L-simulation relating u to v.

We consider the language $\mathcal{L}(X)$ with X being the set of variables. Let Φ be a function from relations in $\mathcal{P}(X \times Y)$ to valuations in $\mathcal{P}(Y)^X$, such that $\Phi(R)(x) = \{y \mid (x, y) \in R\}$. Let Ψ be the function from relations in $\mathcal{P}(Y \times X)$ to valuations in $\mathcal{P}(Y)^X$, such that $\Psi(R)(x) = \Phi(R^\circ)(x) = \{y \mid (y, x) \in R\}$.

Definition 10. *Let $\mathfrak{A} = (X, \alpha)$ and $\mathfrak{B} = (Y, \beta)$ be T-coalgebras. We say that a system of equations $s : X \to \mathcal{L}(X)$ directly expresses the endofunction $\mathcal{F}_L$ if for $Z \subseteq Y \times X$*

$$(\mathfrak{B}, y) \vDash_{\Psi(Z)} s(x) \Leftrightarrow (y, x) \in \mathcal{F}_L(Z).$$

Similarly, we say that a system of equations $s : X \to \mathcal{L}(X)$ conversely expresses the endofunction $\mathcal{F}_L$ if for $Z \subseteq X \times Y$

$$(\mathfrak{B}, y) \vDash_{\Phi(Z)} s(x) \Leftrightarrow (x, y) \in \mathcal{F}_L(Z).$$

Theorem 1. *If s directly expresses $\mathcal{F}_L$, then*

$$(\mathfrak{B}, v) \vDash_s u \text{ iff } (\mathfrak{B}, v) \preceq_L (\mathfrak{A}, u).$$

If s conversely expresses $\mathcal{F}_L$, then

$$(\mathfrak{B}, v) \vDash_s u \text{ iff } (\mathfrak{A}, u) \preceq_L (\mathfrak{B}, v).$$

The idea behind this theorem has been used for some time, and has been given in papers such as [2,14,15]. The presentation in this paper is most similar to a formulation given in [15], which addressed probabilistic simulations in a non-coalgebraic setting. The proofs given in those papers apply to this setting as well. However, we provide a sketch of the proof here to emphasize that it applies to our more general (coalgebraic) setting.

Proof (sketch). Recall that $\mathcal{O}_s$ is a function from $\mathcal{P}(Y)^X$ to itself, such that $\mathcal{O}_s(v)(x) = \{y \mid (\mathfrak{B}, y) \vDash_v s(x)\}$, and this function has the greatest fixed point $GFP(s)$. It follows directly from the definitions that s directly expresses $\mathcal{F}_L$ if and only if the following diagram commutes:

$$
\begin{array}{ccc}
\mathcal{P}(Y \times X) & \xrightarrow{\mathcal{F}_L} & \mathcal{P}(Y \times X) \\
\Psi \downarrow & & \downarrow \Psi \\
\mathcal{P}(Y)^X & \xrightarrow{\mathcal{O}_s} & \mathcal{P}(Y)^X
\end{array}
$$

Similarly, s conversely expresses $\mathcal{F}_L$ if and only if the following commutes:

$$
\begin{array}{ccc}
\mathcal{P}(X \times Y) & \xrightarrow{\mathcal{F}_L} & \mathcal{P}(X \times Y) \\
\Phi \downarrow & & \downarrow \Phi \\
\mathcal{P}(Y)^X & \xrightarrow{\mathcal{O}_s} & \mathcal{P}(Y)^X
\end{array}
$$

Hence, if s directly expresses $\mathcal{F}_L$ then, since the function Ψ is an isomorphism between the lattices of relations in $\mathcal{P}(Y \times X)$ and variable interpretations in $\mathcal{P}(Y)^X$, by [2, Theorem 2.3] it maps the greatest fixed point of $\mathcal{F}_L$ to the greatest fixed point of $\mathcal{O}_s$, that is $GFP(s) = \Psi(GFP(\mathcal{F}_L)) = \Psi(\preceq_L)$. So we get

$$
\begin{aligned}
(\mathfrak{B}, v) \vDash_s u &\Leftrightarrow (\mathfrak{B}, v) \vDash_{GFP(s)} u \\
&\Leftrightarrow v \in GFP(s)(u) \\
&\Leftrightarrow v \in \Psi(\preceq_L)(u) \\
&\Leftrightarrow v \preceq_L u.
\end{aligned}
$$

Similarly, the function Φ is an isomorphism between relations in $\mathcal{P}(X \times Y)$ and variable interpretations in $\mathcal{P}(Y)^X$, and hence if s conversely expresses $\mathcal{F}_L$ it maps the greatest fixed point of $\mathcal{F}_L$ to the greatest fixed point of $\mathcal{O}_s$, that is $GFP(s) = \Phi(GFP(\mathcal{F}_L)) = \Phi(\preceq_L)$. So we get

$$(\mathfrak{B}, v) \vDash_s u \Leftrightarrow (\mathfrak{B}, v) \vDash_{GFP(s)} u$$
$$\Leftrightarrow v \in GFP(s)(u)$$
$$\Leftrightarrow v \in \Phi(\preceq_L)(u)$$
$$\Leftrightarrow u \preceq_L v$$

as required.   □

We are now left with the task to find systems of equations that express $\mathcal{F}_L$ (directly and conversely). The main observation here is that, with the semantics we are using here for the □- and ◇-operators, this is easy: a finite T-coalgebra almost *is* a system of equations!

To be precise, fix a finite T-coalgebra $\mathfrak{A} = (X, \alpha)$. We treat the set X as a set of variables and consider the language $\mathcal{L}(X)$. We define two systems of equations $s_\Box$ and $s_\Diamond$ by setting

$$s_\Box(u) := \Box\alpha(u)$$

and

$$s_\Diamond(u) := \Diamond\alpha(u).$$

We then get the following result:

Lemma 1. *For any lax extension L, where L° is also a lax extension, $s_\Box$ directly expresses $\mathcal{F}_L$, and $s_\Diamond$ conversely expresses $\mathcal{F}_L$.*

Proof. Let $Z \subseteq Y \times X$. Given $x \in X$, note that

$$\vDash_{\Psi(Z)}\!\upharpoonright_{Y \times SPT(\alpha(x))} \subseteq Z \subseteq \vDash_{\Psi(Z)}.$$

Then by **L2** and Observation 4, $\beta(y)(L \vDash_{\psi(Z)})\alpha(x)$ iff $\beta(y)(LZ)\alpha(x)$. Then

$$(\mathfrak{B}, y) \vDash_{\Psi(Z)} s_\Box(x) \Leftrightarrow (\mathfrak{B}, y) \vDash_{\Psi(Z)} \Box\alpha(x)$$
$$\Leftrightarrow \beta(y)(L \vDash_{\Psi(Z)})\alpha(x)$$
$$\Leftrightarrow (\beta(y), \alpha(x)) \in L(Z)$$
$$\Leftrightarrow (y, x) \in \mathcal{F}_L(Z).$$

This proves that $s_\Box$ directly expresses $\mathcal{F}_L$.

To see that $s_\Diamond$ conversely expresses $\mathcal{F}_L$, let $Z \subseteq X \times Y$. Then, as $\Phi(Z) = \Psi(Z^\circ)$,

$$(\mathfrak{B}, y) \vDash_{\Phi(Z)} s_\Diamond(x) \Leftrightarrow (\mathfrak{B}, y) \vDash_{\Psi(Z^\circ)} \Diamond\alpha(x)$$
$$\Leftrightarrow \beta(y)(L^\circ \vDash_{\Psi(Z^\circ)})\alpha(x)$$
$$\Leftrightarrow (\beta(y), \alpha(x)) \in L^\circ(Z^\circ)$$
$$\Leftrightarrow (\alpha(x), \beta(y)) \in L(Z)$$
$$\Leftrightarrow (x, y) \in \mathcal{F}_L(Z).$$

                                                                               □

From Lemma 1 together with Theorem 1, we immediately get our main result:

Theorem 2. *Let $\mathfrak{B} = (Y, \beta)$ be any T-coalgebra. Then, for $u \in X$ and $v \in Y$, relative to the system of equations $s_\square$ we have*

$$(1) \quad (\mathfrak{B}, v) \vDash u \ \text{iff} \ (\mathfrak{B}, v) \preceq_L (\mathfrak{A}, u).$$

Conversely, relative to the system of equations $s_\Diamond$ we have

$$(2) \quad (\mathfrak{B}, v) \vDash u \ \text{iff} \ (\mathfrak{A}, u) \preceq_L (\mathfrak{B}, v).$$

We also get characteristic formulas for various notions of bisimilarity as an easy corollary to this result. Given a lax extension L that extends $\overline{T}$, we use $\sim_L$ as an abbreviation for the simulation relation $\preceq_{B(L)}$, where $B(L)$ is the bisimulator of L.

Corollary 1. *Let $\mathfrak{A} = (X, \alpha)$ be a T-coalgebra and let $s_{\square, \Diamond}$ be the system of equations over X defined by*

$$v \mapsto \square\alpha(v) \wedge \Diamond\alpha(v).$$

Then for $u \in X$ and any pointed T-coalgebra $(\mathfrak{B}, w)$ we have

$$(\mathfrak{B}, w) \vDash_{s_{\square, \Diamond}} u \ \text{iff} \ (\mathfrak{B}, w) \sim_L (\mathfrak{A}, u).$$

Proof. Let ∇ be the "box modality" corresponding to the lax extension $B(L)$. It is easy to see that $s_{\square, \Diamond}$ gives rise to the same operator on the lattice of evaluations in a coalgebra $\mathfrak{B}$ as the system s_∇ defined by

$$v \mapsto \nabla\alpha(v).$$

The corollary now follows from Theorem 2 applied to s_∇. $\square$

Example 3. Consider the $\mathcal{P}$-coalgebra $\mathfrak{A}$ depicted by

$$x \longleftarrow y \longrightarrow z \circlearrowright$$

Then, given that $\square$ is for example the box modality corresponding to L_{sim}, the system of equations $s_\square$ is given by

$$s_\square(x) = \square\emptyset$$
$$s_\square(y) = \square\{x, z\}$$
$$s_\square(z) = \square\{z\}$$

Remark 3. In the case where L is the Barr extension of T (where T preserves weak pullbacks), the system of equations $s_\square$ (equivalently $s_\Diamond$) can viewed as a very simple "T-automaton" in the sense of [18]. Hence [18, Proposition 4.9], which shows that any finite T-coalgebra can be characterized up to bisimilarity by a suitable T-automaton, can be seen as a special instance of Theorem 2.

4.1   Predicate Liftings

The $\Box$ and $\Diamond$ modalities used to obtained characteristic formulas above have the nice feature that the appropriate connection between the formulas and the lax extension L is built directly into the semantics. On the other hand, these modalities are rather abstract. By contrast, modalities based on *predicate liftings* are relatively easy to grasp and are formally closer to the standard modalities used in Hennessy-Milner logic and other modal logics for specification of various kinds of transition systems. In this section we provide conditions on a lax extension L that allow us to derive characteristic systems of equations for L-simulation in the language of predicate liftings. This is very closely related to a recent result by Marti and Venema, appearing first in [10] and later in [12]. The result builds on earlier work by A. Kurz and R. Leal [8], and provides a translation of nabla-style coalgebraic logic corresponding to a lax extension into the logic of predicate liftings. The one subtle difference is that, while Marti and Venema restrict attention to symmetric lax extensions, we are interested also in the non-symmetric case. The non-symmetric case allows us to characterize simulation preorders whereas the symmetric only allows us to characterize behavioral equivalences.

Definition 11. *An n-ary predicate lifting for a set functor T is a natural transformation*

$$\lambda : Q^n \to QT$$

where Q is the contravariant powerset functor[2].

Fix a *finitary* functor $T : \mathbf{Set} \to \mathbf{Set}$ and a lax extension L that extends $\overline{T}$. Given a set V of variables, the language of all predicate liftings Λ for T over the variables V is given by the grammar:

$$\mathcal{L}_\Lambda(V) \ni \varphi ::= x \mid \varphi \wedge \varphi \mid \varphi \vee \varphi \mid \lambda(\varphi, \dots, \varphi),$$

where x ranges over V and λ ranges over predicate liftings. Given a coalgebra $\mathfrak{A} = (X, \alpha)$ and a valuation $v : V \to \mathcal{P}X$, the semantics is given by the usual clauses for variables and Booleans, with the evaluation clause for liftings:

$$(\mathfrak{A}, u) \vDash_v \lambda(\varphi_1, \dots, \varphi_n) \Leftrightarrow \alpha(u) \in \lambda_X(tr^v_\mathfrak{A}(\varphi_1), \dots, tr^v_\mathfrak{A}(\varphi_n)),$$

where, here and from now on, $tr^v_\mathfrak{A} : \mathcal{L}_\Lambda(V) \to QX$ sends a formula φ to its "truth set":

$$tr^v_\mathfrak{A}(\varphi) = \{v \in X \mid (\mathfrak{A}, v) \vDash_v \varphi\}.$$

A *system of equations* is a mapping $s : X \to \mathcal{L}_\Lambda(V)$. Any system of equations s gives rise to an operator $\mathcal{O}_s$ on the lattice of evaluations in $\mathfrak{A}$ in the same way as before; if this operator is always monotone, then we say that the system s is *positive*. In this case the operator always has a greatest fixed point, and we write $(\mathfrak{A}, u) \vDash_s \varphi$ as shorthand for $(\mathfrak{A}, u) \vDash_{GFP(s)} \varphi$, where the evaluation $GFP(s)$ is the greatest fixed point for this operator.

We introduce some notation: given a set X, let $\in_X$ denote the membership relation from X to QX. Consider the following conditions on L:

---

[2] Given a mapping $h : X \to Y$, $Qh : QY \to QX$ is defined by $Qh(Z) = h^{-1}[Z]$.

A1 Given a mapping $f : Z \to X$ and a relation $R \subseteq X \times Y$, we have

$$\widehat{Tf}; LR = L(\widehat{f}; R).$$

A2 Given a relation $R \subseteq X \times Y$ and a mapping $f : Z \to Y$, we have

$$L(R; (\widehat{f})^\circ) = LR; (\widehat{Tf})^\circ.$$

It is shown in [10, Proposition 3.10] and [12, Proposition 5] that these conditions hold for all symmetric lax extensions.

Example 4. The reader can verify that these conditions hold for the lax extension L_{sim} from Example 1.

Observation 6. *If **A1** and **A2** hold for L, then they hold for L° also.*

An immediate consequence of the conditions **A1** and **A2** is the following:

Lemma 2. *If L satisfies **A1** and **A2**, then the mappings $d_X : TQX \to QTX$ defined by*

$$a \mapsto \{b \in TX \mid b(L \in_X)a\}$$

form a distributive law, i.e. they are the components of a natural transformation

$$d : TQ \to QT.$$

The case where L is symmetric is shown is given in [12, Proposition 19]. Below we verify that the equations **A1** and **A2** suffice for the proof to go through.

Proof. Let $h : X \to Y$ be any mapping and let $a \in TQY$. First, note that

$$\in_X; (\widehat{Qh})^\circ = \widehat{h}; \in_Y \qquad\qquad (2)$$

since, for $u \in X$ and $z \in QY$, we have $u \in Qh(z)$ iff $h(u) \in z$. We calculate:

$$
\begin{aligned}
d_X \circ TQh(a) &= \{b \in TX \mid b(L \in_X)TQh(a)\} \\
&= \{b \in TX \mid b(L \in_X); (\widehat{TQh})^\circ a\} \\
&= \{b \in TX \mid b(L(\in_X; (\widehat{Qh})^\circ))a\} &&\text{by } \textbf{A2} \\
&= \{b \in TX \mid b(L(\widehat{h}; \in_Y))a\} &&\text{by } (2) \\
&= \{b \in TX \mid b(\widehat{Th}; (L \in_Y))a\} &&\text{by } \textbf{A1} \\
&= \{b \in TX \mid Th(b)(L \in_Y)a\} \\
&= QTh \circ d_Y(a),
\end{aligned}
$$

and we have proven that $d_X \circ TQh = QTh \circ d_Y$, so that d is a natural transformation. $\square$

Lemma 3. *Suppose L satisfies $\mathbf{A1}$ and $\mathbf{A2}$, and let d be the distributive law determined by L, according to Lemma 2. Let $\mathfrak{A} = (X, \alpha)$ be a coalgebra and $v : V \to \mathcal{P}X$ a valuation. Then for $a \in TX$ and $b \in T\mathcal{L}(V)$, we have*

$$a(L \vDash_v)b \quad \textit{iff} \quad a \in d_X \circ T(tr^v_{\mathfrak{A}})(b).$$

From this point we can simply apply the same techniques that are used in [10,12] to translate ∇-formulas into the language of predicate liftings: since T is finitary it has a presentation as a quotient of a polynomial functor:

$$p : \coprod_{n \in \omega} \Sigma_n \times (-)^n \to T,$$

where each Σ_n is a constant set[3] (see [1] for details). Given $n \in \omega$ and $\sigma_n \in \Sigma_n$, we get a natural transformation $p^{\sigma_n} : (-)^n \to T$ by

$$p^{\sigma_n}_X(u_1, \ldots, u_n) = p_X(\sigma_n, u_1, \ldots, u_n).$$

We will simply write p^σ from now on, letting the index n be made clear from context. We can exploit the presentation p to derive a set of predicate liftings for T:

Definition 12. *Given $\sigma \in \Sigma_n$, define the "Moss lifting" $\mu[\sigma] : Q^n \to QT$ by*

$$(X_1, \ldots, X_n) \mapsto d_X \circ p^\sigma_{QX}(X_1, \ldots, X_n).$$

We now come to the main lemma of this section:

Lemma 4. *Suppose that L satisfies conditions $\mathbf{A1}$ and $\mathbf{A2}$. Let $\mathfrak{A} = (X, \alpha)$ be a finite T-coalgebra. Then there exist systems of equations*

$$s_1, s_2 : X \to \mathcal{L}_\Lambda(X)$$

such that, relative to any coalgebra $\mathfrak{B} = (Y, \beta)$, $\mathcal{O}_{s_1} : \mathcal{P}(Y)^X \to \mathcal{P}(Y)^X$ is the same as $\mathcal{O}_{s_\square}$, and also $\mathcal{O}_{s_2} = \mathcal{O}_{s_\lozenge}$.

Proof. For the first part of the lemma, fix $u \in X$. We have $\alpha(u) \in TX \subseteq T\mathcal{L}(V)$. Since the presentation p is point-wise surjective, there are $x_1, \ldots, x_n \in X$ and $\sigma \in \Sigma_n$ with

$$p^\sigma_X(x_1, \ldots, x_n) = \alpha(u)$$

---

[3] To be concrete, we can take $\Sigma_n = T(n)$, and we can define the action of p_X on $(\sigma, u_1, \ldots, u_n) \in \Sigma_n \times X^n$ by

$$p_X(\sigma, u_1, \ldots, u_n) = Th(\sigma),$$

where $h : n \to X$ is the mapping defined by $i \mapsto u_i$. These details will not be relevant to us, however. All we need to know is that p is a natural transformation, and each of its components is surjective.

Since p^σ is natural and T preserves inclusions we get

$$p^\sigma_{\mathcal{L}(V)}(x_1, \ldots, x_n) = \alpha(u)$$

Let $\mu[\sigma] : Q^n \to QT$ denote the n-ary Moss lifting determined by σ using the distributive law d induced by L, and set

$$s_1(u) = \mu[\sigma](x_1, \ldots, x_n).$$

Then, for $v \in Y, u \in X$ and a valuation v, we get

$\quad (\mathfrak{B}, v) \vDash_v s_\square(u)$
iff $(\mathfrak{B}, v) \vDash_v \square\alpha(u)$
iff $\beta(v)L(\vDash_v)\alpha(u)$
iff $\beta(v) \in d_Y \circ T(tr^v_\mathfrak{B})(\alpha(u))$ \hfill by Lemma 3
iff $\beta(v) \in d_Y \circ T(tr^v_\mathfrak{B})(p^\sigma_{\mathcal{L}(V)}(x_1, \ldots, x_n))$
iff $\beta(v) \in d_Y \circ p^\sigma_{QY}(tr^v_\mathfrak{B}(x_1), \ldots, tr^v_\mathfrak{B}(x_n))$ \hfill by naturality of p^σ
iff $\beta(v) \in \mu[\sigma]_Y(tr^v_\mathfrak{B}(x_1), \ldots, tr^v_\mathfrak{B}(x_n))$
iff $\beta(v) \in \mu[\sigma]_Y(v(x_1), \ldots, v(x_n))$
iff $(\mathfrak{B}, v) \vDash_v \mu[\sigma](x_1, \ldots, x_n)$
iff $(\mathfrak{B}, v) \vDash_v s_1(u).$

It clearly follows that the systems of equations $s_\square$ and s_1 give rise to the same operator on the lattice of valuations in $\mathfrak{B}$.

For the second part of the lemma, we make use of Observation 6 and reason exactly the same way using the distributive law determined by L°. $\qquad\square$

Note that if s_1, s_2 always give rise to the same operators on evaluations as $s_\square, s_\lozenge$, then these systems of equations must be positive! Hence, we get:

Theorem 3. *Suppose that L satisfies conditions **A1** and **A2**. Given a finite T-coalgebra $\mathfrak{A} = (X, \alpha)$, there exist positive systems of equations*

$$s_1, s_2 : X \to \mathcal{L}_\Lambda(X)$$

such that for any $u \in X$ and any pointed T-coalgebra $(\mathfrak{B}, v)$, we have

$$(\mathfrak{B}, v) \vDash_{s_1} u \text{ iff } (\mathfrak{B}, v) \preceq_L (\mathfrak{A}, u)$$

and

$$(\mathfrak{B}, v) \vDash_{s_2} u \text{ iff } (\mathfrak{A}, u) \preceq_L (\mathfrak{B}, v).$$

Proof. Easy corollary from the previous lemma and Theorem 2. $\qquad\square$

5 Applications

In this final section, we provide examples of lax extensions for various functors that give rise to simulations and bisimulations that have been used in the literature. All these examples are taken from the papers [2,15].

Finitary Power Set Functor.

Example 5 (simulations). Consider the following lax extensions for the covariant powerset functor:

$$L_{sim}R := \{(A, B) \in \mathcal{P}_\omega X \times \mathcal{P}_\omega Y \mid \forall a \in A \, \exists b \in B : aRb\},$$
$$L_{rs}R := \{(A, B) \in \mathcal{P}_\omega X \times \mathcal{P}_\omega Y \mid (\forall a \in A \, \exists b \in B : aRb) \, \& \, (A = \emptyset \Rightarrow B = \emptyset)\},$$
$$L_{cs}R := \{(A, B) \in \mathcal{P}_\omega X \times \mathcal{P}_\omega Y \mid A \neq \emptyset \Rightarrow (B \neq \emptyset \, \& \, \forall b \in B \, \exists a \in A : aRb)\}.$$

Recall that L_{sim} was already given in Example 1 and L_{sim}-simulations are ordinary simulations. Also, L_{rs}-simulations are ready simulations and L_{cs}-simulations are conformance simulations. Item (2) of Theorem 2 yields characteristic formulas for each of these simulations.

Example 6 (bisimulation). Let L be one of L_{sim}, L_{rs}, or L_{cs} from Example 5. In each of these cases its bisimulator $B(L)$ is the same, and is given by

$$B(L)R = \{(A, B) \in \mathcal{P}_\omega X \times \mathcal{P}_\omega Y \mid \forall a \in A \, \exists b \in B : aRb, \, \&$$
$$\forall b \in B \, \exists a \in A : aRb\}.$$

Hence by Remark 1, $B(L)$ is the Barr extension $\overline{\mathcal{P}_\omega}$ for the finitary power set functor, and the main theorem gives characteristic formulas for bisimulation.

Example 7 (mutual simulation). Given states $u \in X$ and $v \in Y$ in $\mathcal{P}_\omega$-coalgebras $\mathfrak{A} = (X, \alpha)$ and $\mathfrak{B} = (Y, \beta)$, we say that u and b are mutually simulated, written $(\mathfrak{A}, u) \approx (\mathfrak{B}, v)$, if there is a simulation S from $\mathfrak{A}$ to $\mathfrak{B}$ with uSv and a simulation S' from $\mathfrak{B}$ to $\mathfrak{A}$ with $vS'u$. In other words, $(\mathfrak{A}, u) \approx (\mathfrak{B}, v)$ iff $(\mathfrak{A}, u) \preceq_{L_{sim}} (\mathfrak{B}, v)$ and $(\mathfrak{B}, v) \preceq_{L_{sim}} (\mathfrak{A}, u)$. Given a finite $\mathcal{P}_\omega$-coalgebra $\mathfrak{A} = (X, \alpha)$ and $u \in X$, we want to find a system of equations that allows us to characterize $(\mathfrak{A}, u)$ up to mutual simulation. There is a simple way to obtain such a system of equations from the main theorem. Let $s_\square : u \mapsto \square \alpha(u)$, and $s_\diamond : u \mapsto \diamond \alpha(u)$. Take the disjoint union of X with itself, i.e. the coproduct

$$X + X = (X \times \{0\}) \cup (X \times \{1\})$$

as a new set of variables. Let ι_1 and ι_2 be the left and right insertions of X into this coproduct, and define the system of equations s by setting

- $s(w, 0) = \square(\mathcal{P}_\omega \iota_1(\alpha(w)))$ and
- $s(w, 1) = \diamond(\mathcal{P}_\omega \iota_2(\alpha(w)))$.

Note that $\mathcal{P}_\omega \iota_1(\alpha(w)) \in \mathcal{P}_\omega(X \times \{0\})$, $\mathcal{P}_\omega \iota_1(\alpha(w)) \in \mathcal{P}_\omega(X + X)$, and similarly for $\mathcal{P}_\omega \iota_2(\alpha(w))$, and hence s maps variables in $X + X$ to formulas in the language $\mathcal{L}(X + X)$. With respect to this system of equations, the formula $(u, 0) \wedge (u, 1)$ is a characteristic formula for the pointed coalgebra $(\mathfrak{A}, u)$ w.r.t. mutual simulation. To see this, first let $t_i = s \restriction_{X \times \{i\}}$ for $i = \{0, 1\}$. As $s_\square$ and t_0 are isomorphic, and similarly $s_\diamond$ and t_1, it is easy to see that for any pointed coalgebra $(\mathfrak{B}, v)$, we have

- $(\mathfrak{B}, v) \vDash_{s_\square} u$ iff $(\mathfrak{B}, v) \vDash_{t_0} (u, 0)$,
- $(\mathfrak{B}, v) \vDash_{s_\lozenge} u$ iff $(\mathfrak{B}, v) \vDash_{t_1} (u, 1)$.

Using this and Observation 5, we have that for any pointed coalgebra $(\mathfrak{B}, v)$, we have

- $(\mathfrak{B}, v) \vDash_{s_\square} u$ iff $(\mathfrak{B}, v) \vDash_s (u, 0)$,
- $(\mathfrak{B}, v) \vDash_{s_\lozenge} u$ iff $(\mathfrak{B}, v) \vDash_s (u, 1)$.

It is immediate from this and Theorem 2 (the main theorem) that

$$(\mathfrak{B}, v) \vDash_s (u, 0) \wedge (u, 1) \text{ iff } (\mathfrak{A}, u) \approx (\mathfrak{B}, v)$$

as required.

Finite Probability Functor. Given a partial function $\rho : X \to [0, 1]$, and a subset $B \subseteq X$, let

$$\rho[B] = \sum_{b \in B \cap \mathrm{dom}(\rho)} \rho(b).$$

Let $\mathcal{D}$ be the finite probability functor as given in [13, Example 3.5]: $\mathcal{D}$ maps each set X to the set of partial functions from $\rho : X \to [0, 1]$, such that $\mathrm{dom}(\rho)$ is finite and $\rho[X] = 1$, and maps each function $f : X \to Y$ to $\mathcal{D}f : \mathcal{D}X \to \mathcal{D}Y$ given by

$$((\mathcal{D}f)\rho)(y) = \rho[f^{-1}[\{y\}]] = \sum \{\rho(x) : x \in Supp\,\rho, f(x) = y\}.$$

for each $\rho \in \mathcal{D}X$ and $y \in f[\mathrm{dom}(\rho)]$. Then $\mathcal{D}$ preserves inclusions (this is the reason for ρ being a partial rather than total function). A coalgebra $\alpha : A \to \mathcal{D}A$ corresponds to a Markov chain.

Given a relation $R \subseteq X \times Y$ and $A \subseteq X$, let $R[A] = \{b \mid \exists a \in A : aRb\}$.

Example 8 (Simulation and bisimulation on Markov chains). Let

$$L_{mc}R := \{(p, q) \in \mathcal{D}X \times \mathcal{D}Y \mid \forall C \subseteq X, p[C] \le q[R[C]]\}.$$

Then L is a lax extension of the finite probability functor $\mathcal{D}$.

The lax extension L corresponds to both simulation and bisimulation on Markov chains, and so the main theorem gives characteristic formulas for this relation (simulation and bisimulation are distinguished in variations of these Markov chains such as in [6] as well as with the probabilistic automata in Example 9 below). Furthermore, it is immediate from the equivalence of items 1 and 3 in [15, Lemma 1] that L is in fact just the Barr extension of $\mathcal{D}$.

Finite Non-deterministic Probability Functor. We call the functor $\mathcal{P}_\omega \circ \mathcal{D}$ the *finite nondeterministic probability functor*. A coalgebra for $\mathcal{P}_\omega \circ \mathcal{D}$ corresponds to a *probabilistic automaton* (which is essentially a Markov chain with non-deterministic transitions to distributions).

Example 9 (Simulation on Probabilistic Automata). The finite non-deterministic probability functor has a lax extension

$$L_{pa}R := \{(A, B) \in \mathcal{P}_\omega DX \times \mathcal{P}_\omega DY \mid \forall p \in A, \ \exists q \in B : \forall C \subseteq X, p[C] \leq q[R[C]]\}.$$

Such a lax extension corresponds to simulation (on probabilistic automata), and so we find characteristic formulas for such simulations.

Example 10 (Probabilistic simulation on Probabilistic Automata). Given an element $\mu \in \mathcal{D}DX$, let $\gamma(\mu) = \nu \in \mathcal{D}(X)$, where $\nu(x) = \sum_{\nu' \in \text{dom}(\mu)} \nu'(x)\mu(\nu')$. Then probabilistic simulation (see [16]) is defined by the relation lifting:

$$L_{psim}R := \{(A, B) \in \mathcal{P}_\omega DX \times \mathcal{P}_\omega DY \mid \forall p \in A, \ \exists q \in \mathcal{D}B :$$
$$\forall C \subseteq X, p[C] \leq \gamma(q)[R[C]]\}.$$

It can be checked that this is a lax extension. It is easy to see that L_{psim} is monotone (**L1** holds) and as $L_{pa} \subseteq L_{psim}$, **L3** holds for L_{psim} as well. To see that **L2** holds. Suppose that $A(L_{psim}R)B$ and $B(L_{psim}S)C$, and let $\mu \in A$. Then there exists a $\tilde{\nu} \in \mathcal{D}B$, such that for all $Z \subseteq A, \mu[Z] \leq \gamma(\tilde{\nu})[R[Z]]$. Also, for each $\nu \in Supp\,\tilde{\nu}$, there exists $\tilde{\rho}_\nu \in \mathcal{D}C$, such that for all $Z \subseteq B, \nu[Z] \leq \gamma(\tilde{\rho}_\nu)[S[Z]]$. Now let $\tilde{\sigma} = \sum_{\nu \in Supp\,\tilde{\nu}} \tilde{\nu}(\nu)\tilde{\rho}_\nu$. Then for any $Z \subseteq A$,

$$\mu[Z] \leq \gamma(\tilde{\nu})[R[Z]]$$
$$= \sum_{\nu \in Supp\,\tilde{\nu}} \tilde{\nu}(\nu)\nu[R[Z]]$$
$$\leq \sum_{\nu \in Supp\,\tilde{\nu}} \tilde{\nu}(\nu)\gamma(\tilde{\rho}_\nu)[(R;S)[Z]]$$
$$= \sum_{\nu \in Supp\,\tilde{\nu}} \tilde{\nu}(\nu) \sum_{\rho \in Supp\,\tilde{\rho}_\nu} \tilde{\rho}_\nu(\rho)\rho[(R;S)[Z]]$$
$$= \sum_{\nu \in Supp\,\tilde{\nu}} \sum_{\rho \in Supp\,\tilde{\rho}_\nu} (\tilde{\nu}(\nu)\tilde{\rho}_\nu(\rho)) \cdot \rho[(R;S)[Z]]$$
$$= \sum_{\rho \in Supp\,\tilde{\sigma}} \left(\sum_{\{\nu \mid \rho \in Supp\,\tilde{\rho}_\nu\}} \tilde{\nu}(\nu)\tilde{\rho}_\nu(\rho) \right) \rho[(R;S)[Z]]$$
$$= \gamma(\tilde{\sigma})[(R;S)[Z]].$$

As L_{psim} is a lax extension, Theorem 2 yields characteristic formulas.

Labelled Powerset Functor. Let A be a set of labels, and let $\mathcal{P}_A$ be the functor that maps each object X to $(\mathcal{P}_\omega(X))^A$, and maps each morphism $f : X \to Y$ to $\mathcal{P}_A f : h \mapsto k$, where $k : a \mapsto f[h(a)]$. Such a functor corresponds to a multi-modal Kripke frame.

Example 11 (multi-modal simulation). Let

$$L_{msim}R := \{(h, k) \mid \forall a \in A \forall x \in h(a) \exists y \in k(a) : xRy\}.$$

In other words $L_{msim}R$ consists of all pairs (h, k), such that for all $a \in A$, there is a function $f : h(a) \to k(a)$, whose graph is a subset of R. Since L_{msim} is a lax extension, the main theorem yields a characteristic formula.

Weak Simulation and Bisimulation. Let A be a set of labels and designate $\tau \in A$ to be a "silent action", not to be counted in a weak simulation. We aim to define a lax extension to capture weak simulation of transition systems. Here, we cannot simply work with the labelled powerset functor; the problem is that this functor only catches the "one-step" behaviours, while weak simulation crucially involves iterated behaviour. We will solve this problem by modelling transition systems as coalgebras for a suitable *co-monad*.

It is well known that the forgetful functor from the category of $\mathcal{P}_A$-coalgebras to the category **Set** of sets and mappings has a right adjoint [5], and this adjunction gives rise to a co-monad on **Set**. Here, we shall describe essentially the same co-monad in more concrete terms: let a rooted tree t over a set X be a prefix closed set of strings in $\mathbb{N}^*$; being prefix closed, t must include the empty string ε. An A-labelled rooted tree is a pair (t, λ), where t is a rooted tree and $\lambda : t \to A$ is a labelling function. Let $C_A : \mathbf{Set} \to \mathbf{Set}$ be defined by setting:

- For a set X, $C_A(X)$ is the set of $(X \times A)$-labelled and finitely branching rooted trees, with $\pi_2\lambda(\varepsilon) = \tau$ (as an arbitrary convention).
- For a mapping $h : X \to Y$, $C_A h : C_A X \to C_A Y$ is defined by letting $C_A h$ map a tree (t, λ) to the tree (t, λ') with labelling λ' obtained by the assignment $x \mapsto (h(\pi_1\lambda(x)), \pi_2\lambda(x))$.

Intuitively, $C_A X$ is the set of possible behaviours for a $\mathcal{P}_A$-coalgebra with domain X.

The functor C_A is a co-monad on **Set**. The co-unit $\eta : C_A \to Id_{\mathbf{Set}}$ is defined by letting η_X send a tree $(t, \lambda) \in C_A X$ to $\pi_1(\lambda(\varepsilon)) \in X$. The co-multiplication $\mu : C_A \to C_A \circ C_A$ is defined by letting μ_X send a tree $(t, \lambda) \in C_A X$ to the "tree of trees" (t, λ') in $C_A(C_A(X))$, such that $\lambda' : w \mapsto ((t_w, \lambda_w), \pi_2\lambda(w))$, where

- $t_w = \{v \in \mathbb{N}^* \mid w \cdot v \in t\}$ and
- $\lambda_w : t_w \to X \times A$, where $\lambda_w(v) = \begin{cases} \lambda(w \cdot v) & v \neq \varepsilon \\ (\pi_1\lambda(w), \tau) & v = \varepsilon \end{cases}$.

A labelled transition system can be represented as a coalgebra $\alpha : X \to C_A X$ for this co-monad, meaning that the following diagrams are required to commute:

$$
\begin{array}{ccc}
C_A X & \xrightarrow{\eta_X} & X \\
{\scriptstyle \alpha}\Big\uparrow & \nearrow{\scriptstyle Id_X} & \\
X & &
\end{array}
\qquad\qquad
\begin{array}{ccc}
C_A X & \xrightarrow{\mu_X} & C_A C_A X \\
{\scriptstyle \alpha}\Big\uparrow & & \Big\uparrow{\scriptstyle C_A \alpha} \\
X & \xrightarrow[\alpha]{} & C_A X
\end{array}
$$

Forgetting the co-monad structure of C_A we can just view it as an ordinary set functor, and so it makes sense to speak of a lax extension of C_A. We want to define a lax extension that captures weak simulation between labelled transition systems. Given a set X, a labelled tree $(t, \lambda) \in C_A X$ and a label a, define the relation

$$\xrightarrow{t, \lambda, a} \subseteq X \times X$$

by setting $x \xrightarrow{t, \lambda, a} y$ iff there is an a-labelled edge from x to y in the labelled tree (t, λ), that is, there exists w and $w \cdot n$ in t, such that $\pi_1(\lambda(w)) = x$ and $\lambda(w \cdot n) = (y, a)$. Let $\xrightarrow{t, \lambda, a^*}$ be the transitive reflexive closure of $\xrightarrow{t, \lambda, a}$. For a labelled tree (t, λ), say that a node v is a-reachable from u in (t, λ) if there are nodes u' and v' with

$$u \xrightarrow{t, \lambda, \tau^*} u' \xrightarrow{t, \lambda, a} v' \xrightarrow{t, \lambda, \tau^*} v$$

and denote by $re(t, \lambda, a)$ the set of nodes a-reachable in (t, λ) from the root $\pi_1(\lambda(\varepsilon))$ of t. Then C_A has a lax extension L_{weak} defined, for $R \subseteq X \times Y$, by setting $L_{weak} R$ to be the set of pairs $((t, \lambda), (t', \lambda')) \in C_A X \times C_A Y$ satisfying

$$\forall a \in A \setminus \{\tau\} \; \forall x \in re(t, \lambda, a) \; \exists y \in re(t', \lambda', a) : xRy.$$

This lax extension gives $\Box$- and $\Diamond$-modalities evaluated on C_A-coalgebras as before, and we can derive characteristic formulas for L_{weak}-simulation using Theorem 2. In particular, these formulas will characterize L_{weak}-simulation among coalgebras for C_A as a co-monad, and among these coalgebras L_{weak}-simulation can be taken to model weak simulation in the usual sense. Weak bisimulation is handled by considering the bisimulator of L_{weak}.

References

1. Adámek, J., Gumm, H.P., Trnková, V.: Presentation of set functors: a coalgebraic perspective. J. Logic Comput. **20**(5), 991–1015 (2010)
2. Aceto, L., Ingolfsdottir, A., Levy, P., Sack, J.: Characteristic formulae for fixed-point semantics: a general framework. Math. Struct. Comput. Sci. **22**(02), 125–173 (2012)
3. Baltag, A.: A logic for coalgebraic simulation. Electron. Notes Theor. Comput. Sci. **33**, 42–46 (2000)
4. Barr, M.: Relational algebras. In: MacLane, S., et al. (eds.) Reports of the Midwest Category Seminar IV. Lecture Notes in Mathematics, vol. 137, pp. 39–55. Springer, Heidelberg (1970)
5. Barr, M.: Terminal coalgebras in well-founded set theory. Theor. Comput. Sci. **114**, 299–315 (1993)
6. van Breugel, F., Mislove, M., Ouaknine, J., Worrell, J.: Domain theory, testing and simulation for labelled Markov processes. Theor. Comput. Sci. **333**, 171–197 (2005)
7. Hughes, J., Jacobs, B.: Simulations in coalgebra. Theor. Comput. Sci. **327**(1–2), 71–108 (2004)
8. Kurz, A., Leal, R.: Equational coalgebraic logic. In: Abramsky, S., Mislove, M., Palamidessi, C. (eds.): Proceedings of the 25th Conference on Mathematical Foundations of Programming Semantics (MFPS 2009) Electronic Notes in Theoretical Computer Science, vol. 249, pp. 333–356 (2009)

9. Levy, P.B.: Similarity quotients as final coalgebras. In: Hofmann, M. (ed.) FOS-SACS 2011. LNCS, vol. 6604, pp. 27–41. Springer, Heidelberg (2011)
10. Marti, J.: Relation liftings in coalgebraic modal logic. M.Sc. thesis, Institute for Logic, Language and Computation, University of Amsterdam (2011)
11. Marti, J., Venema, Y.: Lax extensions of coalgebra functors. In: Pattinson, D., Schröder, L. (eds.) CMCS 2012. LNCS, vol. 7399, pp. 150–169. Springer, Heidelberg (2012)
12. Marti, J., Venema, Y.: Lax extensions of coalgebra functors and their logics. J. Comput. Syst. Sci. (2014), to appear
13. Moss, L.: Coalgebraic logic. Ann. Pure Appl. Logic **96**, 277–317 (1999)
14. Müller-Olm, M.: Derivation of characteristic formulae. Electron. Notes Theor. Comput. Sci. **18**, 159–170 (1998)
15. Sack, J., Zhang, L.: A general framework for probabilistic characterizing formulae. In: Kuncak, V., Rybalchenko, A. (eds.) VMCAI 2012. LNCS, vol. 7148, pp. 396–411. Springer, Heidelberg (2012)
16. Segala, R., Lynch, N.: Probabilistic simulations for probabilistic processes. Nord. J. Comput. **2**(2), 250–273 (1995)
17. Thijs, A.: Simulation and fixpoint semantics. Ph.D. thesis, University of Groningen (1996)
18. Venema, Y.: Automata and fixed point logic: a coalgebraic perspective. Inf. Comput. **204**, 637–678 (2006)

Coalgebraic Simulations and Congruences

H. Peter Gumm[✉] and Mehdi Zarrad

Philipps-Universität Marburg, Marburg, Germany
gumm@mathematik.uni-marburg.de

Abstract. In a recent article Gorín and Schröder [3] study λ-simulations of coalgebras and relate them to preservation of positive formulae. Their main results assume that λ is a set of monotonic predicate liftings and their proofs are set-theoretical. We give a different definition of simulation, called strong simulation, which has several advantages:

Our notion agrees with that of [3] in the presence of monotonicity, but it has the advantage, that it allows diagrammatic reasoning, so several results from the mentioned paper can be obtained by simple diagram chases. We clarify the role of λ-monotonicity by showing the equivalence of

- λ is monotonic
- every simulation is strong
- every bisimulation is a (strong) simulation
- every F-congruence is a (strong) simulation.

We relate the notion to bisimulations and F-congruences - which are defined as pullbacks of homomorphisms. We show that

- if λ is a separating set, then each difunctional strong simulation is an F-congruence,
- if λ is monotonic, then the converse is true: if each difunctional strong simulation is an F-congruence, then λ is separating.

1 Introduction

Coalgebraic logic as introduced by D. Pattinson [6] and refined by L. Schröder [9], has been very successful in providing a common framework for quite a variety of modal logics, see for instance [2,5], or [11]. In many cases, the type functor, used to model such coalgebras preserves weak pullbacks, so logical equivalence can be modeled by structural relations called bisimulations. Two states related by a bisimulation are equivalent. In a recent paper Gorín and Schröder have introduced a notion of λ-simulation, where λ is a (set of) predicate lifting(s). Their definition is set theoretical and their proofs are calculational. In all of their results they assumed that all predicate liftings are monotonic.

Here we offer a different notion of simulation, which we call strong simulation. The definition is amenable to diagrammatical reasoning, whose utility we show in a number of proofs. Moreover, we show that under the assumption of monotonicity our definition coincides with that of [3]. Since they used monotony

© IFIP International Federation for Information Processing 2014
M.M. Bonsangue (Ed.): CMCS 2014, LNCS 8446, pp. 118–134, 2014.
DOI: 10.1007/978-3-662-44124-4_7

as a general hypothesis in their work, their results could be proved as well with our definition. We relate our strong simulations to the notion of Aczel-Mendler bisimulation (called F-bisimulation) and to (generalized) congruences.

2   Basic Notions and Preparations

Given a binary relation $R \subseteq A \times B$, let $R^- \subseteq B \times A$ be the converse relation. If $S \subseteq B \times C$ is another relation, then $R \circ S := \{(a, c) \mid \exists b \in B.aRb \wedge bSc\}$ is called the composition of R and S. Obviously, $\circ$ is associative and $(R \circ S)^- = S^- \circ R^-$.

For $R \subseteq A \times A$, a relation *on* a set A, notice that R is transitive iff $R \circ R \subseteq R$. Let $R^\star \subseteq A \times A$ be the reflexive transitive closure of R. The smallest equivalence relation containing R is $R^{eq} := (R \cup R^-)^\star$. It is well known that kernels of maps are equivalence relations, where for a map $f : A \to B$ the kernel is defined as

$$ker\, f := \{(a, a') \mid f(a) = f(a')\},$$

and conversely, any equivalence relation $E \subseteq A \times A$ is the kernel of the projection map $\pi_E : A \to A/E$, sending each element $a \in A$ to a/E, its equivalence class under E. With Δ_A we denote the identity relation on a set A.

2.1   Difunctionality

Difunctional relations are generalizations of equivalence relations, for the case of relations $R \subseteq A \times B$ between possibly different sets. Reflexivity, symmetry and transitivity make no sense for such relations, so a possible generalization is:

Definition 1. *A relation $R \subseteq A \times B$ is called* difunctional, *if it satisfies:*

$$(a_1, b_1), (a_2, b_1), (a_2, b_2) \in R \implies (a_1, b_2) \in R.$$

Immediately from the definition we see ([7]):

Lemma 1. *R is difunctional $\iff$ $R \circ R^- \circ R \subseteq R$ $\iff$ $R^- \circ R \circ R^- \subseteq R^-$ $\iff$ R^- is difunctional. The difunctional closure of a relation R is obtained as $R^d := R \circ (R^- \circ R)^\star = (R \circ R^-)^\star \circ R$.*

Each equivalence relation on A is obviously difunctional. More generally, let $f : A \to C$ and $g : B \to C$ be two maps then we define

$$ker(f, g) := \{(a, b) \in A \times B \mid f(a) = g(b)\}.$$

It is easy to see that $ker(f, g)$ is a difunctional relation, and, in perfect analogy to the situation with equivalence relations, every difunctional relation arises that way:

Lemma 2. *A relation $R \subseteq A \times B$ is difunctional, if and only if there are maps $f : A \to C$, $g : B \to C$ with $R = ker(f, g)$.*

Proof. Let $R \subseteq A \times B$ be difunctional. Let $e_A : A \to A + B$ and $e_B : B \to A + B$ be the canonical inclusions of A and B into their sum. On $A + B$ define

$$\bar{R} := \{(e_A(x), e_B(y)) \mid (x, y) \in R\}.$$

Obviously, $\bar{R}$ is difunctional, so

$$\bar{R} \circ \bar{R}^- \circ \bar{R} \subseteq \bar{R} \text{ and } \bar{R}^- \circ \bar{R} \circ \bar{R}^- \subseteq \bar{R}^-.$$

Moreover, $\bar{R} \circ \bar{R} = \emptyset = \bar{R}^- \circ \bar{R}^-$ by construction. Therefore,

$$E := \Delta_{A+B} \cup \bar{R} \cup \bar{R}^- \cup \bar{R} \circ \bar{R}^- \cup \bar{R}^- \circ \bar{R}$$

is an equivalence relation, since one easily calculates $E \circ E \subseteq E$. Notice that

$$(e_A[A] \times e_B[B]) \cap E = \bar{R}.$$

With the projection $\pi_E : A + B \to (A + B)/E$ it is now easy to calculate:

$$
\begin{aligned}
(x, y) \in ker(\pi_E \circ e_A, \pi_E \circ e_B) &\iff (e_A(x), e_B(y)) \in E \\
&\iff (e_A(x), e_B(y)) \in \bar{R} \\
&\iff (x, y) \in R.
\end{aligned}
$$

Thus, $R = ker(f, g)$ where f and g are constructed as the pushout of the projections $\pi_A^R : R \to A$ and $\pi_B^R : R \to B$.

More generally, if $f : A \to C$ and $g : B \to D$, then any difunctional relation $R \subseteq C \times D$ gives rise to a difunctional relation $ker(f, g)_R := \{(a, b) \mid f(a) \, R \, g(b)\} \subseteq A \times B$.

2.2 Directed Diagrams

Each map $\theta : A \to 2$, where $2 = \{0, 1\}$ is understood as an ordered set, is called a *predicate*. The *carrier* of predicate $\theta : A \to 2$ is the subset

$$[\![\theta]\!] := \{a \in A \mid \theta(a) = 1\}$$

and conversely, every subset $U \subseteq A$ arises as $U = [\![\chi_U]\!]$ from its characteristic function χ_U. We shall often use the same symbol for a predicate and its carrier, such as in $\square : F(2) \to 2$ and $\square \subseteq F(2)$.

It is sometimes convenient to write $a \models \theta$ rather than $\theta(a) = 1$ or $a \in \theta$. Similarly, $A \models \theta$ means that $a \models \theta$ for each $a \in A$.

We say $\theta \implies \psi$ provided $[\![\theta]\!] \subseteq [\![\psi]\!]$. It will be convenient to encode this diagrammatically, where the inclusion is indicated by an upwards arrow, as in

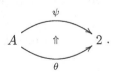

Establishing such a diagram amounts to showing that for any $a \in A$, taking the lower path in the diagram yields a result which is smaller than or equal to the result obtained by the upper path, i.e. $\theta(a) \leq \psi(a)$ for all $a \in A$. We generalize this notation in the following way:

Definition 2. *Given a relation S between sets A and B and predicates $\theta : A \to 2$ and $\psi : B \to 2$, we introduce*

$$\theta \overset{S}{\Longrightarrow} \psi \ :\Longleftrightarrow \ \forall(x,y) \in S. \, (\, x \models \theta \implies y \models \psi \,) \tag{2.1}$$

which may be spelled as "θ *implies* ψ *modulo* S". With our above notation, we can visualize $\theta \overset{S}{\Longrightarrow} \psi$ by the following "upwards-commuting" diagram

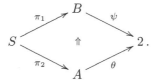

Notice that with $A = B$ and $S = \Delta_A$, we have $\theta \implies \psi$ being the same as $\theta \overset{\Delta_A}{\Longrightarrow} \psi$.

Interpreting a relation $S \subseteq A \times B$ as a map between the powersets $S : \mathbb{P}(A) \to \mathbb{P}(B)$, via $S(U) := \{b \in B \mid \exists a \in U.(a,b) \in S\}$, we could equivalently write:

$$\theta \overset{S}{\Longrightarrow} \psi \iff S(\llbracket \theta \rrbracket) \subseteq \llbracket \psi \rrbracket.$$

This shows that this notation is closely related to the notation of Hoare triples, where the relation S would be given as the semantics of an imperative program. We can immediately gather a number of simple properties inspired by this association. These correspond to the rules of precondition strengthening/postcondition weakening and sequencing:

Lemma 3

1. $\theta' \subseteq \theta \overset{S}{\Longrightarrow} \psi$ *implies* $\theta' \overset{S}{\Longrightarrow} \psi$
2. $\theta \overset{S}{\Longrightarrow} \psi \subseteq \psi'$ *implies* $\theta \overset{S}{\Longrightarrow} \psi'$
3. $\theta \overset{R}{\Longrightarrow} \varphi$ *and* $\varphi \overset{S}{\Longrightarrow} \psi$ *implies* $\theta \overset{R \circ S}{\Longrightarrow} \psi$

Proof. The first two claims can be readily obtained by gluing diagrams where we use the obvious naming conventions for the projections of a relation to its components:

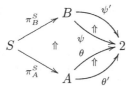

For the third claim, we note that if $R \bowtie S$ is the pullback of π_B^R with π_B^S, then $R \circ S$ is the image obtained by factoring the span $(R \bowtie S, \pi_C^{R \bowtie S}, \pi_C^{R \bowtie S})$ into an epi followed by a mono source:

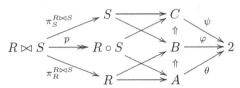

Explicitly, an upwards diagram chase, for instance in the right diagram, would be:

$$\theta \circ \pi_A^{R \circ S} \circ p = \theta \circ \pi_A^R \circ \pi_R^{R \bowtie S}$$
$$\leq \varphi \circ \pi_B^R \circ \pi_R^{R \bowtie S}$$
$$= \varphi \circ \pi_B^S \circ \pi_S^{R \bowtie S}$$
$$\leq \psi \circ \pi_C^S \circ \pi_S^{R \bowtie S}$$
$$= \psi \circ \pi_C^{R \circ S} \circ p.$$

Cancelling the epi p results in $\theta \circ \pi_A^{R \circ S} \leq \psi \circ \pi_C^{R \circ S}$.

3   Functors, Coalgebras and Bisimulations

Let $F : Set \rightarrow Set$ be an endofunctor on the category of sets. We shall write $F(X)$ for the action of F on an object X and Ff for the action of F on a map f.

Typical endofunctors describe set-theoretical constructions, such as *sets, lists, tuples, bags,* etc. In programming they include all generic collection classes such as List<X>, Set<X>, Bag<X> etc. The action of F on a map $f : X \rightarrow Y$ is generically called : map f. It will be useful to keep the following visualization in mind:

- F defines a type of "constructions".
- Elements of $F(X)$ are those "constructions" whose elements are drawn from a set X; we will call them $X - patterns$.
- Given a map $f : X \rightarrow Y$, the map $Ff : F(X) \rightarrow F(Y)$ acts on an X-pattern $p \in F(X)$ by replacing in p each x by $f(x)$.
- A pattern $p \in F(X)$ is finite, if there is a subset $\{x_1, \ldots, x_n\} \subseteq X$ such that $p \in F(\{x_1, \ldots, x_n\})$. In this case, we write $p = p(x_1, \ldots, x_n)$ and we let $p(f(x_1), \ldots, f(x_n))$ denote $(Ff)p(x_1, \ldots, x_n)$.
- In particular, if $\theta : X \rightarrow 2$ is a predicate, then $F\theta$ acts on an element $p \in F(X)$ by replacing in p each x by 1 if $x \models \theta$ and by 0 otherwise.
- If $p = p(x_1, \ldots, x_n)$, then $(F\theta)p(x_1, \ldots, x_n) = p(\theta(x_1), \ldots, \theta(x_n))$ is called a $0 - 1 - pattern$.

If $f : X \rightarrow Y$ is injective and $X \neq \emptyset$, then f is left-invertible, hence Ff is injective, too. F can always be modified just on the empty set and on empty mappings, so that it preserves injectivity for all mappings, including the empty one, see [13]. We therefore assume for the rest of this article, that F preserves all monos.

3.1   Coalgebras

Definition 3. *An F-coalgebra $\mathcal{A} = (A, \alpha)$ consists of a set A and a map $\alpha : A \to F(A)$. A is called the* base set *and α the* structure map. *The functor F is called the* type *of coalgebra $\mathcal{A}$.*

We shall keep F fixed and consider only coalgebras of that given type F.

Definition 4. *A map $\varphi : A \to B$ between two coalgebras $\mathcal{A} = (A, \alpha)$ and $\mathcal{B} = (B, \beta)$ is called a* homomorphism, *if $\beta \circ \varphi = F\varphi \circ \alpha$.*

The functor properties immediately guarantee that the class of all F-coalgebras with homomorphisms as morphisms forms a category Set_F. The forgetful functor $U : Set_F \to Set$ which associates with every coalgebra $\mathcal{A}$ its underlying set A and with every homomorphism its underlying map is known to create and preserve colimits [8], so in particular the category Set_F is cocomplete and colimits have the same underlying set and mappings as the corresponding colimits in Set.

Example 1. Kripke *frames* are coalgebras of type $\mathbb{P}$ where $\mathbb{P}$ is the covariant powerset functor, acting on a map $f : X \to Y$ as $\mathbb{P}f : \mathbb{P}(X) \to \mathbb{P}(Y)$ where $(\mathbb{P}f)(U) := f[U] := \{f(u) \mid u \in U\}$ for any $U \in \mathbb{P}(X)$.

Kripke *structures* come with a fixed set V of atomic properties, so they are modeled as coalgebras of type $\mathbb{P}(-) \times \mathbb{P}(V)$, where the second component is simply a constant. A coalgebra of type $\mathbb{P}(-) \times \mathbb{P}(V)$ is therefore a base set A with a structure map $\alpha : A \to \mathbb{P}(A) \times \mathbb{P}(V)$. Its first component associates to a state $a \in A$ the set of its successors $succ_A(a) := (\pi_1 \circ \alpha)(a)$ and its second component yields the set of all atomic values $val_A(a) := (\pi_2 \circ \alpha)(a)$ which are true for a.

Homomorphisms $\varphi : A \to B$ between Kripke frames, resp. Kripke structures are also known as *bounded morphisms*. They are maps preserving and reflecting successors and atomic values in the following sense: $\varphi[succ_A(a)] = succ_B(\varphi(a))$ and $val_A(a) = val_B(\varphi(a))$.

3.2   Bisimulations

In the structure theory of coalgebras, bisimulations play the role of compatible relations.

Definition 5. *([1]) A* bisimulation *between coalgebras $\mathcal{A}$ and $\mathcal{B}$ is a relation $R \subseteq A \times B$ for which there exists a coalgebra structure $\rho : R \to F(R)$ such that the projections $\pi_A^R : R \to A$ and $\pi_B^R : R \to B$ are homomorphisms.*

Typical bisimulations are graphs of homomorphisms $G(\varphi) := \{(a, \varphi(a)) \mid a \in A\}$. In fact, a map $f : A \to B$ is a homomorphism iff its graph is a bisimulation ([8]). If $R \subseteq A \times B$ is a bisimulation between coalgebras $\mathcal{A}$ and $\mathcal{B}$, then there could be several possible structure maps $\rho : R \to F(R)$ establishing that R is a bisimulation.

The empty relation $\emptyset \subseteq A \times B$ is always a bisimulation and (more generally) the union of bisimulations is a bisimulation, so that bisimulations between $\mathcal{A}$ and $\mathcal{B}$ form a complete lattice with largest element called $\sim_{\mathcal{A},\mathcal{B}}$.

The following proposition will be needed later in the proof of Theorem 3. It shows that bisimulations can be enlarged as long as the structure maps are not affected in the following sense:

Proposition 1. *Let $\mathcal{A}_1$ and $\mathcal{A}_2$ be coalgebras with corresponding structure maps α_1 and α_2. Let $R \subseteq \mathcal{A}_1 \times \mathcal{A}_2$ be a bisimulation and R' an enlargement i.e. $R \subseteq R' \subseteq \ker \alpha_1 \circ R \circ \ker \alpha_2$. Then R' is also a bisimulation.*

Proof. R is a bisimulation, so there exists a structure map $\rho : R \rightarrow F(R)$ with $\alpha_i \circ \pi_i^R = F\pi_i^R \circ \rho$. Let $\iota : R \rightarrow R'$ be the inclusion map, then clearly $\pi_i^R = \pi_i^{R'} \circ \iota$. By assumption, we find for every $(x', y') \in R'$ a pair $(x, y) \in R$ such that $\alpha_1(x) = \alpha_1(x')$ and $\alpha_2(y) = \alpha_2(y')$. The axiom of choice provides for a map $\mu : R' \rightarrow R$ satisfying

$$\alpha_i \circ \pi_i^{R'} \circ \iota \circ \mu = \alpha_i \circ \pi_i^{R'}.$$

We now define $\rho' : R' \rightarrow F(R')$ by $\rho' := F\iota \circ \rho \circ \mu$.

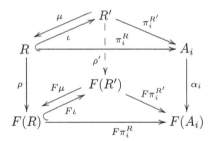

The rest is a simple calculation.

Corollary 1. *Let $\mathcal{A} = (A, \alpha)$ be a coalgebra, then every reflexive relation $R \subseteq \ker \alpha$ is a bisimulation.*

Proof. Since $\Delta \subseteq A$ is always a bisimulation, we have $\Delta \subseteq R \subseteq \ker \alpha = \ker \alpha \circ \ker \alpha = \ker \alpha \circ \Delta \circ \ker \alpha$, because $\ker \alpha$ is transitive.

3.3 Predicate Liftings and Boxes

We denote the contravariant powerset functor by 2^-. Thus 2^X is the set of all subsets of X and a map $f : X \rightarrow Y$ induces a map $2^f : 2^Y \rightarrow 2^X$ via $2^f(V) := f^{-1}[V]$. If we consider the elements of 2^Y as predicates $\tau : Y \rightarrow 2$, we can write $2^f(\tau) = \tau \circ f$, or $2^f = (-) \circ f$.

The classical Kripke style modal logic introduces formulae expressing properties holding for all successors of a point x. If φ is a state formula then $\Box\varphi$ holds at x if φ holds for each successor x' of x. The set of all successors of a point x is $\alpha(x) \in \mathbb{P}(X)$, in the case of Kripke frames. Thus $\Box$ can be understood as lifting a property φ from the base set A to a property $\lambda_A(\varphi) \subseteq \mathbb{P}(A)$, so $x \models \Box\varphi$ iff $\alpha(a)$ satisfies the lifted property $\lambda_A(\varphi)$. Generalizing this observation, Pattinson [6] introduced predicate liftings $\lambda_A : 2^A \to 2^{F(A)}$ as natural transformations between the contravariant powerset functors $2^{(-)}$ and $2^{F(-)}$.

Definition 6. *A predicate lifting λ for F is a natural transformation $\lambda : 2^- \to 2^{F(-)}$ where the latter is the composition of the functor F with 2^-. For each X denote by λ_X its X-component $\lambda_X : 2^X \to 2^{F(X)}$.*

The idea is that every property for elements of a set X is transformed to a property for elements of $F(X)$.

By the Yoneda lemma, such a natural transformation λ is uniquely determined by the action of λ_2 on the input id_2 where $[\![id_2]\!] = \{1\} \subseteq 2$, i.e. by $\lambda_2(id_2) : F(2) \to 2$, which is a predicate on $F(2)$. This was observed in [9]. We shall from now on write $[\lambda]$ or simply $\Box$, if λ is understood, for this predicate.

Conversely, given a predicate $\Box : F(2) \to 2$ on $F(2)$, then $\theta \mapsto \Box \circ F\theta$ defines a predicate transformer, and it is easy to see that id_2 is sent to $\Box$ again.

Intuitively, we think of $\Box \subseteq F(2)$ as a selection of $0 - 1 - patterns$. The map λ_A of the corresponding predicate transformer λ, when applied to $\theta \in 2^A$ takes an A-pattern $p(a_1, \ldots, a_n) \in F(A)$ to 1 if $p(\theta(a_1), \ldots, \theta(a_n)) \in \Box$, and to 0 otherwise.

In this paper we prefer to deal with predicates $\Box : F(2) \to 2$ rather than with predicate transformers $\lambda : 2^{(-)} \to 2^{F(-)}$. Ignoring for a moment the map α, the following figure visualizes the translation between these two views.

$$
\begin{array}{ccccc}
A & \xrightarrow{\theta} & 2 & \xrightarrow{\;\;id\;\;} & 2 \\
{\scriptstyle \alpha}\downarrow & & & & \\
F(A) & \xrightarrow{F\theta} & F(2) & \xrightarrow[\lambda_2(id_2)]{\Box} & 2
\end{array}
$$
$$\underbrace{\hspace{5cm}}_{\lambda_A(\theta)}$$

Let us now consider F-coalgebras $\mathcal{A} = (A, \alpha)$, where $\alpha : A \to F(A)$ is the *structure map*. Every predicate transformer, i.e. every predicate $\Box$ on $F(2)$ defines a modality.

Definition 7. *Given a predicate θ on $\mathcal{A} = (A, \alpha)$, denote by $\Box\theta$ the predicate $\Box \circ F\theta \circ \alpha$, that is for any $a \in A$ we define*

$$a \models \Box\theta :\iff (\Box \circ F\theta \circ \alpha)(a) = 1.$$

3.4 Coalgebraic Modal Logic

Given any choice of predicate liftings, equivalently, any choice of boxes $\Box_i : F(2) \to 2$, $i \in I$, we obtain a logic $\mathcal{L}$ (see [6]) whose formulae are defined inductively by

$$\varphi ::= \top \mid \varphi_1 \vee \varphi_2 \mid \varphi_1 \wedge \varphi_2 \mid \neg\varphi \mid \Box_i\varphi \text{ for each } i \in I$$

A formula is called positive, if it has no occurrence of $\neg$.

Given a coalgebra $\mathcal{A} = (A, \alpha)$ each formula defines a predicate $\varphi_{\mathcal{A}} : A \to 2$, where the propositional connectors have their obvious interpretation and $(\Box_i \varphi)_{\mathcal{A}} := \Box_i \circ (F\varphi_{\mathcal{A}}) \circ \alpha$, which is short for saying

$$a \models \Box_i \varphi : \iff (F\varphi_{\mathcal{A}} \circ \alpha)(a) \in \Box_i.$$

4   Simulations

Given a predicate lifting λ, a λ-simulation S between coalgebras $\mathcal{A} = (A, \alpha)$ and $\mathcal{B} = (B, \beta)$ was defined in [3] as a relation $S \subseteq A \times B$ such that for any $(x, y) \in S$ and any predicate $\theta : A \to 2$ one has $\alpha(x) \models \lambda_A(\theta) \implies \beta(y) \models \lambda_B(S[\theta])$, where $S[\theta]$ is defined as $b \models S[\theta] : \iff \exists a \in A.(aSb \land a \models \theta)$. Most of the results in [3] assume that λ is monotonic, a notion to be discussed in Sect. 4.2. Amongst other things, for instance they prove:

– if λ is monotonic then bisimulations are λ-simulations
– if λ is monotonic, then each λ-simulation preserves positive formulae.

The proofs, in each case, are set theoretical, so it is difficult to see how the notions and results could possibly be lifted to situations beyond set-theoretical categories. Therefore, we introduce a new definition of "strong" simulation which has the advantage that

– proofs are diagrammatical
– monotonicity need not be assumed.

For notational reasons, we shall from now on fix a certain $\Box$ and define simulations relative to that $\Box$. Thus a "simulation" is the same as a λ-simulation from [3] with λ the predicate lifting defined by $\Box$. Next we shall define our new notion of "strong simulation". It will turn out that monotonicity is the property relating simulations with strong simulations, see Theorem 2 below.

4.1   Strong Simulations

A *strong simulation* between coalgebras $\mathcal{A} = (A, \alpha)$ and $\mathcal{B} = (B, \beta)$ is a relation $S \subseteq A \times B$ such that for any predicates $\theta : A \to 2$ and $\psi : B \to 2$ we have

$$\theta \overset{S}{\Longrightarrow} \psi \text{ implies } \Box\theta \overset{S}{\Longrightarrow} \Box\psi.$$

Diagrammatically:

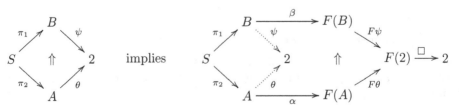

Clearly, every strong simulation is a simulation. This is because the diagram in the premise, above, is trivially satisfied with $\psi = S[\theta]$.

Lemma 4. *Strong simulations are closed under unions and relational composition, i.e. if $R \subseteq A \times B$ and $S \subseteq B \times C$ are strong simulations, then so is $R \circ S \subseteq A \times C$.*

Proof. Closure under unions is easily checked. For closure under relational composition, let $\theta \xRightarrow{R \circ S} \psi$ be given. Obviously, then $\theta \xRightarrow{R} R[\theta]$ and $R[\theta] \xRightarrow{S} \psi$. Assuming that R and S are simulations, we obtain $\Box \theta \xRightarrow{R} \Box R[\theta]$ as well as $\Box R[\theta] \xRightarrow{S} \Box \psi$, so Lemma 3 yields $\Box \theta \xRightarrow{R \circ S} \Box \psi$.

Simulations have a preferred direction. This is emphasized by the following logical fact:

Theorem 1. *Strong simulations preserve positive formulae.*

Proof. Let S be a strong simulation between coalgebras $\mathcal{A}$ and $\mathcal{B}$, and $(x, y) \in S$. By structural induction, we show that for any positive formula ϕ we have: $x \models \phi \implies y \models \phi$, that is we need to show $\phi_{\mathcal{A}} \xRightarrow{S} \phi_{\mathcal{B}}$. The only interesting case is when $\phi = \Box \psi$ with ψ another positive formula. Let $\psi_{\mathcal{A}}$, resp $\psi_{\mathcal{B}}$ be the predicates defined by ψ in $\mathcal{A}$, resp $\mathcal{B}$. By assumption then, $\psi_A \xRightarrow{S} \psi_B$, whence the definition of simulation yields $\Box \psi_A \xRightarrow{S} \Box \psi_B$, hence $(\Box \psi)_{\mathcal{A}} \xRightarrow{S} (\Box \psi)_{\mathcal{B}}$.

By a (strong) *bidirectional simulation* we understand a (strong) simulation S for which S^- is also a simulation. We must be careful not to confuse this with the notion of *bisimulation*.

From Lemmas 1 and 4 we obtain:

Lemma 5. *Let $(S_i)_{i \in I}$ be a family of bidirectional simulations, then their difunctional closure is again a bidirectional simulation.*

4.2   Monotonicity

Definition 8. *A predicate lifting λ is called* monotonic, *if for all sets U, V, A with $U \subseteq V \subseteq A$ one has $\lambda_A(U) \subseteq \lambda_A(V)$. We say that $\Box : F(2) \to 2$ is monotonic, if the predicate lifting given by $\Box$ is monotonic.*

We get the following characterization:

Lemma 6. *$\Box : F(2) \to 2$ is monotonic, iff for any A and any predicates θ, ψ on A with $\theta \implies \psi$, we obtain $\Box \circ F\theta \implies \Box \circ F\psi$.*

Proof. Suppose $\lambda_A = \Box \circ F(-)$ is monotonic, $\theta \implies \psi$ and $\Box \circ F\theta = 1$, that is $\lambda_A(\theta) = 1$. By monotonicity, $\lambda_A(\psi) = 1$, i.e. $\Box \circ F\psi = 1$. Conversely, assume $U \subseteq V \subseteq A$ and $u \in \lambda_A(U)$, where $\lambda_A(U) = [\![\Box \circ F\chi_U]\!]$. Then $\chi_U \implies \chi_V$ and $(\Box \circ F\chi_U)(u) = 1$ whence by assumption $(\Box \circ F\chi_V)(u) = 1$, meaning $u \in \lambda_A(V)$. Thus λ_A is monotonic.

Graphically, monotonicity can be represented as

$$A \overset{\psi}{\underset{\theta}{\Longleftrightarrow}} 2 \quad \Longrightarrow \quad F(A) \overset{F\psi}{\underset{F\theta}{\Longleftrightarrow}} F(2) \overset{\square}{\longrightarrow} 2.$$

The following observation was independently found by L. Schröder and appears in the journal version [10] of [9]. With our diagrammatic notation its proof becomes almost trivial:

Lemma 7. $\square$ *is monotonic if and only if for every ternary pattern* $p(x, y, z)$ *we have that*

$$p(1, 0, 0) \in \square \implies p(1, 1, 0) \in \square.$$

Proof. When $\theta \implies \psi$, we can obtain a joint factorization as $\theta = \chi_{\{x\}} \circ f$ and $\psi = \chi_{\{x,y\}} \circ f$. Thus the above definition of monotonicity reduces to the following implication:

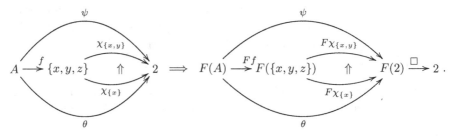

The outer diagrams are upward commutative iff the inner ones are. The one in the premise is automatically upward commutative. Therefore, $\square$ is monotonic, if and only if the inner diagram on the right is upwards commutative.

This means that monotonicity needs only be checked for $\theta = \chi_{\{x\}}$ and $\psi = \chi_{\{x,y\}}$, which translates immediately into the statement $p(1, 0, 0) \in \square \implies p(1, 1, 0) \in \square$ for each $p \in F(\{x, y, z\})$.

Theorem 2. $\square$ *is monotonic iff each simulation is strong.*

Proof. Suppose that $\square$ is monotonic and let S be a simulation between coalgebras $\mathcal{A} = (A, \alpha)$ and $\mathcal{B} = (B, \beta)$. Suppose $\theta \overset{S}{\implies} \psi$, then $S[\theta] \leq \psi$ as shown in the left part of the following figure, where the left inner square trivially commutes. Since S is a simulation we get upwards commutativity of the outer figure with $FS[\theta]$ instead of $F\psi$. Using monotonicity, we get upwards commutativity of the right upper figure and therefore of the whole diagram:

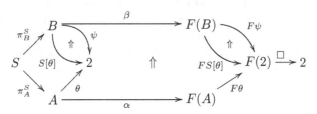

For the converse, consider the identity relation Δ_A on A, which is obviously a simulation, hence it is a strong simulation by assumption. Given any $p \in F(A)$ and $\theta \leq \psi : A \to 2$ we choose the constant coalgebra structure $c_p : A \to F(A)$. Since the left square is upwards commuting, so must be the outer figure. This readily translates into $\square$ being monotonic.

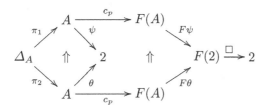

Theorem 3. *The following are equivalent:*

1. $\square$ *is monotonic*
2. *each bisimulation is a simulation*
3. *each bisimulation is a strong simulation*

Proof. (1.→ 3.) Suppose $\square$ is monotonic and $S \subseteq A \times B$ is a bisimulation between coalgebras $\mathcal{A} = (A, \alpha)$ and $\mathcal{B} = (B, \beta)$. Given $\theta \overset{S}{\Longrightarrow} \psi$, the left square is upward commuting. Since $\square$ is monotonic, applying F makes the right hand square (followed by $\square$) upward commuting, too.

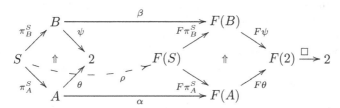

Inserting the bisimulation structure ρ into the picture, an upward diagram chase yields that the outer diagram is upward commuting, too:

$$\square \circ F\theta \circ \alpha \circ \pi_A^S = \square \circ F\theta \circ F\pi_A^S \circ \rho$$
$$\leq \square \circ F\psi \circ F\pi_B^S \circ \rho$$
$$= \square \circ F\psi \circ \beta \circ \pi_B^S$$

which means that S is a strong simulation.

(3 → 2) being trivial, we prove (2→1): By Lemma 7, we need to check monotonicity only for $A = \{x, y, z\}$, $\theta = \chi_{\{x\}}$ and $\psi = \chi_{\{x,y\}}$. Given $p \in F(A)$ with $p(1,0,0) \in \square$, i.e. $(\square \circ F\theta)(p) = 1$, define a coalgebra $\mathcal{A}_p$ on A with constant structure map c_p. By Proposition 1, $R := \Delta_A \cup \{(x, y), (y, x)\}$ is a bisimulation on $\mathcal{A}_p$, and $\psi = R[\theta]$. By hypothesis, R is a simulation, so $\square \circ F\theta \circ c_p \overset{B}{\Longrightarrow} \square \circ F\psi \circ c_p$, in particular,

$$(\Box \circ F\psi)(p) = (\Box \circ F\psi \circ c_p \circ \pi_2)(x, x)$$
$$\geq (\Box \circ F\theta \circ c_p \circ \pi_1)(x, x)$$
$$= (\Box \circ F\theta)(p)$$
$$= 1$$

i.e. $p(1, 1, 0) \in \Box$, as can be read from the following diagram:

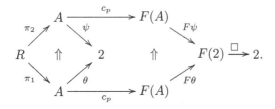

5   Congruences and Separability

5.1   Congruences

In classical examples of coalgebras, such as Kripke structures, deterministic and nondeterministic automata, etc., observational equivalence is definable via bisimulations. The reason is that the corresponding type functors preserve weak pullbacks (see [4]). This in turn has many structural consequences. In particular the largest bisimulation is always the same as the largest congruence relation, where a congruence is defined as the kernel of a homomorphism. Thus a congruence is a relation *on a single* coalgebra. Since we want to study relations between different coalgebras, we have to widen the notion of congruence and therefore introduce the notion of F-congruence. This notion has been studied by Sam Staton under the name *kernel bisimulation* [12]:

Definition 9. *An F-congruence θ between coalgebras $\mathcal{A}$ and $\mathcal{B}$ is the pullback of two homomorphisms $\varphi : \mathcal{A} \to \mathcal{C}$ and $\psi : \mathcal{B} \to \mathcal{C}$:*

$$\theta = ker(\varphi, \psi).$$

Theorem 4. *The following are equivalent:*

1. $\Box$ is monotonic
2. each congruence is a simulation
3. each F-congruence is a strong simulation.

Proof. (1.$\to$3.): An F-congruence $\theta = ker(\varphi, \psi)$ can be obtained as a composition of relations: $\theta = G(\varphi) \circ G(\psi)^-$ where $G(\varphi)$ and $G(\psi)$ are the graphs of φ and ψ. The graphs of homomorphisms are bisimulations ([8]) and the converse of a bisimulation is a bisimulation. Assuming monotonicity of $\Box$, Theorem 3 tells us that they are strong simulations. By Lemma 4, their composition is a strong simulation. In particular, each congruence is a simulation, too. (3.$\to$2) is of course trivial, since each congruence is an F-congruence and each strong simulation is a simulation.

For (2.→1.), assuming that each congruence is a simulation, we can reuse the proof of (3→2) in Theorem 3. This time, we only need to observe that R happens to be a congruence relation, since it is the kernel of the obvious homomorphism from $\mathcal{A}_p = \mathcal{A}_{p(x,y,z)}$ to the constant coalgebra $\mathcal{A}_{p(x,x,z)}$ on $\{x, z\}$.

5.2   Separability

In this section we need to work with a family of boxes $(\Box_i)_{i \in I}$. Such is usually required in order to render coalgebraic modal logic expressive. Separability is usually expressed for the functor and for the boxes separately. A functor is called 2-separable, if for any X and any $p, q \in F(X)$ with $p \neq q$ there is a predicate $\phi : X \to 2$ such $(F\phi)(p) \neq (F\phi)(q)$. Next, we call a family $(\Box_i)_{i \in I}$ of predicate liftings *separating*, if the functor F is 2-separating and the predicates $\Box_i : F(2) \to 2$ combined with the unary boolean operations $\theta : 2 \to 2$ form a mono-source. We can equivalently define this as follows:

Definition 10. $(\Box_i)_{i \in I}$ *is separating if*

$$\forall p \neq q \in F(X). \exists \phi : X \to 2. \exists i \in I. (\Box_i \circ F\phi(p) \neq \Box_i \circ F\phi(q))$$

Theorem 5. *If $(\Box_i)_{i \in I}$ is separating then every difunctional bidirectional strong simulation is an F-congruence.*

Proof. Let S be a difunctional strong simulation between coalgebras $\mathcal{A} = (A, \alpha)$ and $\mathcal{B} = (B, \beta)$ and π_1, π_2 the projections of S. Form the pushout $(P, f : A \to P, g : B \to P)$ of (S, π_1, π_2) in Set. Since S is difunctional, (S, π_1, π_2) is a pullback of f and g in Set. It suffices to show that there exists a coalgebra structure on P so that f and g are homomorphisms. We obtain such a coalgebra structure if we can show that $(FP, Ff \circ \alpha, Fg \circ \beta)$ is a competitor of the pushout (P, f, g) in Set. For this it remains to show : $Ff \circ \alpha \circ \pi_1 = Fg \circ \beta \circ \pi_2$.

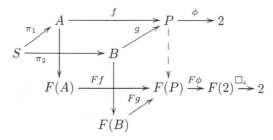

Let $(x, y) \in S$. As $(\Box_i)_{i \in I}$ is separating, it is enough to show that for each $i \in I$ and each $\phi : P \to 2$ we have $Ff \circ \alpha(x) \models \Box_i \theta \iff Fg \circ \beta(y) \models \Box_i \theta$. This we can read from the following diagram:

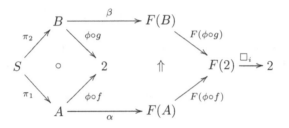

The left square in the diagram commutes, since (P, f, g) is a pushout, in particular it is upward commuting. S being a strong simulation, we obtain $Ff \circ \alpha(x) \models \Box\theta \implies Fg \circ \beta(y) \models \Box\theta$. Since S^- is a strong simulation, too, we similarly have $Fg \circ \beta(y) \models \Box\theta \implies Ff \circ \alpha(x) \models \Box\theta$.

Theorem 6. *If each difunctional simulation is an F-congruence, then $(\Box_i)_{i \in I}$ is separating.*

Proof. Assume $p, q \in FX$ such that $p \models \Box_i\theta \iff q \models \Box_i\theta$ for each $i \in I$ and each $\theta : X \to 2$. We must show $p = q$.

Case 1. $X \neq \emptyset$: On the set X define F-coalgebras $\mathcal{X}_p = (X, c_p)$ and $\mathcal{X}_q = (X, c_q)$, where c_p, resp. c_q, are constant maps with value p, resp. q. Notice that the assumption is then equivalent to saying that Δ_X is a (difunctional) simulation (with respect to each $\Box_i$) between $\mathcal{X}_p$ and $\mathcal{X}_q$. Therefore, by the theorem's premise, Δ_X is an F-congruence. Consequently, there must be homomorphisms $\varphi : \mathcal{X}_p \longrightarrow \mathcal{Z} = (Z, \gamma)$ and $\psi : \mathcal{X}_q \longrightarrow \mathcal{Z}$ with $\Delta_X = Pb(\varphi, \psi)$. This immediately yields $\varphi = \psi$ and φ injective.

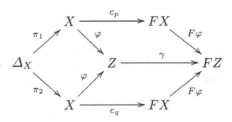

The above diagram commutes, since φ is a homomorphism, so $(F\varphi)(p) = (F\varphi \circ c_p \circ \pi_2)(x) = (F\varphi \circ c_q \circ \pi_2)(x) = (F\varphi)(q)$. Therefore $p = q$ as required.

Case 2. $X = \emptyset$: According to our general assumption, $F\iota : F\emptyset \to F1$ is injective. Thus in order to separate $p, q \in F\emptyset$, it is enough to separate $(F\iota)(p) \in F(1)$ from $(F\iota)(q) \in F(1)$ which is possible due to the previous case.

Corollary 2. *If $\Box$ is monotonic and separating then every difunctional simulation is an F-congruence.*

As a further corollary, we obtain a converse to another result found in [3].

Corollary 3. *Let* $(\Box_i)_{i \in I}$ *be monotonic. Then* $(\Box_i)_{i \in I}$ *are separating and* F *weakly preserves pullbacks if and only if each difunctional simulation is an* F-*bisimulation.*

Proof. The direction from left to right is from [3]. For the converse, suppose that each difunctional simulation is an F-bisimulation. Then by monotony each F-congruence is an F-bisimulation. This is the same as saying that F weakly preserves pullbacks. Similarly, every difunctional simulation is an F-congruence, hence by the above proposition, $(\Box_i)_{i \in I}$ is separating.

6 Conclusion and Further Work

We have given a new definition of coalgebraic simulation, which has the advantage to be amenable to diagrammatic reasoning. We have demonstrated its use with a number of results and related our definition to that of Gorín and Schröder in [3]. In the case where our boxes (respectively predicate liftings) are monotonic, a general assumption in the paper [3], our definition agrees with that of the authors. We have related our simulations to 2-dimensional congruences (so called F-congruences). We suspect that the set of all F-congruences between fixed coalgebras $\mathcal{A}$ and $\mathcal{B}$ forms a complete lattice with the natural ordering. However we were only able to show it under the additional assumption that there exists a set of separating monotonic boxes $(\Box_i)_{i \in I}$. In that case, F-congruences are bidirectional simulations and their supremum is given by difunctional closure. We leave it open whether the existence of a separating set $(\Box_i)_{i \in I}$ is needed.

References

1. Aczel, P., Mendler, N.: A final coalgebra theorem. In: Pitt, D.H., Rydeheard, D.E., Dybjer, P., Pitts, A.M., Poigné, A. (eds.) CTCS 1989. LNCS, vol. 389, pp. 357–365. Springer, Heidelberg (1989)
2. Cîrstea, C., Kurz, A., Pattinson, D., Schröder, L., Venema, Y.: Modal logics are coalgebraic. In: BCS International Academic Conference, pp. 128–140 (2008)
3. Gorín, D., Schröder, L.: Simulations and bisimulations for coalgebraic modal logics. In: Heckel, R. (ed.) CALCO 2013. LNCS, vol. 8089, pp. 253–266. Springer, Heidelberg (2013)
4. Gumm, H.P., Schröder, T.: Types and coalgebraic structure. Algebra Universalis **53**, 229–252 (2005)
5. Myers, R., Pattinson, D., Schröder, L.: Coalgebraic hybrid logic. In: de Alfaro, L. (ed.) FOSSACS 2009. LNCS, vol. 5504, pp. 137–151. Springer, Heidelberg (2009)
6. Pattinson, D.: Coalgebraic modal logic: soundness, completeness and decidability of local consequence. Theor. Comput. Sci. **309**(2–3), 177–193 (2003)
7. Riguet, J.: Relations binaires, fermetures, correspondances de Galois. Bulletin de la Société Mathématique de France **76**, 114–155 (1948)
8. Rutten, J.J.M.M.: Universal coalgebra: a theory of systems. Theor. Comput. Sci. **249**, 3–80 (2000)

9. Schröder, L.: Expressivity of coalgebraic modal logic: the limits and beyond. In: Sassone, V. (ed.) FOSSACS 2005. LNCS, vol. 3441, pp. 440–454. Springer, Heidelberg (2005)
10. Schröder, L.: Expressivity of coalgebraic modal logic: the limits and beyond. Theor. Comput. Sci. **390**(2–3), 230–247 (2008)
11. Schröder, L., Pattinson, D.: Coalgebraic correspondence theory. In: Ong, L. (ed.) FOSSACS 2010. LNCS, vol. 6014, pp. 328–342. Springer, Heidelberg (2010)
12. Staton, S.: Relating coalgebraic notions of bisimulation. Log. Methods Comput. Sci. **7**(1) (2011)
13. Trnková, V.: Some properties of set functors. Comm. Math. Univ. Carol. **10**(2), 323–352 (1969)

Dijkstra Monads in Monadic Computation

Bart Jacobs[✉]

Institute for Computing and Information Sciences (iCIS),
Radboud University Nijmegen, Nijmegen, The Netherlands
bart@cs.ru.nl
http://www.cs.ru.nl/B.Jacobs

Abstract. The Dijkstra monad has been introduced recently for capturing weakest precondition computations within the context of program verification, supported by a theorem prover. Here we give a more general description of such Dijkstra monads in a categorical setting. We first elaborate the recently developed view on program semantics in terms of a triangle of computations, state transformers, and predicate transformers. Instantiations of this triangle for different monads T show how to define the Dijkstra monad associated with T, via the logic involved. Technically, we provide a morphism of monads from the state monad transformation applied to T, to the Dijkstra monad associated with T. This monad map is precisely the weakest precondition map in the triangle, given in categorical terms by substitution.

1 Introduction

A monad is a categorical concept that is surprisingly useful in the theory of computation. On the one hand it describes a form of computation (such as partial, non-deterministic, or probabilistic), and on the other hand it captures various algebraic structures. Technically, the computations are maps in the Kleisli category of the monad, whereas the algebraic structures are described via the category of so-called Eilenberg-Moore algebras. The Kleisli approach has become common in program semantics and functional programming (notably in the language Haskell), starting with the seminal paper [23]. The algebraic structure captured by the monad exists on these programs (as Kleisli maps), technically because the Kleisli category is enriched over the category of algebras.

Interestingly, the range of examples of monads has been extended recently from computation to program logic. So-called Hoare monads [24,29] and Dijkstra monads [28] have been defined in a systematic approach to program verification. Via these monads one describes not only a program but also the associated correctness assertions. These monads have been introduced in the language of a theorem prover, but have not been investigated systematically from a categorical perspective. Here we do so for the Dijkstra monad. We generalise the original definition from [28] and show that a "Dijkstra" monad can be associated with various well-known monads that are used for modelling computations. (The Hoare monad will be mentioned briefly towards the end.)

© IFIP International Federation for Information Processing 2014
M.M. Bonsangue (Ed.): CMCS 2014, LNCS 8446, pp. 135–150, 2014.
DOI: 10.1007/978-3-662-44124-4_8

Since the Dijkstra (and Hoare) monads combine both semantics and logic of programs, we need to look at these two areas in a unified manner. From previous work [13] (see also [12]) a view on program semantics and logic emerged involving a triangle of the form:

$$\mathbf{Log}^{\mathrm{op}} = \left(\begin{array}{c}\text{predicate}\\\text{transformers}\end{array}\right) \xrightleftharpoons[\qquad\top\qquad]{} \left(\begin{array}{c}\text{state}\\\text{transformers}\end{array}\right)$$

$$\begin{array}{c}\nwarrow\ Pred \qquad Stat\ \nearrow\\\left(\text{computations}\right)\end{array}$$

(1)

The three nodes in this diagram represent categories of which only the morphisms are described. The arrows between these nodes are functors, where the two arrows $\rightleftarrows$ at the top form an adjunction. The two triangles involved should commute. In the case where two up-going "predicate" and "state" functors $Pred$ and $Stat$ in (1) are full and faithful, we have three equivalent ways of describing computations. On morphisms, the predicate functor yields what is called substitution in categorical logic, but what amounts to a weakest precondition operation in program semantics.

The upper category on the left is of the form $\mathbf{Log}^{\mathrm{op}}$, where $\mathbf{Log}$ is some category of logical structures. The opposite category $(-)^{\mathrm{op}}$ is needed because predicate transformers operate in the reverse direction, taking a post-condition to a precondition. In this paper we do not expand on the precise logical structure involved (which connectives, which quantifiers, etc. in $\mathbf{Log}$) and simply claim that this 'indexed category' on the left is a model of some predicate logic. The reason is that at this stage we don't need more structure than 'substitution', which is provided by the functoriality of $Pred$.

In a setting of quantum computation this translation back-and-forth $\rightleftarrows$ in (1) is associated with the different approaches of Heisenberg (logic-based, working backwards) and Schrödinger (state-based, working forwards), see e.g. [9]. In certain cases the adjunction $\rightleftarrows$ forms — or may be restricted to — an equivalence of categories, yielding a duality situation. It shows the importance of duality theory in program semantics and logic; this topic has a long history, going back to [1].

Almost all of our examples of computations are given by maps in a Kleisli category of a monad. In this monadic setting, the right-hand-side of the diagram (1) is the full and faithful "comparison" functor $\mathcal{K\ell}(T) \to \mathcal{EM}(T)$, for the monad T at hand. This functor embeds the Kleisli category in the category of (Eilenberg-Moore) algebras. The left-hand-side takes the form $\mathcal{K\ell}(T) \to \mathbf{Log}^{\mathrm{op}}$, and forms an indexed category (or, if you like, a fibration), and thus a categorical model of predicate logic. The monad T captures computations as maps in its Kleisli category. And via the predicate logic in (1) an associated monad is defined (in Sect. 5) that captures predicate transformers. Therefore, this new monad is called a "Dijkstra" monad, following [28].

We list the main points of this paper.

1. The paper explains the unified view on program semantics and logic as given by the above triangle (1) by presenting many examples, involving non-deterministic, partial, linear, probabilistic, and also quantum computation. This involves some new results, like the adjunction for partial computation in (5) in the next section.

2. Additionally, in many of these examples the enriched nature of these categories and functors is shown, capturing some essential compositional aspects of the weakest precondition operation. The role of these enrichments resembles the algebraic effects, see *e.g.* [25]; it goes beyond the topic of the current paper, but definitely deserves further investigation.

3. A necessary step towards understanding the Dijkstra monad is made, by simplifying previous accounts [28] and casting them in proper categorical language.

4. Using this combined view on computations and logic, for the different monad examples T in this paper, an associated "Dijkstra monad" $\mathcal{D}_T$ is defined. This definition depends on the logic **Log** that is used to reason about T, since the monad is defined via a homset in this category **Log**. This logic-based approach goes well beyond the particular logic that is used in the original article [28], where the Dijkstra monad is introduced, since it now also applies to for instance probabilistic computation, in various forms.

5. Once we have the Dijkstra monad $\mathcal{D}_T$ associated with T we define a "map of monads" $\mathfrak{S}_T \Rightarrow \mathcal{D}_T$, where $\mathfrak{S}_T$ is the T-state monad, obtained by applying the state monad transformer to T. This map of monads is precisely the weakest precondition operation (categorically: substitution). This operation that is fundamental in the work of Dijkstra is thus captured neatly in categorical/monadic terms.

6. Finally, a general construction is presented that defines the Dijkstra monad $\mathcal{D}_T$ for an arbitrary monad T on **Sets**. A deeper understanding of the construction requires a systematic account of how the categories "**Log**" in (1) arise in general. This is still beyond current levels of understanding.

We assume that the reader is familiar with the basic concepts of category theory, especially with the theory of monads. The organisation of the paper is as follows: the first three Sects. 2–4 elaborate instances of the triangle (1) for non-deterministic, linear & probabilistic, and quantum computation. Subsequently, Sect. 5 shows how to obtain the Dijkstra monads for the different (concrete) monad examples, and proves that weakest precondition computation forms a map of monads. These examples are generalised in Sect. 6. Finally, Sect. 7 wraps up with some concluding remarks.

2   Non-deterministic and Partial Computation

The powerset operation $\mathcal{P}(X) = \{U \mid U \subseteq X\}$ yields a monad $\mathcal{P}: \textbf{Sets} \to \textbf{Sets}$ with unit $\eta = \{-\}$ given by singletons and multiplication $\mu = \bigcup$ by union. The associated Kleisli category $\mathcal{K}\ell(\mathcal{P})$ is the category of sets and non-deterministic functions $X \to \mathcal{P}(Y)$, which may be identified with relations $R \subseteq X \times Y$. The category $\mathcal{EM}(\mathcal{P})$ of (Eilenberg-Moore) algebras is the category $\mathbf{CL}_\vee$ of complete lattices and join-preserving functions. In this situation diagram (1) takes the form:

$$
\begin{array}{ccc}
(\mathbf{CL}_\wedge)^{\mathrm{op}} & \overset{\cong}{\underset{\longleftarrow}{\longrightarrow}} & \mathbf{CL}_\vee = \mathcal{EM}(\mathcal{P}) \\
& \underset{Pred}{\searrow} \quad \underset{Stat}{\nearrow} & \\
& \mathcal{K}\ell(\mathcal{P}) &
\end{array}
\tag{2}
$$

where $\mathbf{CL}_\wedge$ is the category of complete lattices and meet-preserving maps. The isomorphism $\cong$ arises because each join-preserving map between complete lattices corresponds to a meet-preserving map in the other direction. The upgoing "state" functor $Stat$ on the right is the standard full and faithful functor from the Kleisli category of a monad to its category of algebras. The predicate functor $Pred \colon \mathcal{K}\ell(\mathcal{P}) \to (\mathbf{CL}_\wedge)^{\mathrm{op}}$ on the left sends a set X to the powerset $\mathcal{P}(X)$ of predicates/subsets, as complete lattices; a Kleisli map $f \colon X \to \mathcal{P}(Y)$ yields a map:

$$
\mathcal{P}(Y) \xrightarrow{\ f^* = Pred(f)\ } \mathcal{P}(X) \qquad \text{given by} \qquad (Q \subseteq Y) \longmapsto \{x \mid f(x) \subseteq Q\}.
\tag{3}
$$

In categorical logic, this $Pred(f)$ is often written as f^*, and called a substitution functor. In modal logic one may write it as $\Box_f$. In the current context we also write it as $wp(f)$, since it forms the weakest precondition operation for f, see [4]. Clearly, it preserves arbitrary meets (intersections). It is not hard to see that the triangle (2) commutes.

Interestingly, the diagram (2) involves additional structure on homsets. If we have a collection of parallel maps f_i in $\mathcal{K}\ell(\mathcal{P})$, we can take their (pointwise) join $\bigvee_{i \in I} f_i$. Pre- and post-composition preserves such joins. This means that the Kleisli category $\mathcal{K}\ell(\mathcal{P})$ is enriched over the category $\mathbf{CL}_\vee$. The category $\mathbf{CL}_\vee$ is monoidal closed, and thus enriched over itself. Also the category $(\mathbf{CL}_\wedge)^{\mathrm{op}}$ is enriched over $\mathbf{CL}_\vee$, with joins given by pointwise intersections. Further, the functors in (2) are enriched over $\mathbf{CL}_\vee$, which means that they preserve these joins on posets. In short, the triangle is a diagram in the category of categories enriched over $\mathbf{CL}_\vee$. In particular, the predicate functor is enriched, which amounts to the familiar law for non-deterministic choice in weakest precondition reasoning: $wp(\bigvee_i f_i) = \bigwedge_i wp(f_i)$.

A less standard monad for non-determinism is the *ultrafilter* monad $\mathcal{U} \colon \mathbf{Sets} \to \mathbf{Sets}$. A convenient way to describe it, at least in the current setting, is:

$$
\mathcal{U}(X) = \mathbf{BA}\big(\mathcal{P}(X), 2\big) = \{f \colon \mathcal{P}(X) \to 2 \mid f \text{ is a map of Boolean algebras}\}.
$$

For a finite set X one has $X \overset{\cong}{\to} \mathcal{U}(X)$.

A famous result of [19] says that the category of algebras of $\mathcal{U}$ is the category **CH** of compact Hausdorff spaces (and continuous functions). It yields the following triangle.

$$
\begin{array}{c}
\textit{Spec} = \textit{Hom}(-,2) \\
\mathbf{BA}^{\mathrm{op}} \underset{\textit{Clopen}}{\overset{\top}{\rightleftarrows}} \mathbf{CH} = \mathcal{EM}(\mathcal{U}) \\
{}_{\textit{Pred}} \searrow \qquad \nearrow {}_{\textit{Stat}} \\
\mathcal{K\ell}(\mathcal{U})
\end{array}
\tag{4}
$$

The predicate functor *Pred* sends a set X to the Boolean algebra $\mathcal{P}(X)$ of subsets of X. For a map $f\colon X \to \mathcal{U}(Y)$ we get $f^*\colon \mathcal{P}(Y) \to \mathcal{P}(X)$ by $f^*(Q) = \{x \mid f(x)(Q) = 1\}$. This functor *Pred* is full and faithful, almost by construction.

The precise enrichment in this case is unclear. Enrichment over (compact Hausdorff) spaces, if present, is not so interesting because it does not provide algebraic structure on computations.

We briefly look at the *lift* (or "maybe") monad $\mathcal{L}\colon \mathbf{Sets} \to \mathbf{Sets}$, given by $\mathcal{L}(X) - 1 + X$. Its Kleisli category $\mathcal{K\ell}(\mathcal{L})$ is the category of sets and partial functions. And its (equivalent) category of algebra $\mathcal{EM}(\mathcal{L})$ is the category $\mathbf{Sets_\bullet}$ of pointed sets, $(X, \bullet_X)$, where $\bullet_X \in X$ is a distinguished element; morphisms in $\mathbf{Sets_\bullet}$ are "strict", in the sense that they preserve such points. There is then a situation:

$$
\begin{array}{c}
(\mathbf{ACL}_{\bigvee_\bullet, \wedge})^{\mathrm{op}} \overset{\top}{\rightleftarrows} \mathbf{Sets_\bullet} = \mathcal{EM}(\mathcal{L}) \\
{}_{\textit{Pred}} \searrow \qquad \nearrow {}_{\textit{Stat}} \\
\mathcal{K\ell}(\mathcal{L})
\end{array}
\tag{5}
$$

We call a complete lattice *atomic* if (1) each element is the join of atoms below it, and (2) binary meets $\wedge$ distribute over arbitrary joins $\bigvee$. Recall that an atom a is a non-bottom element satisfying $x < a \Rightarrow x = \bot$. We write $At(L) \subseteq L$ for the subset of atoms. In such an atomic lattice atoms a are completely join-irreducible: for a non-empty index set I, if $a \leq \bigvee_{i \in I} x_i$ then $a \leq x_i$ for some $i \in I$.

The category $\mathbf{ACL}_{\bigvee_\bullet, \wedge}$ contains atomic complete lattices, with maps preserving non-empty joins (written as $\bigvee_\bullet$) and binary meets $\wedge$. Each Kleisli map $f\colon X \to \mathcal{L}(Y) = \{\bot\} \cup Y$ yields a substitution map $f^*\colon \mathcal{P}(Y) \to \mathcal{P}(X)$ by $f^*(Q) = \{x \mid \forall y.\, f(x) = y \Rightarrow Q(y)\}$. This f^* preserves $\wedge$ and non-empty joins $\bigvee_\bullet$. Notice that $f^*(\emptyset) = \{x \mid f(x) = \bot\}$, which need not be empty.

The adjunction $(\mathbf{ACL}_{\bigvee_\bullet, \wedge})^{\mathrm{op}} \rightleftarrows \mathbf{Sets_\bullet}$ amounts to a bijective correspondence:

$$
\frac{L \overset{f}{\longrightarrow} \mathcal{P}(X - \bullet) \qquad \text{in } (\mathbf{ACL}_{\bigvee_\bullet, \wedge})^{\mathrm{op}}}{X \underset{g}{\longrightarrow} \{\bot\} \cup At(L) \qquad \text{in } \mathbf{Sets_\bullet}}
$$

This correspondence works as follows. Given $f: L \to \mathcal{P}(X - \bullet)$ notice that $X = f(\top) = f(\bigvee At(L)) = \bigcup_{a \in At(L)} f(a)$. Hence for each $x \in X$ there is an atom a with $x \in f(a)$. We define $\overline{f}: X \to \{\perp\} \cup At(L)$ as:

$$\overline{f}(x) = \begin{cases} a & \text{if } x \in f(a) - f(\perp) \\ \perp & \text{otherwise.} \end{cases}$$

This is well-defined: if x is both in $f(a) - f(\perp)$ and in $f(a') - f(\perp)$, for $a \neq a'$, then $x \in (f(a) \cap f(a')) - f(\perp) = f(a \wedge a') - f(\perp) = f(\perp) - f(\perp) = \emptyset$.

In the other direction, given $g: X \to \{\perp\} \cup At(L)$, define for $y \in L$,

$$\overline{g}(y) = \{x \in X \mid \exists a \in At(L).\, a \leq y \text{ and } g(x) = a\} \cup \{x \in X - \bullet \mid g(x) = \perp\}.$$

It is not hard to see that this yields a commuting triangle (5), and that the (upgoing) functors are full and faithful.

3   Linear and (sub)Convex Computation

We sketch two important sources for linear and (sub)convex structures.

1. If A is a matrix, say over the real numbers $\mathbb{R}$, then the set of solution vectors v of the associated homogeneous equation $Av = 0$ forms a linear space: it is closed under finite additions and scalar multiplication. For a fixed vector $b \neq 0$, the solutions v of the non-homogeneous equation $Ax = b$ form a convex set: it is closed under convex combinations $\sum_i r_i v_i$ of solutions v_i and "probability" scalars $r_i \in [0, 1]$ with $\sum_i r_i = 1$. Finally, for $b \geq 0$, the solutions v to the inequality $Av \leq b$ are closed under subconvex combinations $\sum_i r_i v_i$ with $\sum_i r_i \leq 1$. These examples typically occur in linear programming.
2. If V is a vector space of some sort, we can consider the space of linear functions $f: V \to \mathbb{R}$ to the real (or complex) numbers. This space is linear again, via pointwise definitions. Now if V contains a unit 1, we can impose an additional requirement that such functions $f: V \to \mathbb{R}$ are 'unital', i.e. satisfy $f(1) = 1$. This yields a convex set of functions, where $\sum_i r_i f_i$ again preserves the unit, if $\sum_i r_i = 1$. If we require only $0 \leq f(1) \leq 1$, making f 'subunital', we get a subconvex set. These requirements typically occur in a setting of probability measures.

Taking (formal) linear and (sub)convex combinations over a set yields the structure of a monad. We start by recalling the definitions of these (three) monads, namely the multiset monad $\mathcal{M}_R$, the distribution monad $\mathcal{D}$, and the subdistribution monad $\mathcal{D}_{\leq 1}$, see [12] for more details. A semiring is given by a set R which carries a commutative monoid structure $(+, 0)$, and also another monoid structure $(\cdot, 1)$ which distributes over $(+, 0)$. As is well-known [11], each such semiring R gives rise to a multiset monad $\mathcal{M}_R: \textbf{Sets} \to \textbf{Sets}$, where:

$$\mathcal{M}_R(X) = \{\varphi: X \to R \mid supp(\varphi) \text{ is finite}\},$$

where $supp(\varphi) = \{x \in X \mid \varphi(x) \neq 0\}$ is the support of φ. Such $\varphi \in \mathcal{M}_R(X)$ may also be written as finite formal sum $\varphi = \sum_i s_i|x_i\rangle$ where $supp(\varphi) = \{x_1, \ldots, x_n\}$ and $s_i = \varphi(x_i) \in R$ is the multiplicity of $x_i \in X$. The "ket" notation $|x\rangle$ for $x \in X$ is just syntactic sugar. The unit of the monad is given by $\eta(x) = 1|x\rangle$ and its multiplication by $\mu(\sum_i s_i|\varphi_i\rangle) = \sum_x (\sum_i s_i \cdot \varphi_i(x))|x\rangle$.

The *distribution* monad $\mathcal{D} \colon \textbf{Sets} \to \textbf{Sets}$ is defined similarly. It maps a set X to the set of finite formal convex combinations over X, as in:

$$\mathcal{D}(X) = \{\varphi \colon X \to [0,1] \mid supp(\varphi) \text{ is finite, and } \sum_x \varphi(x) = 1\}$$
$$= \{r_1|x_1\rangle + \cdots + r_n|x_n\rangle \mid x_i \in X, r_i \in [0,1] \text{ with } \sum_i r_i = 1\}.$$

The unit η and multiplication μ for $\mathcal{D}$ are as for $\mathcal{M}_R$. We consider another variation, namely the *subdistribution* monad $\mathcal{D}_{\leq 1}$, where $\mathcal{D}_{\leq 1}(X)$ contains the formal *subconvex* combinations $\sum_i r_i|x_i\rangle$ where $\sum_i r_i \leq 1$. It has the same unit and multiplication as $\mathcal{D}$.

These three monads $\mathcal{M}_R, \mathcal{D}$ and $\mathcal{D}_{\leq 1}$ are used to capture different kinds of computation, in the style of [23]. Maps (coalgebras) of the form $c \colon X \to \mathcal{M}_R(X)$ capture "multi-computations", which can be written in transition notation as $x \xrightarrow{r} x'$ if $c(x)(x') = r$. This label $r \in R$ can represent the time or cost of a transition. Similarly, the monads $\mathcal{D}$ and $\mathcal{D}_{\leq 1}$ capture probabilistic computation: for coalgebras $c \colon X \to \mathcal{D}(X)$ or $c \colon X \to \mathcal{D}_{\leq 1}(X)$ we can write $x \xrightarrow{r} x'$ if $c(x)(x') = r \in [0,1]$ describes the probability of the transition $x \to x'$.

The category $\mathcal{EM}(\mathcal{M}_R)$ of (Eilenberg-Moore) algebras of the multiset monad $\mathcal{M}_R$ contains the *modules* over the semiring R. Such a module is given by a commutative monoid $M = (M, +, 0)$ together with a scalar multiplication $S \times M \to M$ which preserves $(+, 0)$ in both arguments. More abstractly, if we write **CMon** for the category of commutative monoids, then the semiring R is a monoid in **CMon**, and the category $\textbf{Mod}_R = \mathcal{EM}(\mathcal{M}_R)$ of modules over R is the category $Act_R(\textbf{CMon})$ of R-actions $R \otimes M \to M$ in **CMon**, see also [21, VII§4]. For instance, for the semiring $R = \mathbb{N}$ of natural numbers we obtain $\textbf{CMon} = \mathcal{EM}(\mathcal{M}_\mathbb{N})$ as associated category of algebras; for $R = \mathbb{R}$ or $R = \mathbb{C}$ we obtain the categories $\textbf{Vect}_\mathbb{R}$ or $\textbf{Vect}_\mathbb{C}$ of vector spaces over real or complex numbers; and for the Boolean semiring $R = 2 = \{0,1\}$ we get the category **JSL** of join semi-lattices, since $\mathcal{M}_2$ is the finite powerset monad.

We shall write $\textbf{Conv} = \mathcal{EM}(\mathcal{D})$ for the category of *convex* sets. These are sets X in which for each formal convex sum $\sum_i r_i|x_i\rangle$ there is an actual convex sum $\sum_i r_i x_i \in X$. Morphisms in **Conv** preserve such convex sums, and are often called affine functions. A convex set can be defined alternatively as a barycentric algebra [27], see [10] for the connection. Similarly, we write $\textbf{Conv}_{\leq 1} = \mathcal{EM}(\mathcal{D}_{\leq 1})$ for the category of *subconvex* sets, in which subconvex sums exist.

For linear "multi" computation and computation the general diagram (1) takes the following form, where $\textbf{Mod}_R = \mathcal{EM}(\mathcal{M}_R)$ and $\textbf{Conv} = \mathcal{EM}(\mathcal{D})$.

$$(6)$$

The adjunction $(\mathbf{Mod}_R)^{\mathrm{op}} \rightleftarrows \mathbf{Mod}_R$ is given by the correspondence between homomorphisms $M \to (N \multimap R)$ and $N \to (M \multimap R)$, where $\multimap$ is used for linear function space. The predicate functor $R^{(-)} \colon \mathcal{K}\ell(\mathcal{M}_R) \to (\mathbf{Mod}_R)^{\mathrm{op}}$ sends a set X to the module R^X of functions $X \to R$, with pointwise operations. A Kleisli map $f \colon X \to \mathcal{M}_R(Y)$ yields a map of modules $f^* = R^f \colon R^Y \to R^X$ by $f^*(q)(x) = \sum_y q(y) \cdot f(x)(y)$. Like before, this $f^*(q)$ may be understood as the weakest precondition of the post-condition q. In one direction the triangle commutes: $\mathrm{Hom}(\mathcal{M}_R(X), R) \cong \mathbf{Sets}(X, R) = R^X$ since $\mathcal{M}_R(X)$ is the free module on X. Commutation in the other direction, that is $\mathrm{Hom}(R^X, R) \cong \mathcal{M}_R(X)$ holds for finite sets X. Hence in order to get a commuting triangle we should restrict to the full subcategory $\mathcal{K}\ell_{\mathbb{N}}(\mathcal{M}_R) \hookrightarrow \mathcal{K}\ell(\mathcal{M}_R)$ with objects $n \in \mathbb{N}$, considered as n-element set.

Now let R be a *commutative* semiring. The triangle (6) is then a diagram enriched over $\mathbf{Mod}_R$: the categories, functors, and natural transformations involved are all enriched. Indeed, if the semiring R is commutative, then so is the monad $\mathcal{M}_R$, see e.g. [12]; this implies that $\mathbf{Mod}_R$ is monoidal closed, and in particular enriched over itself. Similarly, the Kleisli category $\mathcal{K}\ell(\mathcal{M}_R)$ is then enriched over $\mathbf{Mod}_R$.

In the probabilistic case one can choose to use a logic with classical predicates (subsets, or characteristic functions) $\{0, 1\}^X$ or 'fuzzy predicates' $[0, 1]^X$. These options are captured in the following two triangles.

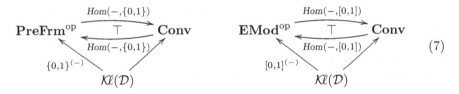

$$(7)$$

The adjunctions both come from [12]. The one on the left is investigated further in [20]. It uses the category $\mathbf{PreFrm}$ of preframes: posets with directed joins and finite meets, distributing over these joins, see [16]. Indeed, for a Kleisli map $f \colon X \to \mathcal{D}(Y)$ we have a substitution functor $f^* \colon \mathcal{P}(Y) \to \mathcal{P}(X)$ in $\mathbf{PreFrm}$ given by $f^*(Q) = \mathrm{wp}(f)(Q) = \{x \in X \mid \mathrm{supp}(f(x)) \subseteq Q\}$. This f^* preserves directed joins because the support of $f(x) \in \mathcal{D}(Y)$ is finite.

The homsets $\mathbf{PreFrm}(X, Y)$ of preframe maps $X \to Y$ have finite meets $\wedge, \top$, which can be defined pointwise. As a result, these homsets are convex sets, in a trivial manner: a sum $\sum_i r_i h_i$ is interpreted as $\bigwedge_i h_i$, where we implicitly assume that $r_i > 0$ for each i. With this in mind one can check that the triangle on the left in (7) is enriched over $\mathbf{Conv}$. It yields the rule $\mathrm{wp}(\sum_i r_i f_i)(Q) = \bigcap_i \mathrm{wp}(f_i)(Q)$.

The situation on the right in (7) requires more explanation. We sketch the essentials. A *partial commutative monoid* (PCM) is a given by a set M with a partial binary operation $\varotimes \colon M \times M \to M$ which is commutative and associative, in a suitable sense, and has a zero element $0 \in M$. One writes $x \perp y$ if $x \varotimes y$ is defined. A morphism $f \colon M \to N$ of PCMs satisfies: $x \perp x'$ implies $f(x) \perp f(x')$, and then $f(x \varotimes x') = f(x) \varotimes f(x')$. This yields a category which we shall write as $\mathbf{PCMon}$.

The unit interval $[0,1]$ is clearly a PCM, with $r \oslash r'$ defined and equal to $r + r'$ if $r + r' \leq 1$. With its multiplication operation this $[0,1]$ is a monoid in the category **PCMon**, see [14] for details. We define a category **PCMod** $= Act_{[0,1]}(\textbf{PCMon})$ of *partial commutative modules*; its objects are PCMs M with an action $[0,1] \times M \to M$, forming a homomorphism of PCMs in both coordinates. These partial commutative modules are thus like vector spaces, except that their addition is partial and their scalars are probabilities in $[0,1]$.

Example 1. Consider the set of *partial* functions from a set X to the unit interval $[0,1]$. Thus, for such a $f \colon X \rightharpoonup [0,1]$ there is an output value $f(x) \in [0,1]$ only for $x \in X$ which are in the domain $dom(f) \subseteq X$. Obviously, one can define scalar multiplication $r \bullet f$, pointwise, without change of domain. We take the empty function — nowhere defined, with empty domain — as zero element. Consider the following two partial sums that turn these partial functions into a partial commutative module.

One way to define a partial sum $\oslash$ is to define $f \perp g$ as $dom(f) \cap dom(g) = \emptyset$; the sum $f \oslash g$ is defined on the union of the domains, via case distinction.

A second partial sum $f \oslash' g$ is defined if for each $x \in dom(f) \cap dom(g)$ one has $f(x) + g(x) \leq 1$. For those x in the overlap of domains, we define $(f \oslash' g)(x) = f(x) + g(x)$, and elsewhere $f \oslash' g$ is f on $dom(f)$ and g on $dom(g)$.

An *effect algebra* (see [5,7]) is a PCM with for each element x a unique complement $x^{\perp}$ satisfying $x \oslash x^{\perp} = 1 = 0^{\perp}$, together with the requirement $1 \perp x \Rightarrow x = 0$. In the unit interval $[0,1]$ we have $r^{\perp} = 1 - r$. In Example 1 for both the partial sums $\oslash$ and $\oslash'$ one does *not* get an effect algebra: in the first case there is not always an $f^{\perp}$ with $f \oslash f^{\perp} = 1$, where 1 is the function that is everywhere defined and equal to 1. For $\oslash'$ there is $f^{\perp}$ with $f \oslash' f^{\perp}$, but $f^{\perp}$ need not be unique. E.g. the function 1 has both the empty function and the everywhere 0 function as complement. We can adapt this example to an effect algebra by considering only partial functions $X \rightharpoonup (0,1]$, excluding 0 as outcome.

A map of effect algebras f is a map of PCMs satisfying $f(1) = 1$. This yields a subcategory **EA** $\hookrightarrow$ **PCMon**. An *effect module* is at the same time an effect algebra and a partial commutative module. We get a subcategory **EMod** $\hookrightarrow$ **PCMod**. By "homming into $[0,1]$" one obtains an adjunction **EMod**^{op} $\rightleftarrows$ **Conv**, see [12] for details. The resulting triangle on the right in (7) commutes in one direction, since $\textbf{Conv}(\mathcal{D}(X), [0,1]) \cong [0,1]^X$. In the other direction one has $\textbf{EMod}([0,1]^X, [0,1]) \cong \mathcal{D}(X)$ for finite sets X.

In [26] it is shown that each effect module is a convex set. The proof is simple, but makes essential use of the existence of orthocomplements $(-)^{\perp}$. In fact, the category **EMod** is enriched over **Conv**. Even stronger, the triangle on the right in (7) is enriched over **Conv**. This yields $wp(\sum_i r_i f_i) = \sum_i r_i wp(f_i)$.

There are two variations on the distribution monad $\mathcal{D}$ that are worth pointing out. The first one is the expectation monad $\mathcal{E}(X) = \textbf{EMod}([0,1]^X, [0,1])$ introduced in [15] (and used for instance in [2] for probabilistic program semantics). It can be seen as a probabilistic version of the ultrafilter monad from the previous section. For a finite set one has $\mathcal{E}(X) \cong \mathcal{D}(X)$. The category of algebras

$\mathcal{EM}(\mathcal{E})$ contains the convex compact Hausdorff spaces, see [15]. This monad $\mathcal{E}$ gives rise to a triangle as on the left below, see [15] for details.

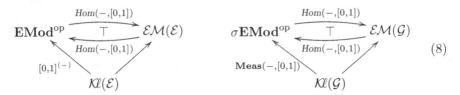

$$(8)$$

The triangle on the right captures continuous probabilistic computation, via the Giry monad $\mathcal{G}$ on the category **Meas** of measurable spaces. This is elaborated in [13]. The category $\sigma\mathbf{EMod}$ contains effect modules in which countable ascending chains have a join. Both these triangles commute, and are enriched over convex sets.

We continue with the category $\mathbf{Conv}_{\leq 1} = \mathcal{EM}(\mathcal{D}_{\leq 1})$ of subconvex sets. We now get a triangle of the form:

$$
\begin{array}{c}
\mathbf{GEMod}^{\mathrm{op}} \xrightarrow[\;Hom(-,[0,1])\;]{\overset{Hom(-,[0,1])}{\top}} \mathbf{Conv}_{\leq 1} = \mathcal{EM}(\mathcal{D}_{\leq 1}) \\
{}_{[0,1]^{(-)}}\searrow \qquad \nearrow \\
\mathcal{K\ell}(\mathcal{D}_{\leq 1})
\end{array}
\qquad (9)
$$

We need to describe the category **GEMod** of generalised effect modules. First, a generalised effect algebra, according to [5], is a partial commutative monoid (PCM) in which $x \oslash y = 0 \Rightarrow x = y = 0$ and $x \oslash z = y \oslash z \Rightarrow x = y$ hold. In that case one can define a partial order $\leq$ in the usual way. We obtain a full subcategory $\mathbf{GEA} \hookrightarrow \mathbf{PCMon}$. In fact we have $\mathbf{EA} \hookrightarrow \mathbf{GEA} \hookrightarrow \mathbf{PCMon}$, since a generalised effect algebra is not an effect algebra, but a more general 'topless' structure: a generalized effect algebra with a top element 1 is an effect algebra.

One can now add multiplication with scalars from $[0,1]$ to generalised effect algebras, like for partial commutative modules. But we require more, namely the existence of subconvex sums $r_1 x_1 \oslash \cdots \oslash r_n x_n$, for $r_i \in [0,1]$ with $\sum_i r_i \leq 1$. As noted before, such sums exist automatically in effect algebras, but this is not the case in generalised effect algebra with scalar multiplication, as the first structure in Example 1 illustrates. Thus we define a full subcategory $\mathbf{GEMod} \hookrightarrow \mathbf{PCMod}$, where objects of **GEMod** are at the same time partial commutative modules and generalised effect algebras, with the additional requirement that all subconvex sums exist. Summarising, we have the following diagram of 'effect' structures, where the bottom row involves scalar multiplication.

$$
\begin{array}{ccc}
\mathbf{EA} & \hookrightarrow \mathbf{GEA} & \hookrightarrow \mathbf{PCMon} \\
\uparrow & \uparrow & \uparrow \\
\mathbf{EMod} & \hookrightarrow \mathbf{GEMod} & \hookrightarrow \mathbf{PCMod}
\end{array}
$$

Once we know what generalized effect modules are, it is easy to see that 'homming into $[0,1]$' yields the adjunction in (9). Moreover, this diagram (9) is enriched over $\mathbf{Conv}_{\leq 1}$, so that weakest precondition wp preserves subconvex sums of Kleisli maps (programs).

4 Quantum Computation, Briefly

In this section we wish to point out that the triangle (1) applies beyond the monadic setting. For instance, quantum computation, modelled via the category $\mathbf{Cstar}_{\mathrm{PU}}$ of C^*-algebras (with unit) and positive, unital maps, one obtains a triangle:

$$
\begin{array}{c}
\mathbf{EMod}^{\mathrm{op}} \underset{Hom(-,[0,1])}{\overset{Hom(-,[0,1])}{\rightleftarrows}} \mathbf{Conv} \\
{}_{Pred}\searrow \qquad \swarrow {}_{Stat} \\
(\mathbf{Cstar}_{\mathrm{PU}})^{\mathrm{op}}
\end{array}
\tag{10}
$$

The predicate functor sends a C^*-algebra A to the unit interval $[0,1]_A \subseteq A$ of "effects" in A, where $[0,1]_A = \{a \in A \mid 0 \leq a \leq 1\}$. This functor is full and faithful, see [8]. On the other side, the state functor sends a C^*-algebra A to the (convex) set of its states, given by the homomorphisms $A \to \mathbb{C}$. This diagram is enriched over convex sets. A similar setting of states and effects, for Hilbert spaces instead of C^*-algebras, is used in [3] for a quantum precondition calculus.

In [8] it was shown that *commutative* C^*-algebras, capturing the probabilistic, non-quantum case, can be described as a Kleisli category. It is unclear if the non-commutative, proper quantum, case can also be described via a monad.

5 Dijkstra Monad Examples

In [28] the "Dijkstra" monad is introduced, as a variant of the "Hoare" monad from [24]. It captures weakest precondition computations for the state monad $X \mapsto (S \times X)^S$, where S is a fixed collection of states (the heap). Here we wish to give a precise description of the Dijkstra monad, for various concrete monads T.

For the powerset monad $\mathcal{P}$, a first version of the Dijkstra monad, following the description in [28], yields $\mathfrak{D}_{\mathcal{P}} \colon \mathbf{Sets} \to \mathbf{Sets}$ defined as:

$$
\mathfrak{D}_{\mathcal{P}}(X) = \mathcal{P}(S)^{\mathcal{P}(S \times X)},
\tag{11}
$$

where S is again a fixed set of states. Thus, an element $w \in \mathfrak{D}_{\mathcal{P}}(X)$ is a function $w \colon \mathcal{P}(S \times X) \to \mathcal{P}(S)$ that transforms a postcondition $Q \in \mathcal{P}(X \times S)$ into a precondition $w(Q) \in \mathcal{P}(S)$. The post-condition is a binary predicate, on both an output value from X and a state from S; the precondition is a unary predicate, only on states.

In this first version (11) we simply take *all* functions $\mathcal{P}(S \times X) \to \mathcal{P}(S)$. But in the triangle (2) we see that predicate transformers are maps in $\mathbf{CL}_\wedge$, *i.e.* are meet-preserving maps between complete lattices. Hence we now properly (re)define $\mathfrak{D}_\mathcal{P}$ as the set of meet-preserving functions:

$$\mathfrak{D}_\mathcal{P}(X) \stackrel{\text{def}}{=} \mathbf{CL}_\wedge\Big(\mathcal{P}(S \times X), \mathcal{P}(S)\Big) = \big(\mathbf{CL}_\wedge\big)^{\text{op}}\Big(\mathrm{Pred}(S), \mathrm{Pred}(S \times X)\Big) \quad (12)$$

This is indeed a monad, following [28], with unit and multiplication:

$$\eta(x) = \lambda Q.\,\{s \mid (s, x) \in Q\} \qquad \mu(H) = \lambda Q.\,H\big(\{(s, h) \mid s \in h(Q)\}\big).$$

We introduce some notation ($\mathfrak{S}$, *i.e.* fraktur S) for the result of applying the state transformer monad to an arbitrary monad (see *e.g.* [18]).

Definition 1. *For a monad $T \colon \mathbf{Sets} \to \mathbf{Sets}$ and for a fixed set (of "states") S, the T-state monad $\mathfrak{S}_T$ is defined as:*

$$\mathfrak{S}_T(X) = T(S \times X)^S = \mathcal{K}\ell(T)\big(S, S \times X\big).$$

For the record, its unit and multiplication are given by:

$$x \longmapsto \lambda s \in S.\,\eta(s, x) \quad \text{and} \quad H \longmapsto \mu \circ T(\lambda(s, h).\,h(s)) \circ H,$$

where η, μ are the unit and multiplication of T.

Proposition 1. *There is a map of monads $\mathfrak{S}_\mathcal{P} \Rightarrow \mathfrak{D}_\mathcal{P}$ from the $\mathcal{P}$-state monad to the $\mathcal{P}$-Dijkstra monad (12), with components:*

$$\mathfrak{S}_\mathcal{P}(X) = \mathcal{K}\ell(\mathcal{P})\big(S, S \times X\big) \xrightarrow{\;\sigma_X\;} \big(\mathbf{CL}_\wedge\big)^{\text{op}}\big(\mathrm{Pred}(S), \mathrm{Pred}(S \times X)\big) = \mathfrak{D}_\mathcal{P}(X)$$

given by substitution/weakest precondition:

$$\sigma_X(f) = \mathrm{Pred}(f) = f^* = \mathrm{wp}(f) = \lambda Q \in \mathcal{P}(S \times X).\,\{s \mid f(s) \subseteq Q\},$$

following the description from (3).

Proof. We have to check that substitution is natural in X and commutes with the units and multiplications. This is easy; for instance:

$$\begin{aligned}
(\sigma \circ \eta^{\mathfrak{S}})(x)(Q) = \big(\eta^{\mathfrak{S}}(x)\big)^*(Q) &= \{s \mid \eta^{\mathfrak{S}}(x)(s) \subseteq Q\} \\
&= \{s \mid \eta^{\mathcal{P}}(s, x) \subseteq Q\} \\
&= \{s \mid \{(s, x)\} \subseteq Q\} \\
&= \{s \mid (s, x) \in Q\} = \eta^{\mathfrak{D}}(x)(Q). \qquad \square
\end{aligned}$$

At this stage the generalisation of the Dijkstra monad for other monads — with an associated logic as in (1) — should be clear. For instance, for the multiset

$\mathcal{M}_R$ and (sub)distribution monad $\mathcal{D}, \mathcal{D}_{\leq 1}$ we use the triangles in (6), (7) and (9) to define associated Dijkstra monads:

$$
\begin{aligned}
\mathfrak{D}_{\mathcal{M}_R}(X) &= \mathbf{Mod}_R\Big(Pred(S \times X), Pred(S)\Big) = \mathbf{Mod}_R\Big(R^{S \times X}, R^S\Big) \\
\mathfrak{D}_{\mathcal{D}}(X) &= \mathbf{EMod}\Big(Pred(S \times X), Pred(S)\Big) = \mathbf{EMod}\Big([0,1]^{S \times X}, [0,1]^S\Big) \quad (13) \\
\mathfrak{D}_{\mathcal{D}_{\leq 1}}(X) &= \mathbf{GEMod}\Big(Pred(S \times X), Pred(S)\Big) = \mathbf{GEMod}\Big([0,1]^{S \times X}, [0,1]^S\Big)
\end{aligned}
$$

Then there is the following result, analogously to Proposition 1. The proofs involve extensive calculations but are essentially straightforward.

Proposition 2. *For the multiset, distribution, and subdistribution monads $\mathcal{M}_R, \mathcal{D}$, and $\mathcal{D}_{\leq 1}$ there are maps of monads given by substitution:*

$$
\mathfrak{S}_{\mathcal{M}_R} \xLongrightarrow{(-)^*} \mathfrak{D}_{\mathcal{M}_R} \qquad \mathfrak{S}_{\mathcal{D}} \xLongrightarrow{(-)^*} \mathfrak{D}_{\mathcal{D}} \qquad \mathfrak{S}_{\mathcal{D}_{\leq 1}} \xLongrightarrow{(-)^*} \mathfrak{D}_{\mathcal{D}_{\leq 1}}
$$

from the associated state monads to the associated Dijkstra monads (13). □

The Dijkstra monad associated with the expectation monad $\mathcal{E}$ is the same as for the distribution monad $\mathcal{D}$. Hence one gets a map of monads $\mathfrak{S}_{\mathcal{E}} \Rightarrow \mathfrak{D}_{\mathcal{D}}$, with substitution components:

$$
\mathfrak{S}_{\mathcal{E}}(X) = \mathcal{E}(S \times X)^S = \mathbf{EMod}\Big([0,1]^{S \times X}, [0,1]\Big)^S
$$

$$
\downarrow (-)^*
$$

$$
\mathbf{EMod}\Big([0,1]^{S \times X}, [0,1]^S\Big) = \mathfrak{D}_{\mathcal{D}}(X)
$$

where $f^*(q)(s) = f(s)(q)$. Details are left to the reader.

6   Dijkstra's Monad, Beyond Examples

In the end it remains a bit unsatisfactory to see only particular instances of what we called a Dijkstra monad $\mathfrak{D}_T$. Below we offer a more general description, even though it is not the definitive story. For convenience we restrict ourselves to monads on **Sets**.

So let $T: \mathbf{Sets} \to \mathbf{Sets}$ be an arbitrary monad. As observed in (an exercise in) [11], each (fixed) Eilenberg-Moore algebra $\omega: T(\Omega) \to \Omega$ determines an adjunction $\mathbf{Sets}^{\mathrm{op}} \rightleftarrows \mathcal{EM}(T)$, via functors $\Omega^{(-)}: \mathbf{Sets}^{\mathrm{op}} \to \mathcal{EM}(T)$ and $Hom(-, \omega): \mathcal{EM}(T) \to \mathbf{Sets}^{\mathrm{op}}$. It makes sense to require that the algebra ω is a cogenerator in $\mathcal{EM}(T)$, making the unit of the adjunction injective, but this is not needed in general. The adjunction can be generalised to strong monads on monoidal categories with equalisers, but that is not so relevant at this stage.

With this adjunction we can form a triangle of the form:

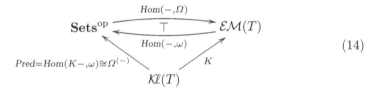

$$(14)$$

The induced predicate functor *Pred* is defined on a Kleisli map $f\colon X \to T(Y)$ as:

$$\Omega^Y \ni q \longmapsto \left(X \xrightarrow{f} T(Y) \xrightarrow{T(q)} T(\Omega) \xrightarrow{\omega} \Omega \right).$$

Appropriate restrictions of this adjunction may give rise to more suitable triangles, like in (2) and (4)–(9). How to do this restriction in a systematic manner is unclear at this stage.

But what we can do is define for a fixed set of states S, a Dijkstra monad, namely:

$$\mathfrak{D}_T(X) = \mathbf{Sets}^{\mathrm{op}}\big(Pred(S), Pred(S \times X)\big) = \mathbf{Sets}\big(\Omega^{S \times X}, \Omega^S\big). \qquad (15)$$

There is a unit $\eta_X\colon X \to \mathfrak{D}_T(X)$, namely $\eta_X(x)(q)(s) = q(s, x)$, and a multiplication $\mu_X\colon (\mathfrak{D}_T)^2(X) \to \mathfrak{D}_T(X)$ given by $\mu(H)(q) = H\big(\lambda(t, k).\, k(q)(t)\big)$.

In this general situation we can define a map of monads $\sigma\colon \mathfrak{S}_T \Rightarrow \mathfrak{D}_T$, where $\mathfrak{S}_T$ is the T-state monad $X \mapsto T(S \times X)^S$ from Definition 1. This σ has components $\sigma_X\colon T(S \times X)^S \to \mathbf{Sets}(\Omega^{S \times X}, \Omega^S)$ given by weakest precondition: $\sigma_X(f) = Pred(f) = f^* = wp(f)\colon \Omega^{S \times X} \to \Omega^S$.

Thus, in this purely set-theoretic setting we can define for an arbitrary monad T an associated Dijkstra monad $\mathfrak{D}_T$ as in (15), together with a 'weakest precondition' map of monads $\mathfrak{S}_T \Rightarrow \mathfrak{D}_T$. However, the general formulation (15) does not take into account that predicate transformers preserve certain logical structure, as in the concrete examples in Sect. 5.

We conclude with two more observations.

1. In the triangle (14) there are two functors $\mathcal{K}\ell(T) \to \mathcal{E}\mathcal{M}(T)$, namely the comparison functor K and $L = Hom(-, \Omega) \circ Pred = \mathbf{Sets}(\Omega^{(-)}, \Omega)$. There is a natural transformation $\tau\colon K \Rightarrow L$ with components:

$$\tau_X(u)(p) = \big(\omega \circ T(p)\big)(u) \quad \text{where } u \in K(X) = T(X) \text{ and } p \in \Omega^X.$$

The triangle (14) commutes in both directions if this τ is an isomorphism.
2. By composing the two adjunctions $\mathbf{Sets} \rightleftarrows \mathcal{E}\mathcal{M}(T) \rightleftarrows \mathbf{Sets}^{\mathrm{op}}$ in (14) one obtains a composite adjunction, which yields another monad T_ω on $\mathbf{Sets}$, namely:

$$T_\omega(X) = \big(U \circ \Omega^{(-)} \circ Hom(-, \omega) \circ F\big)(X) \cong \mathbf{Sets}(\Omega^X, \Omega).$$

This is what Lawvere [17] calls the dual monad; a similar construction occurs for instance in [6, Sect. 5]. There is in this case a map of monads $T \Rightarrow T_\omega$.

7   Concluding Remarks

The triangle-based semantics and logic that was presented via many examples forms the basis for (a) several versions of the Dijkstra monad, associated with different monads T, and (b) a description of the weakest precondition operation as a map of monads. There are many issues that remain to be investigated.

- We have concentrated on Dijkstra monads $\mathcal{D}$, but there is also the Hoare monad $\mathfrak{H}$, see [24,29]. It may be described explicitly as:

$$\mathfrak{H}(X) = \coprod_{P \subseteq S} \coprod_{Q \subseteq S \times X \times S} \{f \colon P \to X \times S \mid \forall s \in S. Q(s, f(s))\},$$

  where S is the set of states. It would be nice to extend this Hoare construction also to other monads than powerset.
- As already mentioned in the beginning, we only scratch the surface when it comes to the enrichment involved in the examples. This also requires further investigation, especially in connection with the algebraic effects approach, see e.g. [25], or the (enriched) monad models of [22].

Acknowledgements. Thanks to Sam Staton, Mathys Rennela, and Bas Westerbaan for their input & feedback.

References

1. Abramsky, S.: Domain theory in logical form. Ann. Pure Appl. Logic **51**(1/2), 1–77 (1991)
2. Barthe, G., Grégoire, B., Zanella Béguelin, S.: Formal certification of code-based cryptographic proofs. In: Principles of Programming Languages, pp. 90–101. ACM Press (2009)
3. D'Hondt, E., Panangaden, P.: Quantum weakest preconditions. Math. Struct. Comput. Sci. **16**(3), 429–451 (2006)
4. Dijkstra, E., Scholten, C.: Predicate Calculus and Program Semantics. Springer, Berlin (1990)
5. Dvurečenskij, A., Pulmannová, S.: New Trends in Quantum Structures. Kluwer Academic Publishers, Dordrecht (2000)
6. Egger, J., Møgelberg, R.E., Simpson, A.: Linearly-used continuations in the enriched effect calculus. In: Ong, L. (ed.) FOSSACS 2010. LNCS, vol. 6014, pp. 18–32. Springer, Heidelberg (2010)
7. Foulis, D.J., Bennett, M.K.: Effect algebras and unsharp quantum logics. Found. Phys. **24**(10), 1331–1352 (1994)
8. Furber, R., Jacobs, B.: From Kleisli categories to commutative C^*-algebras: probabilistic Gelfand duality. In: Heckel, R., Milius, S. (eds.) CALCO 2013. LNCS, vol. 8089, pp. 141–157. Springer, Heidelberg (2013)
9. Heinosaari, T., Ziman, M.: The Mathematical Language of Quantum Theory. From Uncertainty to Entanglement. Cambridge University Press, Cambridge (2012)
10. Jacobs, B.: Convexity, duality and effects. In: Calude, C.S., Sassone, V. (eds.) TCS 2010. IFIP AICT, vol. 323, pp. 1–19. Springer, Heidelberg (2010)

11. Jacobs, B.: Introduction to coalgebra. Towards mathematics of states and observations. Book, version 2 (2012, in preparation)
12. Jacobs, B.: New directions in categorical logic, for classical, probabilistic and quantum logic. See arxiv.org/abs/1205.3940 (2014)
13. Jacobs, B.: Measurable spaces and their effect logic. In: Logic in Computer Science. IEEE, Computer Science Press (2013)
14. Jacobs, B., Mandemaker, J.: Coreflections in algebraic quantum logic. Found. Phys. **42**(7), 932–958 (2012)
15. Jacobs, B., Mandemaker, J.: The expectation monad in quantum foundations. In: Jacobs, B., Selinger, P., Spitters, B. (eds.) Quantum Physics and Logic (QPL) 2011. Electronic Proceedings in Theoretical Computer Science, vol. 95, pp. 143–182 (2012)
16. Johnstone, P., Vickers, S.: Preframe presentations present. In: Carboni, A., Pedicchio, M.C., Rosolini, G. (eds.) Como Conference on Category Theory. Lecture Notes in Mathematics, vol. 1488, pp. 193–212. Springer, Berlin (1991)
17. Lawvere, F.: Ordinal sums and equational doctrines. In: Eckman, B. (ed.) Seminar on Triples and Categorical Homology Theory. Lecture Notes in Mathematics, vol. 80, pp. 141–155. Springer, Berlin (1969)
18. Liang, S., Hudak, P., Jones, M.: Monad transformers and modular interpreters. In: Principles of Programming Languages, pp. 333–343. ACM Press (1995)
19. Manes, E.: A triple-theoretic construction of compact algebras. In: Eckman, B. (ed.) Seminar on Triples and Categorical Homology Theory. Lecture Notes in Mathematics, vol. 80, pp. 91–118. Springer, Berlin (1969)
20. Maruyama, Y.: Categorical duality theory: with applications to domains, convexity, and the distribution monad. In: Ronchi Della Rocca, S. (ed.) Computer Science Logic. Leibniz International Proceedings in Informatics, pp. 500–520 (2013)
21. Mac Lane, S.: Categories for the Working Mathematician. Springer, Berlin (1971)
22. Møgelberg, R.E., Staton, S.: Linearly-used state in models of call-by-value. In: Corradini, A., Klin, B., Cîrstea, C. (eds.) CALCO 2011. LNCS, vol. 6859, pp. 298–313. Springer, Heidelberg (2011)
23. Moggi, E.: Notions of computation and monads. Inf. Comput. **93**(1), 55–92 (1991)
24. Nanevski, A., Morrisett, G., Shinnar, A., Govereau, P., Birkedal, L.: Ynot: dependent types for imperative programs. In: International Conference on Functional Programming (ICFP). ACM SIGPLAN Notices, pp. 229–240 (2008)
25. Plotkin, G., Power, J.: Computational effects and operations: an overview. In: Proceedings of the Workshop on Domains VI. Electronic Notes in Theoretical Computer Science, vol. 73, pp. 149–163. Elsevier, Amsterdam (2004)
26. Pulmannová, S., Gudder, S.: Representation theorem for convex effect algebras. Commentat. Math. Univ. Carol. **39**(4), 645–659 (1998)
27. Stone, M.: Postulates for the barycentric calculus. Ann. Math. **29**, 25–30 (1949)
28. Swamy, N., Weinberger, J., Schlesinger, C., Chen, J., Livshits, B.: Verifying higher-order programs with the Dijkstra monad. In: Proceedings of the 34th ACM SIGPLAN Conference on Programming Language Design and Implementation (PLDI), pp. 387–398. ACM (2013)
29. Swierstra, W.: A Hoare logic for the state monad. In: Berghofer, S., Nipkow, T., Urban, C., Wenzel, M. (eds.) TPHOLs 2009. LNCS, vol. 5674, pp. 440–451. Springer, Heidelberg (2009)

Categories of Coalgebras with Monadic Homomorphisms

Wolfram Kahl[(✉)]

McMaster University, Hamilton, ON, Canada
kahl@cas.mcmaster.ca

Abstract. Abstract graph transformation approaches traditionally consider graph structures as algebras over signatures where all function symbols are unary.

Attributed graphs, with attributes taken from (term) algebras over arbitrary signatures do not fit directly into this kind of transformation approach, since algebras containing function symbols taking two or more arguments do not allow component-wise construction of pushouts. We show how shifting from the algebraic view to a coalgebraic view of graph structures opens up additional flexibility, and enables treating term algebras over arbitrary signatures in essentially the same way as unstructured label sets. We integrate substitution into our coalgebra homomorphisms by identifying a factoring over the term monad, and obtain a flexible framework for graphs with symbolic attributes. This allows us to prove that pushouts can be constructed for homomorphisms with unifiable substitution components.

We formalised the presented development in Agda, which crucially aided the exploration of the complex interaction of the different functors, and enables us to report all theorems as mechanically verified.

1 Introduction

In computer science, algebras are used in two different rôles:

- "Algebras providing datatype" are the concern in particular of the field of algebraic specifications [EM85, BKL+91, BM04]: The carrier sets of an algebra are datatypes, and the operations are available as some kind of executable function. Frequently, the carrier sets are so large that one would not consider to keep all their elements simultaneously available in some data structure.
- "Algebras as data" are most obviously the topic of the algebraic approach to graph transformation [CMR+97, EHK+97, EEPT06] which derives its name from the fact that it considers graphs as algebras.

In *attributed graphs*, the two views come together: An attributed graph is first of all a graph, that is an algebra that is considered in its whole as a piece of data, but (some of) its items may be assigned attributes which are elements of some datatypes provided by an attribute algebra, which is normally not considered in its whole as a piece of data. Although an attributed graph can be

© IFIP International Federation for Information Processing 2014
M.M. Bonsangue (Ed.): CMCS 2014, LNCS 8446, pp. 151–167, 2014.
DOI: 10.1007/978-3-662-44124-4_9

considered as a single algebra, implementation considerations alone already dictate a separation into a graph structure part and an attribution part. As far as attributed graphs are to be transformed via the algebraic approach to graph transformation, theoretical reasons contribute to this separation: for graph structures, considered as unary algebras, the pushouts of their homomorphisms can be calculated component-wise and independent of the presence of operations between the different sorts, while in the presence of non-unary operations, this is no longer the case. With more-than-unary operations, even a pushout of finite algebras can become infinite, so that calculations of these pushouts is in general not feasible. There is also typically little motivation to consider non-trivial pushouts of attribute algebras, since most transformation concepts for attributed graphs expect the transformation results to be attributed over the same attribute datatypes. An exception to this consideration are symbolic attributes, which can easily be drawn from term algebras over different variable sets during different stages of transformation.

In the context of the algebraic approach to graph transformation, graph structures have traditionally been presented as unary algebras [Löw90, CMR+97]. However, as such they are the intersection between algebras and coalgebras, and in this paper we show how more general coalgebras are useful in modelling graph features, in particular symbolic attribution. Therefore, we define our graph structures not via algebraic signatures, but via coalgebraic signatures, and integrate label types and term type constructors for attributes into the coalgebraic result types.

For example, the following is a signature for directed hypergraphs where each hyperedge has a sequence of source nodes and a sequence of target nodes, and each node is labelled with an element of the constant set L:

$$
\begin{aligned}
\mathsf{sigDHG} := \langle\ &\textbf{sorts: } \mathsf{N}, \mathsf{E} \\
&\textbf{ops: } \mathsf{src} : \mathsf{E} \to \mathsf{List}\ \mathsf{N} \\
&\qquad\ \mathsf{trg} : \mathsf{E} \to \mathsf{List}\ \mathsf{N} \\
&\qquad\ \mathsf{nlab} : \mathsf{N} \to L \\
\rangle&
\end{aligned}
$$

While constant sets like L are perfectly standard as results in coalgebras, modelling labelled graphs as algebras always has to employ the trick of declaring the label sets as additional sorts, and then consider the subcategory that has algebras with a fixed choice for these label sets, and morphisms that map them only with the identity. Similarly, list-valued source and target functions are frequently considered for algebraic graph transformation, but with ad-hoc definitions for morphisms and custom proofs of their properties.

In contrast, declaring these features via a coalgebra signature such as sigDHG makes the generic theory of coalgebras available, which immediately produces the standard homomorphism definition for directed hypergraphs considered as sigDHG structures, without any necessity for ad-hoc treatment of the label type or the list structure.

Even more interesting is the use of coalgebras for symbolically attributed graphs, where morphisms are required to also contain substitutions for attribute variables; the main contribution of this paper is to formulate the beginnings of a coalgebraic approach to corresponding categories of symbolically attributed graphs.

After discussing related work in the next section, we provide some more detailed motivation for moving to coalgebras. We quickly fix our categorical notation in Sect. 3 and explain basics of (co)algebras in Sect. 4. We show more complex graphs structures in Sect. 5, and discuss the limitations of using standard coalgebra homomorphisms. In Sect. 6 we show how a factoring of the coalgebra functor over a monad allows us to replace the morphisms underlying the coalgebra homomorphisms with Kleisli arrows, enabling typical applications of symbolic attributes where instantiation of variables via substitution is required and the variable set may be modified by transformations. We show that this general factoring accommodates a natural formalisation of term graphs as monadic coalgebras. Refining this factoring in Sect. 7 for a general class of structures encompassing in particular common kinds of symbolically attributed graphs, we show that pushouts in that setting can be constructed from unifications for the substitution components of the homomorphisms.

The whole theoretical development has been formalised in the dependently typed programming language and proof checker Agda2 [Nor07] on top of the basic category formalisations provided by [Kah11, Kah14]. The Agda source code for this development is available on-line[1]. For not disrupting the flow of the presentation, we just add a check mark "✓" to statements for which a formalised version has been mechanically checked by Agda.

2   Related Work

Löwe *et al.* [LKW93] started to consider attributed graphs in the context of the algebraic approach to graph transformation; they propose working with a tri-partitioned signature, with a unary graph structure part, an arbitrary attribute signature, and a set of unary attribution operators connecting the two. Rewriting uses the single pushout approach. Without discussing the issue in depth, they propose to add sorts of attribute carriers that are deleted and re-created for relabelling. König and Kozioura [KK08] follow the approach of [LKW93], but impose a rigid organisation of unlabelled nodes, and labelled hyperedges with label-conforming attribution.

In the double-pushout approach, Heckel *et al.* [HKT02] treat data algebra carriers as graph nodes, with graph edges to them allowed, but data algebra function symbols are not part of the graph. Their attribution edges from graph nodes to data nodes are equivalent to the attribute carriers of [LKW93]. The data part is kept constant during transformation. Rule graphs are attributed over a term algebra with a fixed set of variables. The E-graphs of [EPT04] allows also attribute edges starting from edges. The algebra integration of [HKT02] is

---

[1] URL: http://relmics.mcmaster.ca/RATH-Agda/

strengthened from a commuting square to a pullback, which is used for showing the equivalence of categories of typed attributed graphs over type graph *ATG* with categories of algebras over a derived signature *AGSIG(ATG)*, where each type graph item is turned into a sort. For the symbolic graphs of [OL10], the Σ-algebra is not integrated into the graph structure, but only connected to it via constraints: A symbolic graph is an E-graph over a sorted variable set together with a set of formulae that may refer to constants drawn from the Σ-algebra.

While all the approaches presented so far worked with total algebras throughout, the relabelling DPO graph transformations of Habel and Plump [HP02, Plu09] use partially labelled interface graphs. Rule side images of unlabelled interface nodes are unlabelled as well, and natural pushouts (that are also pullbacks) with injective matching are used for rewriting. In [PS04], rule schemas are introduced to get around the fixed label sets of [HP02]; these rule schemas are rules that are labelled over a term algebra. A different approach to relabelling is that of Rebout [RFS08], which employs a special mechanism for relabelling via "computations" in the left-hand side of the rule.

For general theory of coalgebra, we refer to Rutten's overview article [Rut00]. The part of the coalgebra literature that deals with combining algebras and coalgebras is probably closest to our current endeavour; one approach considers separate algebraic and coalgebraic structures in the same carriers, for example Kurz and Hennicker's "Institutions for Modular Coalgebraic Specifications" [KH02]. A further generalisation are "dialgebras" [Hag87, PZ01], which have a single carrier X, and operations $f_i : F_i X \to G_i X$, where both F_i and G_i are polynomial functors.

Pardo studies the combination of corecursion with monads [Par98], using as an essential tool natural transformations for distribution of the monad over the functor; his "monadic coalgebras" are defined by an operation of type $A \to \mathcal{M} (\mathcal{F} A)$, which is the opposite functor composition to the one we use in Sect. 6. Capretta's survey [Cap11] covers coalgebras in functional programming and type theory; like Pardo, also Capretta concentrates on coinduction and infinite structures.

3   Category Notation

We assume familiarity with the basics of category theory; for notation, we write "$f : A \to B$" to declare that morphism f goes from object A to object B, and use ";" as the associative binary *forward composition* operator that maps two morphisms $f : A \to B$ and $g : B \to C$ to $(f \,;g) : A \to C$. The identity morphism for object A is written $\mathbb{I}_A$.

We assign ";" higher priority than other binary operators, and assign unary operators higher priority than all binary operators.

The category of sets and functions is denoted by *Set*.

A *functor* $\mathcal{F}$ from one category to another maps objects to objects and morphisms to morphisms respecting the structure generated by $\to$, $\mathbb{I}$, and composition; we denote functor application by juxtaposition both for objects, $\mathcal{F} A$, and for morphisms, $\mathcal{F} f$. Although we use forward composition of morphisms, we use backward composition "$\_ \circ \_$" for functors, with $(\mathcal{G} \circ \mathcal{F}) A = \mathcal{G} (\mathcal{F} A)$, and may even omit parentheses and just write $\mathcal{G}\mathcal{F}A$.

A *bifunctor* is a functor where the source is a product category. An important example is the coproduct bifunctor $+ : \mathcal{C} \times \mathcal{C} \to \mathcal{C}$ for a category $\mathcal{C}$ with a choice of coproducts. Functors with more than two arguments are handled similarly.

The double-pushout (DPO) approach to high-level rewriting [CMR+97], uses transformation rules that are spans $L \xleftarrow{l} G \xrightarrow{r} R$ in an appropriate category between the left-hand side L, gluing object G, and right-hand side R. A direct transformation step from object A to object B via such a rule is given by a double pushout diagram, where m is called the match:

$$
\begin{array}{ccccc}
L & \xleftarrow{\ l\ } & G & \xrightarrow{\ r\ } & R \\
\Big\downarrow{m} & & \Big\downarrow{h} & & \Big\downarrow{n} \\
A & \xleftarrow{\ a\ } & H & \xrightarrow{\ b\ } & B
\end{array}
$$

4 Algebras and Coalgebras

The category-theoretic definitions of algebras and coalgebras are simple: Given a (unary) functor $\mathcal{F}$,

- an $\mathcal{F}$-algebra $A = (C_A, f_A)$ is an object C_A together with a morphism $f_A : \mathcal{F}\, C_A \to C_A$
- an $\mathcal{F}$-coalgebra $A = (C_A, f_A)$ is an object C_A together with a morphism $f_A : C_A \to \mathcal{F}\, C_A$.

The algebraic approach to graph transformation was named for its understanding of graphs as algebras — unlabelled graphs are conventionally presented as algebras over the the following signature:

$$
\mathsf{sigGraph} := \langle\ \mathbf{sorts:}\ \mathsf{N}, \mathsf{E}
$$
$$
\mathbf{ops:}\ \mathsf{src} : \mathsf{E} \to \mathsf{N}
$$
$$
\mathsf{trg} : \mathsf{E} \to \mathsf{N}\ \ \rangle
$$

(From now on, we assume the product bifunctor $\times$, the coproduct bifunctor $+$, the terminal object $\mathbb{1}$, and the initial object $\mathbb{O}$ to be given.) The functor giving rise to graphs as algebras is a functor on the product category $Set \times Set$ since there is more than one sort:

$$
\mathcal{F}_{\mathsf{sigGraph-alg}}\ (N\ ,\ E)\ =\ ((E+E)\ ,\ \mathbb{O})\ ,
$$

since there is an isomorphism mapping functions $E+E \to N$ to pairs of functions $(E \to N) \times (E \to N)$, and there is only one "empty" function in $\mathbb{O} \to E$.

This functor can be constructed systematically from the signature above, and the signatures for which this systematic procedure works are called "algebraic":

– An *algebraic signature* has only single sort symbols as *result* types.

Dually, a coalgebra functor can be constructed systematically for the following:

– An *coalgebraic signature* has only single sort symbols as *argument* types.

Obviously, sigGraph is also a coalgebraic signature, and the functor giving rise to graphs as coalgebras is the following:

$$\mathcal{F}_{\mathsf{sigGraph}} \ (N \ , \ E) \ = \ (\mathbb{1} \ , \ (N \times N))$$

For algebras, one frequently considers only *polynomial functors*, that is, functors constructed from $+$, $\times$, and $\mathbb{1}$. For coalgebras, more varied functors are the norm, and many more complicated kinds of graphs can easily be characterised via coalgebraic signatures, for example:

– Node-labelled graphs are often presented with signature $\mathsf{sigNLG_1}$ for some node label set L — note that $\mathsf{sigNLG_1}$ is not an algebraic signature, since L is not a sort symbol:

$\mathsf{sigNLG_1}\langle$ **sorts:** N, E **ops:** src : E → N trg : E → N nlab : N → L $\rangle$	$\mathsf{sigNLG_2}\langle$ **sorts:** N, E, L **ops:** src : E → N trg : E → N nlab : N → L $\rangle$

Although this is easily fixed, see $\mathsf{sigNLG_2}$ which introduces an additional sort L, this comes at the cost of considering the label set a *part of the graph*, while usually one may want to consider it as fixed. The category of $\mathsf{sigNLG_2}$-structures admits morphisms that change labels, and encompasses as subcategories images of the categories of $\mathsf{sigNLG_1}$-structures for different choices of the interpretation of L.

However, both $\mathsf{sigNLG_1}$ and $\mathsf{sigNLG_2}$ are coalgebraic signatures, which shows that the coalgebraic view has advantages even when dealing with very simple graph structures. For a fixed node set L, coalgebras over the functor for $\mathsf{sigNLG_1}$ form exactly the category of graphs with node labels drawn from L, without any complications:

$$\mathcal{F}_{\mathsf{sigNLG_1}} \ (N, E) \ = \ (L \ , \ N \times N)$$

– sigDHG, already mentioned in the introduction, is a signature for directed hypergraphs where each hyperedge has a sequence of source nodes and a sequence of target nodes, and each node is labelled with an element of the constant set L:

$$\mathsf{sigDHG} := \langle \ \textbf{sorts:} \ \mathsf{N}, \mathsf{E}$$
$$\textbf{ops:} \ \mathsf{src} : \mathsf{E} \to \mathsf{List} \ \mathsf{N}$$
$$\mathsf{trg} : \mathsf{E} \to \mathsf{List} \ \mathsf{N}$$
$$\mathsf{nlab} : \mathsf{N} \to L \ \ \ \rangle$$

Writing List for the list functor[2], the functor corresponding to sigDHG is again a functor between product categories because of the two sorts:

$$F_{\mathsf{sigDHG}} \ (N \ E) \ = \ (L \ , \ ((\mathsf{List} \ N) \times (\mathsf{List} \ N)))$$

In general, we assume a language of functor symbols (with arity), and a *signature* introduces first, after "**sorts:**", a list of *sort symbols*, and then, after "**ops:**", a list of *function symbols* (or *operation symbols*), and for each operation symbol, an argument type expression and a result type expression (separated by "→") each built from the functor symbols and the sort symbols.

In sigDHG, we used the unary functor symbol List and the zero-ary functor symbol L — we will not make any notational distinction between functor symbols and their interpretation as functors.

5   Limitations of Standard Coalgebra Homomorphisms

For a different situation consider edge-attributed graphs, with symbolic attributes taken from the term algebra $T_\Sigma \ \mathsf{V}$ over some term signature Σ and with variables from the variable carrier set for sort V:

$$\mathsf{sigAG}_\Sigma := \langle \ \mathbf{sorts:} \ \mathsf{N}, \mathsf{E}, \mathsf{V}$$
$$\mathbf{ops:} \ \ \mathsf{src} : \mathsf{E} \rightarrow \mathsf{N}$$
$$\mathsf{trg} : \mathsf{E} \rightarrow \mathsf{N}$$
$$\mathsf{attr} : \mathsf{E} \rightarrow T_\Sigma \ \mathsf{V} \ \ \rangle$$

The resulting homomorphism concept only allows renaming of variables:

Fact 5.1. A sigAG_Σ-coalgebra homomorphism $F : G_1 \rightarrow G_2$ consists of three mappings $F_\mathsf{N} : \mathsf{N}_1 \rightarrow \mathsf{N}_2$ and $F_\mathsf{E} : \mathsf{E}_1 \rightarrow \mathsf{E}_2$ and $F_\mathsf{V} : \mathsf{V}_1 \rightarrow \mathsf{V}_2$ satisfying the following conditions:

$$F_\mathsf{E} \ ; \mathsf{src}_2 = \mathsf{src}_1 \ ; F_\mathsf{N}$$
$$F_\mathsf{E} \ ; \mathsf{trg}_2 = \mathsf{trg}_1 \ ; F_\mathsf{N}$$
$$F_\mathsf{E} \ ; \mathsf{attr}_2 = \mathsf{attr}_1 \ ; T_\Sigma \ F_\mathsf{V} \qquad\qquad \square$$

DPO rewriting in this category therefore has to rely on deletion and re-creation of attribute carrying edges to implement relabelling, like the approaches of [LKW93,KK08]. In addition we also lack the ability to instantiate rules via variable substitution as part of the morphism concept, and might therefore be tempted to add such instantiation outside the DPO rewriting framework, as in [PS04].

Another example where the coalgebra category is unsatisfactory are term graphs, where each node is either a variable (of sort V), or an inner node (of sort N) that has a label (from set L) and a list of successors, which can be either variables or other nodes:

---

[2]   Note that List A can be defined as the initial algebra of the functor $L_A \ Y = \mathbb{1} + A \times Y$.

$$\text{sig}\,\mathsf{TG} := \langle\ \textbf{sorts:}\ \mathsf{V}, \mathsf{N}$$
$$\textbf{ops:}\ \mathsf{lab} : \mathsf{N} \to L$$
$$\mathsf{suc} : \mathsf{N} \to \mathsf{List}\ (\mathsf{N} + \mathsf{V})\ \rangle$$

The resulting standard homomorphism concept also has $F_{\mathsf{V}} : \mathsf{V}_1 \to \mathsf{V}_2$ and therefore does not allow mapping of variables to inner nodes:

Fact 5.2. A sig TG-coalgebra homomorphism $F : G_1 \to G_2$ consists of two mappings $F_{\mathsf{N}} : \mathsf{N}_1 \to \mathsf{N}_2$ and $F_{\mathsf{V}} : \mathsf{V}_1 \to \mathsf{V}_2$ satisfying the following conditions:

$$F_{\mathsf{N}}\,;\mathsf{lab}_2 = \mathsf{lab}_1$$
$$F_{\mathsf{N}}\,;\mathsf{suc}_2 = \mathsf{suc}_1\,;\mathsf{List}(F_{\mathsf{N}} + F_{\mathsf{V}}) \qquad\qquad \square$$

In the resulting category, pushout complements exist only in very special cases, and the resulting DPO rewriting concept does not correspond to any useful term graph rewriting concept.

6   Monadic Coalgebra Morphisms

We now introduce a more powerful morphism concept to remedy these shortcomings. We first show how the homomorphism concepts for sigAG_Σ-coalgebras and for sig TG-coalgebras can be "fixed" to allow substitution, and then extract the general pattern behind this class of "fixes".

6.1   Substituting Attributed Graph Homomorphisms

If we want to allow substitutions in morphisms between sigAG_Σ-coalgebras, we also have to adapt the morphism conditions to take the substituted variables inside the image terms of the attribution function into account:

Definition 6.1. We define the category AG_Σ to have sigAG_Σ-coalgebras as objects, and a morphism $F : G_1 \to G_2$ consists of three mappings typed as shown to the left, satisfying the conditions shown to the right:

$$F_{\mathsf{N}} : \mathsf{N}_1 \to \mathsf{N}_2 \qquad\qquad F_{\mathsf{E}}; \mathsf{src}_2 = \mathsf{src}_1; F_{\mathsf{N}}$$
$$F_{\mathsf{E}} : \mathsf{E}_1 \to \mathsf{E}_2 \qquad\qquad F_{\mathsf{E}}; \mathsf{trg}_2 = \mathsf{trg}_1; F_{\mathsf{N}}$$
$$F_{\mathsf{V}} : \mathsf{V}_1 \to \mathcal{T}_\Sigma\, \mathsf{V}_2 \qquad\qquad F_{\mathsf{E}}; \mathsf{attr}_2 = \mathsf{attr}_1; \mathcal{T}_\Sigma\, F_{\mathsf{V}}; \mu_{\mathcal{T}_\Sigma}$$

where $\mu_{\mathcal{T}_\Sigma} : \forall X\,.\,\mathcal{T}_\Sigma(\mathcal{T}_\Sigma\, X) \to \mathcal{T}_\Sigma\, X$ is the canonical "term flattening" function that turns two-level nested terms into one-level terms. $\qquad\qquad \square$

It is not hard to verify that this category is well-defined $\checkmark$ — the key to the proof is to recognise that the F_{V} components are substitutions and compose via Kleisli composition of the term monad.

The category AG_Σ of course does not have all pushouts, since pushout construction for the F_{V} components involves term unification, which is not always defined.

6.2 "Substituting" Term Graph Homomorphisms

For term graphs, we just want to allow variables to be mapped also to inner nodes, and therefore adapt the type of F_V accordingly. The resulting adaptation in the commutativity condition for suc affects only the argument of the List functor:

Definition 6.2. We define the category TG to have sig TG-coalgebras as objects, and a morphism $F : G_1 \to G_2$ consists of two mappings

$$F_N : N_1 \to N_2$$
$$F_V : V_1 \to N_2 + V_2$$

satisfying the following conditions:

$$F_N ; lab_2 = lab_1$$
$$F_N ; suc_2 = suc_1 ; List((F_N + F_V)) ; \mu_{(N_2+)}$$

where $\mu_{(N+)} : \forall X . (N + (N + X)) \to (N + X)$ is the canonical flattening function for nested alternatives with N.   □

This time, we are dealing with a parameterised monad, namely $(N+)$, which maps any X to $N + X$, where the parameter N is instantiated with the respective carrier of that sort. Composition of the V components of $F : T_1 \to T_2$ and $G : T_2 \to T_3$ is defined accordingly:

$$(F ; G)_V = F_V ; (G_N + G_V) ; \mu_{(N_3+)}$$

Again, the resulting category is well-defined. ✓

6.3 Generalised Coalgebra Morphisms

For obtaining the general shape of such "monadic coalgebra morphisms", inspection of the signatures shows that each of the functors underlying these kinds of coalgebras not only contains a primitive monad (the term monad for attributed graphs, and an "alternative monad" for term graphs), but even can be factored over a monad on the relevant product category.

Since the signature sigAG$_\Sigma$ has three sorts, the underlying category is the triple product $Set \times Set \times Set$, with triples of sets as objects.

The coalgebra functor for sigAG$_\Sigma$ is then

$$\mathcal{G}_{sigAG_\Sigma} (N, E, V) = (\mathbb{1}, (N \times N \times \mathcal{T}_\Sigma V), \mathbb{1}) \ ,$$

mapping node and edge set to the terminal object $\mathbb{1}$ since no operations take nodes or edges as arguments; since there are three operations taking edges as arguments, $\mathcal{G}_{sigAG_\Sigma}$ produces as "edge component" of its result type the Cartesian product $N \times N \times \mathcal{T}_\Sigma V$ consisting of the target sets of the three operations.

We can decompose this as $\mathcal{G}_{\text{sigAG}_\Sigma} = \mathcal{F}_{\text{sigAG}_\Sigma} \circ \mathcal{M}_{\text{sigAG}_\Sigma}$, where:

$$\mathcal{M}_{\text{sigAG}_\Sigma} (N, E, V) = (N, E, T_\Sigma\, V)$$
$$\mathcal{F}_{\text{sigAG}_\Sigma} (N, E, T) = (\mathbb{1}, (N \times N \times T), \mathbb{1})$$

Since $\mathcal{M}_{\text{sigAG}_\Sigma}$ is the product of twice the identity monad with the term monad T_Σ, it is obviously a monad. ✓

Analogously, the coalgebra functor for sigTG is

$$\mathcal{G}_{\text{sigTG}} (N, V) = (L \times \text{List}\ (N + V), \mathbb{1})\ ,$$

mapping the variable set to the terminal object $\mathbb{1}$ since no operations take variables as arguments. We can decompose this as $\mathcal{G}_{\text{sigTG}} = \mathcal{F}_{\text{sigTG}} \circ \mathcal{M}_{\text{sigTG}}$, where:

$$\mathcal{M}_{\text{sigTG}} (N, V) = (N, (N + V))$$
$$\mathcal{F}_{\text{sigTG}} (N, S)\ = (L \times \text{List}\ S, \mathbb{1})$$

It is straightforward to prove that $\mathcal{M}_{\text{sigTG}}$ is a monad constructed as a "dependent product monad". ✓

In general, we define:

Definition 6.3. Given a monad $\mathcal{M}$ and an endofunctor $\mathcal{F}$ over a category $\mathcal{C}$, an *$\mathcal{M}$-$\mathcal{F}$-coalgebra* is a coalgebra over the functor $\mathcal{F} \circ \mathcal{M}$, that is, a pair (A, op_A) consisting of

- an object A of $\mathcal{C}$, and
- a morphism $\text{op}_A : A \to \mathcal{F}\,(\mathcal{M}\ A)$

A *raw $\mathcal{M}$-$\mathcal{F}$-coalgebra homomorphism* from (A, op_A) to (B, op_B) is a morphism from A to B in the Kleisli category of $\mathcal{M}$; raw morphism composition is Kleisli composition. □

One might expect that we obtain just coalgebras over the Kleisli category of $\mathcal{M}$. However, the following complications hold:

- $\mathcal{F}$ does in general not give rise to a functor over the Kleisli category of $\mathcal{M}$.
- If a natural transformation from $\mathcal{F} \circ \mathcal{M}$ to $\mathcal{M} \circ \mathcal{F}$ exists, an endofunctor on the Kleisli category that coincides with $\mathcal{F}$ on objects can be constructed — however, no such a natural transformation exists for sigAG$_\Sigma$. ✓
- Constructing an endofunctor on $\mathcal{C}$ from an endofunctor on the Kleisli category would require transformations to "extract from the monad $\mathcal{M}$" which can be natural neither on $\mathcal{C}$ nor on the Kleisli category. ✓

In addition, not all raw $\mathcal{M}_{\text{sigAG}_\Sigma}$-$\mathcal{F}_{\text{sigAG}_\Sigma}$-coalgebra homomorphisms satisfy the conditions we listed above for monadic sigAG$_\Sigma$ coalgebra homomorphisms — we need to identify an appropriate subcategory of the Kleisli category.

From the material we have, we can easily construct the following two morphisms:

$$\begin{array}{rcl} f\,;\mathcal{M}\ \text{op}_B & : & A \to \mathcal{M}\mathcal{F}\mathcal{M}B \\ \text{op}_A\,;\mathcal{F}\mathcal{M}f & : & A \to \mathcal{F}\mathcal{M}\mathcal{M}B \end{array}$$

"Obviously", we can complete this to a commutativity condition using a natural transformation constructed from the return η and the join μ transformations of the monad $\mathcal{M}$, namely:

$$\mathcal{F}\mu;\eta \;:\; \mathcal{F}\circ\mathcal{M}\circ\mathcal{M} \Rightarrow \mathcal{M}\circ\mathcal{F}\circ\mathcal{M}$$

For the category AG_Σ of Definition 6.1, this condition is unfortunately only satisfied by morphisms where F_V only renames variables ✓, which defeats our intentions. This problem is actually not even due to the choice of $\mathcal{F}\mu;\eta$, but to the choice of direction, since it arises for every natural transformation from $\mathcal{F}\circ\mathcal{M}\circ\mathcal{M}$ to $\mathcal{M}\circ\mathcal{F}\circ\mathcal{M}$. Therefore, using distribution transformations from $\mathcal{F}\circ\mathcal{M}$ to $\mathcal{M}\circ\mathcal{F}$ as used by Pardo [Par98] is not an option either.

We found that natural transformations from $\mathcal{M}\circ\mathcal{F}\circ\mathcal{M}$ to $\mathcal{F}s\circ\mathcal{M}$ "work" when combined with a join on the other side:

Definition 6.4. For an endofunctor $\mathcal{F}$ and a monad $(\mathcal{M},\eta,\mu)$ on $\mathcal{C}$, an $\mathcal{M}$-$\mathcal{F}$-*distrjoin* transformation is a natural transformation $\xi : \mathcal{M}\circ\mathcal{F}\circ\mathcal{M} \Rightarrow \mathcal{F}\circ\mathcal{M}$ for which the following properties hold:

- $\eta;\xi = \mathbb{I}$
- $\mu;\xi = \mathcal{M}\xi;\xi$
- $\mathcal{M}\mathcal{F}\mu;\xi = \xi;\mathcal{F}\mu$ □

Definition 6.5. Given an endofunctor $\mathcal{F}$ and a monad $(\mathcal{M},\eta,\mu)$ on $\mathcal{C}$, and also an $\mathcal{M}$-$\mathcal{F}$-distrjoin transformation ξ, an $\mathcal{M}$-$\mathcal{F}$-*coalgebra homomorphism* from (A,op_A) to (B,op_B) is a morphism $f : A \to MB$ making the following diagram commute:

Theorem 6.6. $\mathcal{M}$-$\mathcal{F}$-coalgebras with such $\mathcal{M}$-$\mathcal{F}$-coalgebra homomorphisms form a category. ✓ □

The instantiations of this for edge-attributed graphs (sigAG_Σ) and for term graphs (sigTG) are appropriate:

Theorem 6.7. The $\mathcal{M}_{\mathsf{sigAG}_\Sigma}$-$\mathcal{F}_{\mathsf{sigAG}_\Sigma}$-coalgebra category is equivalent to the category AG_Σ Definition 6.1. ✓ □

Theorem 6.8. The $\mathcal{M}_{\mathsf{sigTG}}$-$\mathcal{F}_{\mathsf{sigTG}}$-coalgebra category is equivalent to the category TG of Definition 6.2. ✓ □

6.4   Pushouts of Monadic Coalgebra Morphisms

It is well-known that the Kleisli category over the term monad $\mathcal{T}$ does not have all pushouts, and that existence of pushouts essentially corresponds to unifiability.

Therefore we cannot expect the category of monadic sigAG_Σ coalgebras to have all pushouts. Nevertheless, since pushouts are the key ingredient of the categoric approach to graph transformation, an interesting question is whether pushouts in the Kleisli category of $\mathcal{M}$ give rise to pushouts in the $\mathcal{M}$-$\mathcal{F}$-coalgebra category. It is easy to see that this is not the case for term graphs; however, it does hold for symbolically edge-attributed graphs, and in this section we explore the question how to prove this from the point of view of $\mathcal{M}$-$\mathcal{F}$-coalgebra categories without considering decompositions of $\mathcal{M}$ and $\mathcal{F}$.

A pushout for a span $B \xleftarrow{F} A \xrightarrow{G} C$ is a completion $B \xrightarrow{H} D \xleftarrow{K} C$ to a commuting square that is "minimal" in the sense that every other candidate completion factors uniquely over it (via U).

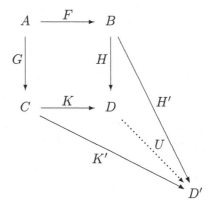

Assuming such a pushout in the Kleisli category of $\mathcal{M}$, for constructing the operation op_D of the target coalgebra we need to choose appropriate H' and K' such that the U can be used to construct op_D and prove its pushout property in the category of $\mathcal{M}$-$\mathcal{F}$-coalgebras.

Without assuming additional (natural) transformations, we can only choose the following:

$$H' = \mathsf{op}_B \, ; \mathcal{F}(\mathcal{M}H \, ; \mu) \, ; \eta$$
$$K' = \mathsf{op}_C \, ; \mathcal{F}(\mathcal{M}K \, ; \mu) \, ; \eta$$

However, commutativity $F \, \mathbin{\raise.1ex\hbox{$_\circ$}} H' = G \, \mathbin{\raise.1ex\hbox{$_\circ$}} K'$ in the Kleisli category can only be shown assuming additional (natural) transformations and/or laws, which however are not available for the setting of sigAG_Σ — there, this commutativity does not hold. ✓ The essential reason for this is that $\mathcal{F}$ maps V to $\mathbb{1}$, while V is also the only component that has a non-trivial monad. Commutativity fails for the V components due to the fact that op_B and op_C map these to $\mathbb{1}$, while $\mathcal{M}$ for the V components is the term monad $\mathcal{T}_\Sigma$:

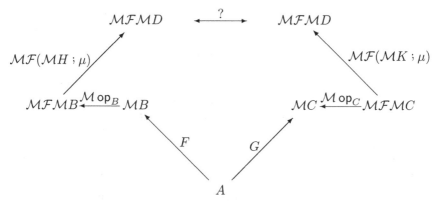

Due to this property of the operators in sigAG_Σ, commutativity will actually fail for any definition of H' of the shape "$H' = \mathsf{op}_B\,;\,\ldots$". We solve this problem in the next section by restriction to more specialised versions of $\mathcal{M}$ and $\mathcal{F}$, which allows us to "patch" H' and K' so as to avoid this conflict.

7   Monadic Product Coalgebras

We now specialise the $\mathcal{M}$-$\mathcal{F}$-coalgebras of Sect. 6.3 in a way that still generalises the setup for symbolically edge-attributed graphs there, while also allowing a pushout construction.

The most general shape we have been able to identify for this are "monadic product coalgebras" over a product category $\mathcal{C}_1 \times \mathcal{C}_2$, defined in the following setting (which we will assume for the remainder of this section): Let $\mathcal{C}_1$ and $\mathcal{C}_2$ be two categories; let $\mathcal{M}$ be a monad on $\mathcal{C}_2$, and $\mathcal{F}$ a functor from $\mathcal{C}_1 \times \mathcal{C}_2$ to $\mathcal{C}_1$.

In terms of coalgebraic signatures this implements the restriction that sorts mentioned as monad arguments do not occur as source sorts of operators, and that the monad must not depend on sorts that do occur as source sorts of operators. This restriction is satisfied by all simple kinds of symbolically attributed graphs where the monad is typically a term monad, is applied only to sets of free variables, and these variables do not otherwise participate in the graph structure.

Definition 7.1. An $\mathcal{M}$-$\mathcal{F}$-*product-coalgebra* A is a triple $(A_1, A_2, \mathsf{op}_A)$ consisting of

- an object A_1 of $\mathcal{C}_1$, and
- an object A_2 of $\mathcal{C}_2$, and
- a morphism $\mathsf{op}_A : A_1 \to \mathcal{F}\,(A_1,\,\mathcal{M}\,A_2)$

A $\mathcal{M}$-$\mathcal{F}$-*product-coalgebra homomorphism* f from $(A_1, A_2, \mathsf{op}_A)$ to $(B_1, B_2, \mathsf{op}_B)$ is a pair (f_1, f_2) consisting of a $\mathcal{C}_1$-morphism f_1 from A_1 to B_1 and a morphism f_2 from A_2 to B_2 in the Kleisli category of $\mathcal{M}$ such that

$$f_1\,;\mathsf{op}_B = \mathsf{op}_A\,;\mathcal{F}\,(f_1,\,\mathcal{M}\,f_2\,;\mu)\ .$$

Morphism composition is composition of the corresponding product category. □

This morphism composition is well-defined ✓, and induces a category ✓.

Now let $\mathcal{M}_0$ be the product monad of the identity monad on $\mathcal{C}_1$ and $\mathcal{M}$, and define $\mathcal{F}_0$ as endofunctor on $\mathcal{C}_1 \times \mathcal{C}_2$ by:

$$\mathcal{F}_0(X_1, X_2) = (\mathcal{F}(X_1, X_2), \mathbb{1})$$

With these definitions, the $\mathcal{M}_0$-$\mathcal{F}_0$-distrjoin transformation (see Definition 6.4) has identities of $\mathcal{C}_1$ and terminal morphisms of $\mathcal{C}_2$ as its two components $\checkmark$, which allows us to identify monadic product coalgebras as a special case of the monadic coalgebras of Sect. 6.3:

Theorem 7.2. The category of $\mathcal{M}$-$\mathcal{F}$-product-coalgebra homomorphisms is equivalent to the category of $\mathcal{M}_0$-$\mathcal{F}_0$-coalgebra homomorphisms. $\checkmark$ □

In addition, the more fine-grained structure of monadic product coalgebras allows us to circumvent the problems we encountered in Sect. 6.4.

Let $\mathbb{K}$ be the Kleisli category of $\mathcal{M}_0$. Since $\mathcal{M}_0$ is a product monad, pushouts in $\mathbb{K}$ are calculated component-wise, that is, they consist of a pushout in $\mathcal{C}_1$ and a pushout in the Kleisli category of $\mathcal{M}$.

Theorem 7.3. Let a span $B \xleftarrow{F} A \xrightarrow{G} C$ of $\mathcal{M}$-$\mathcal{F}$-product-coalgebra homomorphisms be given, and a cospan $(B_1, B_2) \xrightarrow{H} (D_1, D_2) \xleftarrow{K} (C_1, C_2)$ in $\mathbb{K}$ that is a pushout for the Kleisli morphisms underlying F and G. Then (D_1, D_2) can be extended to a $\mathcal{M}$-$\mathcal{F}$-product-coalgebra $D = (D_1, D_2, \mathsf{op}_D)$ such that $B \xrightarrow{H} D \xleftarrow{K} C$ is a pushout for $B \xleftarrow{F} A \xrightarrow{G} C$ in the $\mathcal{M}$-$\mathcal{F}$-product-coalgebra category. $\checkmark$

Proof sketch: The first step is the construction of a cospan

$$(B_1, B_2) \xrightarrow{H'} D' \xleftarrow{K'} (C_1, C_2)$$

in $\mathbb{K}$ such that the first component of the universal morphism $U : (D_1, D_2) \to D'$ from the $\mathbb{K}$ pushout can be used as op_D.

The first constituent of D' therefore must be the target of op_D, so we define:

$$D' = (\qquad \mathcal{F}(D_1, \mathcal{M}\ D_2)\qquad , D_2)$$
$$H' = (\ \mathsf{op}_B\ ; \mathcal{F}(H_1, \mathcal{M}\ H_2\ ; \mu)\ , H_2)$$
$$K' = (\ \mathsf{op}_C\ ; \mathcal{F}(K_1, \mathcal{M}\ K_2\ ; \mu)\ , K_2)$$

The second constituent of D' is inherited from (D_1, D_2), which allows us to use the universality of the original $\mathbb{K}$ pushout when proving universality of the resulting $\mathcal{M}$-$\mathcal{F}$-product-coalgebra pushout.   □

Together with the two equivalences of categories of Theorems 7.2 and 6.7, pushouts for edge-attributed graphs essentially reduce to unification for their variable components (if we choose an underlying category that has pushouts, such as *Set*):

Corollary 7.4. A span $B \xleftarrow{F} A \xrightarrow{G} C$ in the $\mathcal{M}_{\mathsf{sigAG}}$-$\mathcal{F}_{\mathsf{sigAG}}$-coalgebra category over Set for edge-attributed graphs (as sigAG_Σ structures) has a pushout if F_V and G_V, as substitutions, have a pushout. $\checkmark$ □

8   Conclusion and Outlook

We have shown how the additional flexibility of coalgebraic signatures enables us to integrate label types and symbolic attribute types into graph structure signatures, and then modified the coalgebraic homomorphism concept via integration of Kleisli arrows to achieve the flexibility necessary to allow substitutions as part of our morphisms. We showed that several seemingly plausible formalisations for this do not model the intended applications, and arrived at a simple factoring setup (Sect. 6.3) that additionally encompasses a natural formalisation of term graphs. For symbolically attributed graphs fitting into the pattern of monadic product coalgebras (Sect. 7), we showed that pushouts can be obtained where the substitution components of their homomorphisms are unifiable.

Without the support of our mechanised formalisation in Agda, the mentioned failures of inappropriate formalisations, and also the successful proofs reported in the paper would have been extremely hard to arrive at with comparable confidence.

Since pushouts do not necessarily exist in Kleisli categories, an important question is whether general results for appropriate classes of restricted monomorphisms can be obtained. In this context, the "guarded monads" of Ghani *et al.* [GLDM05] might be useful, since a guarded monad essentially can be represented as $\mathsf{Id} + \mathcal{N}$ for some functor $\mathcal{N}$, and the identity component can be used to lift monomorphisms from the base category into the Kleisli category.

The ultimate goal of this work is a fully verified implementation of monadic coalgebra transformation that can be instantiated in particular for the transformation of symbolically attributed graph structures.

Acknowledgements. I am grateful to the anonymous referees for their constructive comments.

References

[BKL+91] Bidoit, M., Kreowski, H.-J., Lescanne, P., Orejas, F., Sannella, D. (eds.): Algebraic System Specification and Development. LNCS, vol. 501. Springer, Heidelberg (1991)

[BM04] Bidoit, M., Mosses, P.D. (eds.): Casl User Manual. LNCS, vol. 2900. Springer, Heidelberg (2004). (With Chapters by Mossakowski, T., Sannella, D., Tarlecki, A.)

[Cap11] Capretta, V.: Coalgebras in functional programming and type theory. Theoret. Comput. Sci. **412**, 5006–5024 (2011)

[CMR+97] Corradini, A., Montanari, U., Rossi, F., Ehrig, H., Heckel, R., Löwe, M.: Algebraic approaches to graph transformation, Part I: basic concepts and double pushout approach. In: [Roz97], Chap. 3, pp. 163–245

[CEKR02] Corradini, A., Ehrig, H., Kreowski, H.-J., Rozenberg, G. (eds.): Graph Transformation. Lecture Notes in Computer Science, vol. 2505. Springer, Heidelberg (2002)

[EM85] Ehrig, H., Mahr, B.: Fundamentals of Algebraic Specification 1: Equations and Initial Semantics. Springer, New York (1985)

[EHK+97] Ehrig, H., Heckel, R., Korff, M., Löwe, M., Ribeiro, L., Wagner, A., Corradini, A.: Algebraic approaches to graph transformation, Part II: single pushout approach and comparison with double pushout approach. In: [Roz97], Chap. 4, pp. 247–312

[EPT04] Ehrig, H., Prange, U., Taentzer, G.: Fundamental theory for typed attributed graph transformation. In: [PPBE04], pp. 161–177

[EEPT06] Ehrig, H., Ehrig, K., Prange, U., Taentzer, G.: Fundamentals of Algebraic Graph Transformation. Springer, Heidelberg (2006)

[GLDM05] Ghani, N., Lüth, C., De Marchi, F.: Monads of coalgebras: rational terms and term graphs. Math. Struct. Comput. Sci. **15**, 433–451 (2005)

[HP02] Habel, A., Plump, D.: Relabelling in graph transformation. In: [CEKR02], pp. 135–147

[Hag87] Hagino, T.: A Categorical Programming Language. Ph.D. thesis, Edinburgh University (1987)

[HKT02] Heckel, R., Küster, J. M., Taentzer, G.: Confluence of typed attributed graph transformation systems. In: [CEKR02], pp. 161–176

[Kah11] Kahl, W.: Dependently-typed formalisation of relation-algebraic abstractions. In: de Swart, H. (ed.) RAMICS 2011. LNCS, vol. 6663, pp. 230–247. Springer, Heidelberg (2011)

[Kah14] Kahl, W.: Relation-Algebraic Theories in Agda – RATH-Agda-2.0.1. Mechanically checked Agda theories available for download, with 456 pages literate document output (2014). http://RelMiCS.McMaster.ca/RATH-Agda/

[KK08] König, B., Kozioura, V.: Towards the verification of attributed graph transformation systems. In: Ehrig, H., Heckel, R., Rozenberg, G., Taentzer, G. (eds.) ICGT 2008. LNCS, vol. 5214, pp. 305–320. Springer, Heidelberg (2008)

[KH02] Kurz, A., Hennicker, R.: On institutions for modular coalgebraic specifications. Theoret. Comput. Sci. **280**, 69–103 (2002)

[Löw90] Löwe, M.: Algebraic Approach to Graph Transformation Based on Single Pushout Derivations. Technical report 90/05, TU Berlin (1990)

[LKW93] Löwe, M., Korff, M., Wagner, A.: An algebraic framework for the transformation of attributed graphs. In: Sleep, M., Plasmeijer, M., van Eekelen, M. (eds.) Term Graph Rewriting: Theory and Practice, pp. 185–199. Wiley, New York (1993)

[Nor07] Norell, U.: Towards a Practical Programming Language Based on Dependent Type Theory. Ph.D. thesis, Department of Computer Science and Engineering, Chalmers University of Technology (2007)

[OL10] Orejas, F., Lambers, L.: Delaying constraint solving in symbolic graph transformation. In: Ehrig, H., Rensink, A., Rozenberg, G., Schürr, A. (eds.) ICGT 2010. LNCS, vol. 6372, pp. 43–58. Springer, Heidelberg (2010)

[Par98] Pardo, A.: Monadic corecursion - definition, fusion laws and applications. ENTCS **11**, 105–139 (1998)

[PPBE04] Parisi-Presicce, F., Bottoni, P., Engels, G. (eds.): ICGT 2004. LNCS, vol. 3256. Springer, Heidelberg (2004)

[PS04] Plump, D., Steinert, S.: Towards graph programs for graph algorithms. In: [PPBE04], pp. 128–143

[Plu09] Plump, D.: The graph programming language GP. In: Bozapalidis, S., Rahonis, G. (eds.) CAI 2009. LNCS, vol. 5725, pp. 99–122. Springer, Heidelberg (2009)

[PZ01] Poll, E., Zwanenburg, J.: From algebras and coalgebras to dialgebras. ENTCS **44**, 289–307 (2001)

[RFS08] Rebout, M., Féraud, L., Soloviev, S.: A unified categorical approach for attributed graph rewriting. In: Hirsch, E.A., Razborov, A.A., Semenov, A., Slissenko, A. (eds.) Computer Science – Theory and Applications. LNCS, vol. 5010, pp. 398–409. Springer, Heidelberg (2008)

[Roz97] Rozenberg, G. (ed.) Handbook of Graph Grammars and Computing by Graph Transformation, Foundations, vol. 1. World Scientific, Singapore (1997)

[Rut00] Rutten, J.J.: Universal coalgebra: a theory of systems. Theoret. Comput. Sci. **249**, 3–80 (2000)

Lifting Adjunctions to Coalgebras
to (Re)Discover Automata Constructions

Henning Kerstan[1 (✉)], Barbara König[1], and Bram Westerbaan[2]

[1] Universität Duisburg-Essen, Essen, Germany
{henning.kerstan,barbara_koenig}@uni-due.de
[2] Radboud Universiteit Nijmegen, Nijmegen, The Netherlands
bram.westerbaan@cs.ru.nl

Abstract. It is a well-known fact that a nondeterministic automaton can be transformed into an equivalent deterministic automaton via the powerset construction. From a categorical perspective this construction is the right adjoint to the inclusion functor from the category of deterministic automata to the category of nondeterministic automata. This is in fact an adjunction between two categories of coalgebras: deterministic automata are coalgebras over **Set** and nondeterministic automata are coalgebras over **Rel**. We will argue that this adjunction between coalgebras originates from a canonical adjunction between **Set** and **Rel**.

In this paper we describe how, in a quite generic setting, an adjunction can be lifted to coalgebras, and we compare some sufficient conditions. Then we illustrate this technique in length: we recover several constructions on automata as liftings of basic adjunctions including determinization of nondeterministic and join automata, codeterminization, and the dualization of linear weighted automata. Finally, we show how to use the lifted adjunction to check behavioral equivalence.

1 Introduction

Coalgebra offers a general framework for specifying transition systems with various branching types. Given a functor $F \colon \mathbf{Set} \to \mathbf{Set}$, describing the branching type of the transition system, an F-coalgebra is a function $c \colon X \to FX$, where X represents the state set and c the transition function. Depending on the choice of F, one can describe labeled, nondeterministic, probabilistic or various other types of branching and it is possible to combine several of them. Coalgebras come with a natural notion of behavioral equivalence, and coalgebra homomorphisms can be seen as functional bisimulations, mapping states to equivalent states.

In recent years, it has also become customary to study coalgebras in categories different from **Set**. There are several reasons, for instance, one can impose an algebraic structures on the states, or one can work in presheaf categories in order to model name passing [7]. Particularly relevant to this paper is the study of coalgebras in Kleisli categories, where a monad offers a way to model side-effects that can also be understood as implicit branching, different from the explicit branching that can be modeled directly in **Set**. Such coalgebras have

© IFIP International Federation for Information Processing 2014
M.M. Bonsangue (Ed.): CMCS 2014, LNCS 8446, pp. 168–188, 2014.
DOI: 10.1007/978-3-662-44124-4_10

for instance been studied in [8], where nondeterministic automata are specified as coalgebras in **Rel**, the category of sets and relations, which is isomorphic to the Kleisli category of the powerset monad on **Set**. The behavioral equivalence induced by such coalgebras in **Rel** is indeed trace (or language) equivalence, as desired, and not bisimilarity.

When studying coalgebras in various categories a natural question to ask is how to transform such coalgebras from one representation into another. Our motivating examples come from the world of deterministic and nondeterministic automata where various forms of determinization can be seen as functors which map coalgebras living in one category, into coalgebras living in another category. For instance, nondeterministic automata living in **Rel** can be transformed into deterministic automata in **Set** via the powerset construction. In the other direction, a deterministic automaton in **Set** can be trivially regarded as a nondeterministic automaton in **Rel**. It turns out that the transformations together form an adjunction between categories of coalgebras where the powerset construction is the right adjoint. In the same vein various other determinization-like constructions arise as adjunctions.

In the following we will first show under which circumstances adjunctions on categories can be lifted to adjunctions on coalgebras. Part of the answer was already given by Hermida and Jacobs [9] and we extend their characterization by giving another, equivalent, condition. Then we study several examples in detail, especially various forms of automata. Apart from the well-known deterministic and nondeterministic automata, we consider codeterministic automata (also known as átomata, see [6], or backwards-deterministic automata) and deterministic join automata, i.e. automata that have an algebraic structure on the states, allowing to take the join of a given set of states. Such automata live in the category of join semilattices **JSL**, which is the Eilenberg-Moore category of the powerset monad on **Set** (whereas **Rel** is the Kleisli category of the powerset monad) and have already been considered in [17]. In total we consider four different adjunctions between such automata.

In addition we consider an adjunction in the realm of linear weighted automata, where we take up an example from [3], transforming input into output linear weighted automata (and vice versa).

In order to explain what these adjunctions really mean in terms of behavioral equivalence, we study a general notion of behavioral equivalence for arbitrary categories. We first observe that the final coalgebra, if it exists, is preserved by right adjoints and hence can be "inherited" from coalgebras living in a different category. Furthermore we show how queries on behavioral equivalence can be translated to equivalent queries on coalgebras in another category. This reflects the well-known construction of determinizing a nondeterministic automaton before answering questions about language equivalence.

2   Theoretical Background – Lifting Adjunctions

Within this section we are first going to present a short, motivating example which introduces our approach. Then we will recall some basic definitions from the theory of adjunctions (mainly to introduce our notation) and start to develop our theory which is then summarized in our main theoretical result. The result itself is not very surprising and, in fact, was discovered already earlier by C. Hermida and B. Jacobs [9] in a different setting (we will compare our approach with their result) and can be obtained using standard (2-)categorical methods [12]. However, the focus of our work is *not just the theory itself* but we are more interested in *how this theory helps to understand* certain (algorithmic) constructions on automata by *applying* it to various types of automata, modeled as coalgebras.

2.1   Motivating Example

Consider the following (non-commutative) diagram of functors where the bottom part is a (canonical) adjunction (see below for a definition) between **Set** and **Rel**.

Let $\mathcal{A}$ be an alphabet, i.e. a finite set of labels. It is known (and we will also recall this in Sect. 3.1) that the coalgebras for the functor $2 \times (\_)^{\mathcal{A}}$ on **Set** are the deterministic automata (**DA**) and the coalgebras for the functor $\mathcal{A} \times \_ + 1$ on **Rel** are the nondeterministic automata (**NDA**). (For final coalgebra semantics of **NDA** see [8].) We aim at finding the functors $\overline{L}$, $\overline{R}$ (dashed arrows on top) that form an adjunction which is a *lifting* of the original adjunction and we will see that for this particular example everything works out as planned and the lifted right adjoint $\overline{R}$ "performs" the well-known powerset construction to determinize an **NDA**.

2.2   Adjunctions

We recall some basics from category theory [2,14] to fix our notation.

Definition 1 (Adjunction, Adjoint Functors). *Let **C** and **D** be categories. An* adjunction *between **C** and **D** consists of a functor $L \colon \mathbf{C} \to \mathbf{D}$, called* left adjoint, *a functor $R \colon \mathbf{D} \to \mathbf{C}$, called* right adjoint *and two natural transformations $\eta \colon 1_{\mathbf{C}} \Rightarrow RL$, called* unit, *and $\varepsilon \colon LR \Rightarrow 1_{\mathbf{D}}$, called* counit, *satisfying*

$$\varepsilon L \circ L\eta = 1_L, \qquad and \qquad R\varepsilon \circ \eta R = 1_R. \tag{1}$$

We denote such an adjunction by $\langle L \dashv R, \eta, \varepsilon \rangle \colon \mathbf{C} \to \mathbf{D}$.

A functor $L\colon \mathbf{C} \to \mathbf{D}$ $[R\colon \mathbf{D} \to \mathbf{C}]$ is a *left adjoint [right adjoint]* if it is the left adjoint [right adjoint] of some adjunction $\langle L \dashv R, \eta, \varepsilon \rangle : \mathbf{C} \to \mathbf{D}$. One can prove that L $[R]$ determines the other parts of the adjunction unique up to isomorphism. Since this is not trivial, we will always give the full adjunction.

There are special cases of adjunctions which have their own names.

Definition 2 (Equivalence and Duality of Categories). *We call an adjunction $\langle L \dashv R, \eta, \varepsilon \rangle : \mathbf{C} \to \mathbf{D}$ an* equivalence *if both η and ε are natural isomorphisms. Whenever such an equivalence exists, the categories $\mathbf{C}$ and $\mathbf{D}$ are called* equivalent. *Similarly, we say that the categories $\mathbf{C}$ and $\mathbf{D}$ are* dually equivalent *if there is an equivalence $\langle L \dashv R, \varepsilon, \eta \rangle : \mathbf{C} \to \mathbf{D}^{\mathrm{op}}$.*

Let us now consider our first example of an adjunction (which is well-known in the literature and also the adjunction from our motivating example).

Example 1. Let $L\colon \mathbf{Set} \to \mathbf{Rel}$ be the inclusion functor from $\mathbf{Set}$ to $\mathbf{Rel}$ mapping each set X to itself and each function $f\colon X \to Y$ to the corresponding relation $f\colon X \leftrightarrow Y$. Moreover, let $R\colon \mathbf{Rel} \to \mathbf{Set}$ be the functor which maps each set X to its powerset $\mathbf{2}^X$ and each relation $f\colon X \leftrightarrow Y$ to the function $Rf\colon \mathbf{2}^X \to \mathbf{2}^Y$ given by $(Rf)(S) = \{y \in Y \mid \exists x \in S : \langle x, y \rangle \in f\}$ for every $S \in \mathbf{2}^X$. We obtain an adjunction $\langle L \dashv R, \eta, \varepsilon \rangle : \mathbf{Set} \to \mathbf{Rel}$ where the unit $\eta\colon 1_{\mathbf{Set}} \Rightarrow \mathbf{2}^-$ is given by all the functions $\eta_X\colon X \to \mathbf{2}^X, \eta_X(x) = \{x\}$ for every set X and the counit $\varepsilon\colon \mathbf{2}^- \Rightarrow 1_{\mathbf{Rel}}$ consists of all the relations $\varepsilon_X\colon \mathbf{2}^X \leftrightarrow X$ where $\langle S, x \rangle \in \varepsilon_X \iff x \in S$ for every set X.

2.3   Lifting an Adjunction to Coalgebras

The theory of coalgebras [10, 16] has proven to be an appropriate tool for modeling and analyzing various types of transition systems. We will examine several of the standard examples (deterministic, nondeterministic, codeterministic and linear weighted automata) in detail in the following sections and hence will not provide any examples here. Besides these examples also probabilistic automata [18] and even arbitrary labeled Markov processes [15] or probabilistic transition systems [13] can be seen as coalgebras in suitable categories.

Recently the coalgebraic treatment of automata has provided new views on algorithms for minimization and (co)determinization [1, 3, 4]. In these works the authors make use of certain adjunctions of categories to obtain the minimization of (various kinds of) automata.

We shall hereafter try to find and analyze a common and generic pattern on how we can make use of an adjunction $\langle L \dashv R, \eta, \varepsilon \rangle : \mathbf{C} \to \mathbf{D}$ to reason about and to find constructions (algorithms) on automata modeled as coalgebras. For that purpose let us fix two endofunctors $F\colon \mathbf{C} \to \mathbf{C}$ and $G\colon \mathbf{D} \to \mathbf{D}$ and look at the (non-commutative) diagram of functors on the next page where $U\colon \mathbf{Coalg}\,(F) \to \mathbf{C}$ and $V\colon \mathbf{Coalg}\,(G) \to \mathbf{D}$ are the forgetful functors mapping a coalgebra to its carrier and a coalgebra homomorphism to the underlying arrow.

The question we are interested in is whether we can in some canonical way obtain the functors $\overline{L}$ and $\overline{R}$ as indicated by the dashed lines such that they form an adjunction which "arises" from the initial adjunction. A precise definition for this is given below. In several cases such adjoint functors transform coalgebras in a way that we (re)discover algorithmic con-

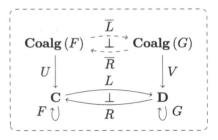

structions on the modeled automata and we will back this hypothesis by the examples given in the following sections.

Definition 3 (Lifting). *Let* $\mathbf{C}$ *and* $\mathbf{D}$ *be categories,* $F\colon \mathbf{C} \to \mathbf{C}$, $G\colon \mathbf{D} \to \mathbf{D}$ *be endofunctors and* $U\colon \mathbf{Coalg}\,(F) \to \mathbf{C}$ *and* $V\colon \mathbf{Coalg}\,(G) \to \mathbf{D}$ *be the forgetful functors mapping a coalgebra to its carrier and a coalgebra morphism to the underlying arrow. Let* $\langle L \dashv R, \eta, \varepsilon\rangle \colon \mathbf{C} \to \mathbf{D}$ *be an adjunction.*

1. *We call a functor* $\overline{L}\colon \mathbf{Coalg}\,(F) \to \mathbf{Coalg}\,(G)$ $\big[\,\overline{R}\colon \mathbf{Coalg}\,(G) \to \mathbf{Coalg}\,(F)\,\big]$ *a lifting of* L $[R]$ *if it satisfies the equality* $V\overline{L} = LU$ $\big[\,U\overline{R} = RV\,\big]$.
2. *We call an adjunction* $\langle \overline{L} \dashv \overline{R}, \overline{\eta}, \overline{\varepsilon}\rangle \colon \mathbf{Coalg}\,(F) \to \mathbf{Coalg}\,(G)$ *a lifting of* $\langle L \dashv R, \eta, \varepsilon\rangle \colon \mathbf{C} \to \mathbf{D}$ *if* $\overline{L}$ *is a lifting of* L, $\overline{R}$ *is a lifting of* R *and we have* $U\overline{\eta} = \eta$ *and* $V\overline{\varepsilon} = \varepsilon$.

Although this definition is straightforward it has one setback: it does not tell us how to construct a lifted adjunction. Let us therefore introduce a method for handling this. If we had a natural transformation $\alpha\colon LF \Rightarrow GL$ it is not hard to see that we obtain a functor $\overline{L}\colon \mathbf{Coalg}\,(F) \to \mathbf{Coalg}\,(G)$ by defining

$$\overline{L}\left(X \xrightarrow{\ c\ } FX\right) = \left(LX \xrightarrow{\ Lc\ } LFX \xrightarrow{\ \alpha_X\ } GLX\right), \quad \overline{L}f = Lf \qquad (2)$$

for all F-coalgebras $c\colon X \to FX$ and all F-coalgebra homomorphisms f and analogously, given a natural transformation $\beta\colon RG \Rightarrow FR$ we can define a functor $\overline{R}\colon \mathbf{Coalg}\,(G) \to \mathbf{Coalg}\,(F)$ by

$$\overline{R}\left(Y \xrightarrow{\ d\ } GY\right) = \left(RY \xrightarrow{\ Rd\ } RGY \xrightarrow{\ \beta_Y\ } FRY\right), \quad \overline{R}g = Rg \qquad (3)$$

for all G-coalgebras $d\colon Y \to GY$ and all G-coalgebra homomorphisms g. By definition these functors are liftings and thus the only remaining question is whether we obtain a lifting of the adjunction. The equation $U\overline{\eta} = \eta$ can be spelled out as the requirement that for all F-coalgebras $c\colon X \to FX$ the arrow $\eta_X\colon X \to RLX$ is an F-coalgebra homomorphism $c \to \overline{RL}c$ and likewise the equation $V\overline{\varepsilon} = \varepsilon$ translates to the requirement that for every G-coalgebra $d\colon Y \to GY$ the arrow $\varepsilon_Y\colon LRY \to Y$ is a G-coalgebra homomorphism $\overline{LR}d \to d$. This is the case iff the outer rectangles of the following two diagrams commute.

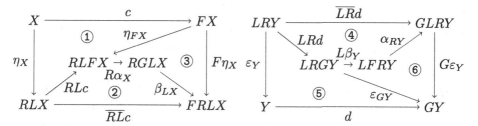

These diagrams certainly commute if their inner parts commute: ① commutes because η is a natural transformation, ② by definition of $\overline{RLc}$ and commutativity of ③ is equivalent to $F\eta_X = \beta_{LX} \circ R\alpha_X \circ \eta_{FX}$. Moreover, ④ commutes by definition of $\overline{LRd}$ and ⑤ because ε is a natural transformation. Finally, the commutativity of ⑥ is equivalent to $\varepsilon_{GY} = G\varepsilon_Y \circ \alpha_{RY} \circ L\beta_Y$.

With these observations at hand it is easy to spell out a sufficient condition for the existence of a lifting which we will do in the following theorem.

Theorem 1 (Lifting an Adjunction to Coalgebras). *Let $F\colon \mathbf{C} \to \mathbf{C}$ and $G\colon \mathbf{D} \to \mathbf{D}$ be endofunctors and $\langle L \dashv R, \eta, \varepsilon \rangle \colon \mathbf{C} \to \mathbf{D}$ be an adjunction. There is a lifting $\langle \overline{L} \dashv \overline{R}, \overline{\eta}, \overline{\varepsilon} \rangle \colon \mathbf{Coalg}\,(F) \to \mathbf{Coalg}\,(G)$ of the adjunction if one of the following equivalent conditions is fulfilled.*

(i) There are two natural transformations $\alpha\colon LF \Rightarrow GL$ and $\beta\colon RG \Rightarrow FR$ satisfying the following equalities.

$$F\eta = \beta L \circ R\alpha \circ \eta F \tag{4}$$

$$\varepsilon G = G\varepsilon \circ \alpha R \circ L\beta \tag{5}$$

(ii) There is a natural isomorphism $\beta\colon RG \Rightarrow FR$. [9, 2.15 Corollary]

If (i) holds, the adjoint mate $\alpha^{\bullet}$ of α, which is defined as

$$\alpha^{\bullet} := RG\varepsilon \circ R\alpha R \circ \eta FR \tag{6}$$

is the inverse of β. Conversely, if (ii) holds we can define α as the adjoint mate $(\beta^{-1})^{\bullet}$ of β^{-1} which is defined as

$$(\beta^{-1})^{\bullet} = \varepsilon GL \circ L\beta^{-1}L \circ LF\eta. \tag{7}$$

In both cases $\overline{L}$ and $\overline{R}$ are defined by (2) and (3).

By the observations from above it should be quite clear, that (i) is sufficient for a lifting to exist. The fact that the second condition (ii) of this theorem is also sufficient for the existence of a lifting is due to a result by C. Hermida and B. Jacobs [9, 2.15 Corollary]. They derive this as a "by-product" from a quite generic result in 2-categories using the fact that coalgebras are certain inserters in the 2-category **CAT** of categories, functors and natural transformations. Thus in order to prove the theorem we just have to show that (i) and (ii) are equivalent using the provided definitions of β^{-1} (6) and α (7).

Proof. (of Theorem 1).

(i) $\Rightarrow$ (ii): The equations $\beta_Y \circ \alpha_Y^\bullet = 1_{FRY}$ and $\alpha_Y^\bullet \circ \beta_Y = 1_{RGY}$ are equivalent to commutativity of the outer rectangles of the following diagrams.

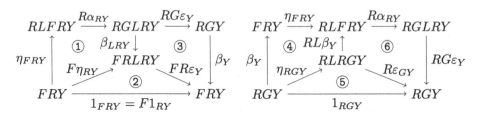

The diagrams commute because their inner parts commute: For the left diagram ① is (4) applied to $X = RY$, ② is F applied to the second unit-counit equation (1) and ③ is the natural transformation diagram for β. For the right diagram we observe that ④ is the natural transformation diagram for η, ⑤ is the second unit-counit equation (1) applied to GY and ⑥ is R applied to (5). Thus β is indeed a natural isomorphism with inverse $\alpha^\bullet$.

(ii) $\Rightarrow$ (i): We have to show that α defined by (7) satisfies (4) and (5).

(4): Let X be an arbitrary **C**-object. Then $F\eta_X = \beta_{LX} \circ R\alpha_X \circ \eta_{FX}$ holds if and only if $\beta_{LX}^{-1} \circ F\eta_X = R\varepsilon_{GLX} \circ RL\beta_{LX}^{-1} \circ RLF\eta_X \circ \eta_{FX}$ holds which in turn is equivalent to commutativity of the outer part of the following diagram.

$$
\begin{array}{ccccccc}
FX & \xrightarrow{\ F\eta_X\ } & FRLX & \xrightarrow{\ \beta_{LX}^{-1}\ } & RGLX & & \\
{\scriptstyle \eta_{FX}}\downarrow & ① & {\scriptstyle \eta_{FRLX}}\downarrow & ② & {\scriptstyle \eta_{RGLX}}\downarrow\ ③ & \searrow^{1_{RGLX}} & \\
RLFX & \xrightarrow[RLF\eta_X]{} & RLFRLX & \xrightarrow[RL\beta_{LX}^{-1}]{} & RLRGLX & \xrightarrow[R\varepsilon_{GLX}]{} & RGLX
\end{array}
$$

① and ② commute because η is a natural transformation from $1_{\mathbf{C}}$ to RL, functors preserve inverses and ③ is the second unit-counit equation (1) applied to GLX.

(5): Let Y be an arbitrary **D**-object. Then: $\varepsilon_{GY} = G\varepsilon_Y \circ \alpha_{RY} \circ L\beta_Y$ holds if and only if $\varepsilon_{GY} \circ L\beta_Y^{-1} = G\varepsilon_Y \circ \left(\varepsilon_{GLRY} \circ L\beta_{LRY}^{-1} \circ LF\eta_{RY}\right)$ holds which in turn is equivalent to commutativity of the outer part of the following diagram.

$$
\begin{array}{ccccccc}
LFRY & \xrightarrow{\ LF\eta_{RY}\ } & LFRLRY & \xrightarrow{\ L\beta_{LRY}^{-1}\ } & LRGLRY & \xrightarrow{\ \varepsilon_{GLRY}\ } & GLRY \\
& {\scriptstyle 1_{LFRY}}\searrow\ ④ & \downarrow{\scriptstyle LFR\varepsilon_Y} & ⑤ & \downarrow{\scriptstyle LRG\varepsilon_Y} & ⑥ & \downarrow{\scriptstyle G\varepsilon_Y} \\
& & LFRY & \xrightarrow[L\beta_Y^{-1}]{} & LRGY & \xrightarrow[\varepsilon_{GY}]{} & GY
\end{array}
$$

④ commutes by applying LF to the second unit-counit equation (1) , ⑤ due to the fact that β^{-1} is a natural transformation and ⑥ because ε is a natural transformation. $\square$

Remark 1. We immediately make the following observations about Theorem 1.

(a) Due to the fact that **Coalg** $(F) \cong$ **Alg** (F^{op}), where F^{op} is the opposite functor to F, we can apply the theorem to obtain liftings to algebras.
(b) If $\langle L \dashv R, \eta, \varepsilon \rangle$ is an equivalence [a dual equivalence] of categories then $\langle \overline{L} \dashv \overline{R}, \eta, \varepsilon \rangle$ is an equivalence [a dual equivalence] of categories.

3 Nondeterministic Automata and Determinization

Within this section we will first shortly recall how deterministic (**DA**), nondeterministic (**NDA**) and codeterministic (**CDA**) automata can be modeled as coalgebras in suitable categories. We will then consider adjunctions between these categories and apply our theorem to obtain a lifting. Via this we will recover the (co)determinization of an automaton via the powerset construction.

The content of this (and the following) section can be summarized in the diagram of categories and functors below. While **DA** live in **Set**, **NDA** can be seen as arrows in **Rel** and **CDA** as arrows in **Set**^{op} (see Sect. 3.1). Furthermore in Sect. 4 we will in addition consider deterministic join automata (**DJA**) which live in the category of complete join semilattices (**JSL**), see Sects. 4.1 and 4.2. Between these categories of coalgebras there are four adjunctions, which will be treated in the following sections.

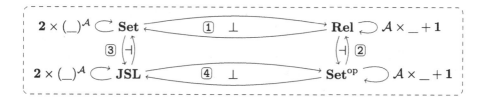

For the rest of this and the following section let $\mathcal{A}$ denote an alphabet, i.e. a finite set of labels. In a coalgebraic treatment of labeled transition systems, one usually omits initial states and the state spaces are not required to be finite.

3.1 Automata as Coalgebras

Deterministic Automata. In the category **Set** of sets and functions, deterministic automata can be modeled as coalgebras for the functor $\mathbf{2} \times (\_)^{\mathcal{A}}$. We can represent a deterministic automaton with states X and alphabet $\mathcal{A}$ as a coalgebra $c\colon X \to \mathbf{2} \times X^{\mathcal{A}}$ where each state $x \in X$ is mapped to a tuple $\langle o, s \rangle$ in which the output flag $o \in \{0,1\}$ determines whether x is final (if and only if $o = 1$) and the successor function $s\colon \mathcal{A} \to X$ determines for each letter $a \in \mathcal{A}$ the unique a-successor $s(a) \in X$ of the state x. We thus define the category of deterministic automata and automata morphisms to be **DA** := **Coalg** $\left(\mathbf{2} \times (\_)^{\mathcal{A}} \colon \mathbf{Set} \to \mathbf{Set}\right)$.

Nondeterministic Automata. Given a set X of states, we model a nondeterministic automaton by a coalgebra for the functor $\mathcal{A} \times \_ + 1$ in **Rel**. Given a coalgebra $c \colon X \leftrightarrow \mathcal{A} \times X + 1$, each state $x \in X$ is in relation with $\checkmark$ if and only if it is a final state. For any letter $a \in \mathcal{A}$ the $y \in X$ such that $\langle x, \langle a, y \rangle \rangle \in c$ are the a-successor(s) (one, multiple or none) of x. We thus define the category of nondeterministic automata and automata morphisms to be $\mathbf{NDA} := \mathbf{Coalg}\,(\mathcal{A} \times \_ + 1 \colon \mathbf{Rel} \to \mathbf{Rel})$.

Codeterministic Automata. Given a set X of states, a codeterministic (backwards deterministic) automaton (**CDA**) is given by a function $c \colon \mathcal{A} \times X + 1 \to X$ where $c(\checkmark) \in X$ is the unique final state and for each pair $\langle a, x \rangle \in \mathcal{A} \times X$ the unique a-predecessor of x is $c(\langle a, x \rangle)$. Hence we can model them as coalgebras for the functor $\mathcal{A} \times X + 1$ on $\mathbf{Set}^{\mathrm{op}}$ and define the category of codeterministic automata and their morphisms to be $\mathbf{CDA} = \mathbf{Coalg}\,(\mathcal{A} \times \_ + 1 \colon \mathbf{Set}^{\mathrm{op}} \to \mathbf{Set}^{\mathrm{op}})$.

Note that $\mathbf{Set}^{\mathrm{op}}$ is equivalent to $\mathbf{CABA}$, the category of all complete atomic boolean algebras, with boolean algebra homomorphisms. So instead of thinking of these automata as codeterministic, one could think of them as deterministic automata with a rich algebraic structure on the states, even richer than the deterministic join automata introduced in Sect. 4.

Codeterministic automata are studied in [6] under the name àtomata. An example automaton is shown in Fig. 1 (right).

3.2 Determinization of Nondeterministic Automata

Let us reconsider the adjunction ① between **Set** and **Rel** which we presented in Example 1. We aim at applying Theorem 1 to this adjunction to get a lifting and claim that this will yield the well known powerset construction to determinize nondeterministic automata.

Recall that, by Theorem 1 (ii), for such a lifting to exist it is sufficient to define a natural isomorphism $\beta \colon 2^{\mathcal{A} \times \_ + 1} \Rightarrow 2 \times (2^-)^{\mathcal{A}}$. Thus we define for every set X the function $\beta_X \colon 2^{\mathcal{A} \times X + 1} \to 2 \times (2^X)^{\mathcal{A}}$ via

$$\beta_X(S) = \langle \chi_S(\checkmark), (s \colon \mathcal{A} \to 2^X, s(a) = \{x \in X \mid \langle a, x \rangle \in S\}) \rangle \qquad (8)$$

for every $S \in 2^{\mathcal{A} \times X + 1}$. Here χ_S is the characteristic function[1] of S. The inverse function $\beta_X^{-1} \colon 2 \times (2^X)^{\mathcal{A}} \to 2^{\mathcal{A} \times X + 1}$ is given by

$$\beta_X^{-1}(\langle o, s \rangle) := \{\checkmark \mid o = 1\} \cup \bigcup_{a \in \mathcal{A}} \{a\} \times s(a) \qquad (9)$$

for every $\langle o, s \rangle \in 2 \times (2^X)^{\mathcal{A}}$. By Theorem 1 (ii) we obtain a lifting. We calculate the natural transformation $\alpha \colon 2 \times (\_)^{\mathcal{A}} \Rightarrow \mathcal{A} \times \_ + 1$ by using (7) to obtain for every set X the relation $\alpha_X \colon 2 \times X^{\mathcal{A}} \leftrightarrow \mathcal{A} \times X + 1$ given by

---

[1] Given a set X, the characteristic function $\chi_S \colon X \to \{0, 1\}$ is defined for any subset $S \subseteq X$ by $\chi_S(x) = 1$ iff $x \in S$ and $\chi_S(x) = 0$ otherwise.

$$\alpha_X = \left\{ \left\langle \langle 1, s \rangle, \checkmark \right\rangle, \left\langle \langle o, s \rangle, \langle a, s(a) \rangle \right\rangle \,\middle|\, o \in \mathbf{2}, s \in X^{\mathcal{A}}, a \in \mathcal{A} \right\}. \tag{10}$$

With these preparations at hand we can now construct the lifted functors. The new left adjoint $\overline{L} \colon \mathbf{DA} \to \mathbf{NDA}$ maps a **DJA** $c \colon X \to \mathbf{2} \times X^{\mathcal{A}}$ to the **NDA** $\overline{L}(c) \colon X \leftrightarrow \mathcal{A} \times X + \mathbf{1}$ which is given by[2]

$$\overline{L}(c) = \left\{ \langle x, \checkmark \rangle \,\middle|\, \begin{matrix} x \in X \\ \pi_1(c(x)) = 1 \end{matrix} \right\} \cup \left\{ \langle x, \langle a, \pi_2(c(x))(a) \rangle \rangle \,\middle|\, \begin{matrix} x \in X \\ a \in \mathcal{A} \end{matrix} \right\} \tag{11}$$

which is simply the same automaton, but interpreted as a nondeterministic one.

The lifted right adjoint $\overline{R} \colon \mathbf{NDA} \to \mathbf{DA}$ maps a nondeterministic automaton $d \colon Y \leftrightarrow \mathcal{A} \times Y + \mathbf{1}$ to the deterministic automaton $\overline{R}(d) \colon \mathbf{2}^Y \to \mathbf{2} \times (\mathbf{2}^Y)^{\mathcal{A}}$. A state of this new automaton is just a set of states $Q \in \mathbf{2}^Y$ of the original automaton. For each such Q the tuple $\overline{R}(d)(Q) = \langle o, s \colon \mathcal{A} \to \mathbf{2}^Y \rangle$ is given as follows: We have $o = 1$ if and only if there is a $q \in Q$ such that $\langle q, \checkmark \rangle \in d$, i.e. Q is final if and only if one of the original states in Q is final. Moreover, the a-successor of Q is determined by $s(a) = \{y \in Y \mid \exists q \in Q \colon \langle q, \langle a, y \rangle \rangle \in d\}$ which we can easily identify to be exactly the definition of the transition function of the usual powerset automaton construction.

3.3   Codeterminization of Nondeterministic Automata

Let us now consider adjunction ② between **Rel** and $\mathbf{Set}^{\mathrm{op}}$ which we automatically obtain by dualizing the adjunction between **Set** and **Rel** from Example 1 and the fact that **Rel** is a self-dual category. This adjunction has already been considered in [1,8].

The left adjoint $L \colon \mathbf{Rel} \to \mathbf{Set}^{\mathrm{op}}$ maps a set X to its powerset $\mathbf{2}^X$ and a relation $f \colon X \leftrightarrow Y$ to the function $Lf \colon \mathbf{2}^X \leftarrow \mathbf{2}^Y$ given by, for every $S \in \mathbf{2}^Y$, $(Lf)(S) = \{x \in X \mid \exists y \in S \colon \langle x, y \rangle \in f\}$. The right adjoint $R \colon \mathbf{Set}^{\mathrm{op}} \to \mathbf{Rel}$ is the inclusion, i.e. it maps a set X to itself and a function $f \colon X \leftarrow Y$ to the corresponding relation $f \colon X \leftrightarrow Y$. The unit η consists of all the relations $\eta_X \colon X \leftrightarrow \mathbf{2}^X$ defined via $\langle x, S \rangle \in \varepsilon_X$ iff $x \in S$ and the counit ε is given by all the functions $\varepsilon_X \colon \mathbf{2}^X \leftarrow X$ mapping each $x \in X$ to the singleton set $\{x\}$.

We proceed to define a lifting as before by specifying a natural isomorphism $\beta \colon \mathcal{A} \times \_ + \mathbf{1} \Rightarrow \mathcal{A} \times \_ + \mathbf{1}$. The obvious choice is to let β_X be the identity relation on $\mathcal{A} \times X + \mathbf{1}$ which indeed yields a natural isomorphism. Moreover, using (7) we obtain the natural transformation $\alpha \colon \mathbf{2}^{\mathcal{A} \times \_ + \mathbf{1}} \Rightarrow \mathcal{A} \times \mathbf{2}^- + \mathbf{1}$ where for each set X the function $\alpha_X \colon \mathbf{2}^{\mathcal{A} \times X + \mathbf{1}} \leftarrow \mathcal{A} \times \mathbf{2}^X + \mathbf{1}$ is given by $\alpha(\checkmark) = 1$ and $\alpha(\langle a, S \rangle) = \{a\} \times S$ for every $\langle a, S \rangle \in \mathcal{A} \times \mathbf{2}^X$.

The lifted left adjoint $\overline{L} \colon \mathbf{NDA} \to \mathbf{CDA}$ performs codeterminization: Given an **NDA** $c \colon X \leftrightarrow \mathcal{A} \times X + \mathbf{1}$ we obtain a **CDA** $\overline{L}(c) \colon \mathbf{2}^X \leftarrow \mathcal{A} \times \mathbf{2}^X + \mathbf{1}$ where $\checkmark$ is mapped to the set $\{x \in X \mid \langle x, 1 \rangle \in c\}$, i.e. the unique final state of the new automaton is the set of all final states of the original automaton. Given a set of states $S \in \mathbf{2}^X$ and a letter $a \in A$ the a-predecessor of S is the set

---

[2] $\pi_1 \colon \mathbf{2} \times X^{\mathcal{A}} \to \mathbf{2}$ and $\pi_2 \colon \mathbf{2} \times X^{\mathcal{A}} \to X^{\mathcal{A}}$ are the projections of the product.

$$\overline{L}(c)(\langle a, S \rangle) = \{x \in X \mid \exists y \in S : \langle x, \langle a, y \rangle \rangle \in c\} \tag{12}$$

containing all the a-predecessors of the states in S. We conclude that the new automaton is indeed a codeterministic automaton which is (language) equivalent to the original one.

While in the previous example the lifted left adjoint was trivial (as was the original left adjoint), in this case we obtain a trivial lifted right adjoint $\overline{R}$: **CDA** $\rightarrow$ **NDA**: it "interprets" a **CDA** d: $Y \leftarrow \mathcal{A} \times Y + 1$ as **NDA**, i.e. as the corresponding relation $\overline{R}(d)$: $Y \leftrightarrow \mathcal{A} \times Y + 1$.

4   Deterministic Join Automata

We will now try to take a different perspective to look at powerset automata instead of just considering them to be determinized nondeterministic automata. In order to do that we briefly recall the notion of complete join semilattices and the corresponding category.

4.1   Complete Join Semilattices

A complete join semilattice is a partially ordered set X such that for every (possibly infinite) set $S \in 2^X$ there is a least upper bound, called *join* and denoted by $\sqcup S$. If Y is another join semilattice we call a function $f : X \rightarrow Y$ *join-preserving* if, for all S, it satisfies $f(\sqcup S) = \sqcup \{f(s) \mid s \in S\}$. The join semilattices and the join-preserving functions form a category which we will denote by **JSL**. It is isomorphic to the Eilenberg-Moore category for the powerset monad on set.

If we equip the set $\mathbf{2} = \{0, 1\}$ with the partial order $0 \leq 1$, we get a complete join semilattice with $\sqcup \emptyset = 0$, $\sqcup \{0\} = 0$, $\sqcup \{1\} = 1$ and $\sqcup \mathbf{2} = 1$.

The product of two join semilattices X and Y is the cartesian product of the base sets equipped with a partial order given by $\langle x_1, y_1 \rangle \leq \langle x_2, y_2 \rangle$ if and only if $x_1 \leq y_1$ and $y_1 \leq y_2$ for all $x_1, x_2 \in X$, $y_1, y_2 \in Y$ and analogously, given a set X we can equip $X^{\mathcal{A}}$ with a partial order based on a given one on X by defining for $f, g \in X^{\mathcal{A}}$ that $f \leq g$ if and only if $f(a) \leq g(a)$ for every $a \in \mathcal{A}$. This is a complete join semilattice if X is one. Given a subset $F \subseteq X^{\mathcal{A}}$ its join is given by $\sqcup F : \mathcal{A} \rightarrow X, \sqcup F(a) = \sqcup \{f(a) \mid f \in F\}$.

4.2   Deterministic Join Automata

We can interpret a coalgebra $c : X \rightarrow \mathbf{2} \times X^{\mathcal{A}}$ for the functor $\mathbf{2} \times (\_)^{\mathcal{A}}$ on **JSL** as a deterministic automaton just as we did before on **Set**. Since the arrows of **JSL** are join-preserving functions, such an automaton possesses a certain additional property. Given a set $S \in 2^X$ of states, we know that there is a *join-state* $\sqcup S$. By the join-preserving property of the transition function c we know that $c(\sqcup S) = \sqcup \{c(x) \mid x \in S\}$ and we conclude that $\sqcup S$ is final if and only if one of the states $x \in S$ is final and moreover any transition of an $x \in S$ can be "simulated" (see below) by $\sqcup S$. We define the category **DJA** := **Coalg** $\left(\mathbf{2} \times (\_)^{\mathcal{A}} : \textbf{JSL} \rightarrow \textbf{JSL}\right)$ and call its objects *deterministic join automata*.

Example 2. Take a look at Fig. 1. If we equip the set $X = \{\bot, u, x, y, \top\}$ with the partial order given by the Hasse diagram on the left we obtain a complete join semilattice. The diagram in the middle shows a deterministic join automaton on this join semilattice. Note that the join of two final states is again final and that for every pair of states and alphabet symbol a, the a-successor of the join of the states is the join of the a-successors. This implies a general property of **DJA** that for every subset of states there exists a state accepting the union of the languages of the given states.

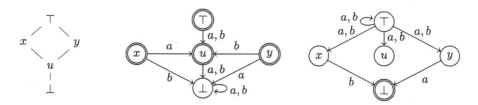

Fig. 1. Hasse diagram of a complete join semilattice, a deterministic join automaton and its codeterminization (from left to right)

4.3   From Deterministic Automata to Deterministic Join Automata

We will now consider adjunction ③ in order to transform **DJA** into **DA**s and vice versa. Given a conventional **DA**, a suitable algorithm to obtain a **DJA** is (again) the powerset construction. We will see that this is a reasonable construction in the sense that it arises from an adjunction between **Set** and **JSL**. As said before, the category of complete join semilattices is equivalent to the Eilenberg-Moore category for the powerset monad on **Set**. The theory of adjunctions gives us a generic construction of an adjunction which we will now spell out in details.

The left adjoint $L\colon \mathbf{Set} \to \mathbf{JSL}$ maps any set X to $\mathbf{2}^X$ which is partially ordered by set inclusion. The join operation is set theoretic union and it is easy to see that we indeed obtain a complete join semilattice. Any function $f\colon X \to Y$ is mapped to its image map $f[.]\colon \mathbf{2}^X \to \mathbf{2}^Y$ which we can easily identify as join- (i.e. union-)preserving function. The right adjoint $R\colon \mathbf{JSL} \to \mathbf{Set}$ takes a complete join semilattice to its base set and forgets about the order and the join operation. Analogously, a join-preserving function is just considered as a function. The unit of the adjunction is given, for every set X, by the function $\eta_X\colon X \to \mathbf{2}^X, \eta_X(x) = \{x\}$. The counit consists of the join-preserving functions $\varepsilon_{\langle Y, \sqcup \rangle}\colon \langle \mathbf{2}^Y, \cup \rangle \to \langle Y, \sqcup \rangle$ mapping each set $S \in \mathbf{2}^Y$ to its join $\sqcup S$ in Y.

In order to obtain the lifting we define $\beta_{\langle X, \sqcup \rangle}\colon \mathbf{2} \times X^{\mathcal{A}} \to \mathbf{2} \times X^{\mathcal{A}}$ to be the identity function on $\mathbf{2} \times X^{\mathcal{A}}$ for every join semilattice $\langle X, \sqcup \rangle$ which obviously yields a natural isomorphism β. Using (7) we construct the natural transformation α where for each set X the join preserving function $\alpha_X\colon \mathbf{2}^{\mathbf{2} \times X^{\mathcal{A}}} \to \mathbf{2} \times (\mathbf{2}^X)^{\mathcal{A}}$ is given by, for every $S \in \mathbf{2}^{\mathbf{2} \times X^{\mathcal{A}}}$,

$$\alpha_X(S) = \left\langle \bigsqcup \{o \mid \langle o, s \rangle \in S\}, \bigsqcup \{s \mid \langle o, s \rangle \in S\} \right\rangle . \tag{13}$$

The lifted left adjoint $\overline{L}\colon \mathbf{DA} \to \mathbf{DJA}$ performs the powerset construction on a deterministic automaton: For a deterministic automaton $c\colon X \to \mathbf{2} \times X^{\mathcal{A}}$ the deterministic join automaton $\overline{L}(c)\colon \mathbf{2}^X \to \mathbf{2} \times (\mathbf{2}^X)^{\mathcal{A}}$ is given by, for all $S \in \mathbf{2}^X$,

$$\overline{L}(c)(S) = \left\langle \bigsqcup \{\pi_1(c(x)) \mid x \in S\}, \bigsqcup \{\pi_2(c(x)) \mid x \in S\} \right\rangle . \tag{14}$$

The lifted right adjoint $\overline{R}\colon \mathbf{DJA} \to \mathbf{DA}$ takes a $\mathbf{DJA}$ and interprets it as $\mathbf{DA}$ by forgetting about its join property.

4.4   Codeterminization of Deterministic Join Automata

Finally, we will describe an unusual construction translating $\mathbf{DJA}$ into $\mathbf{CDA}$, based on adjunction ④. It is unusual, since the unit of the adjunction (which must be join-preserving) maps every element to the complement (!) of its upward-closure (more details are given below).

The left adjoint $L\colon \mathbf{JSL} \to \mathbf{Set}^{\mathrm{op}}$ maps any join semilattice $\langle X, \sqcup \rangle$ to its base set X and each join-preserving function $f\colon \langle X, \sqcup \rangle \to \langle Y, \sqcup \rangle$ to the $\mathbf{Set}^{\mathrm{op}}$-arrow

$$Lf\colon X \leftarrow Y, \quad Lf(y) = \bigsqcup \{x \in X \mid f(x) \sqsubseteq y\} . \tag{15}$$

The right adjoint $R\colon \mathbf{Set}^{\mathrm{op}} \to \mathbf{JSL}$ maps a set X to its powerset $\mathbf{2}^X$ equipped with the subset order, i.e. to the join semilattice $\langle \mathbf{2}^X, \cup \rangle$ and each $\mathbf{Set}^{\mathrm{op}}$-arrow $f\colon X \leftarrow Y$ to the reverse image $f^{-1}[.]\colon \mathbf{2}^X \to \mathbf{2}^Y$. The unit of this adjunction is given by the join-preserving functions

$$\eta_{\langle X, \sqcup \rangle}\colon \langle X, \sqcup \rangle \to \langle \mathbf{2}^X, \cup \rangle, \quad x \mapsto \overline{\uparrow x} = \{x' \in X \mid x' \not\sqsupseteq x\} \tag{16}$$

for every join semilattice $\langle X, \sqcup \rangle$ and the counit is given by, for every set X,

$$\varepsilon_X\colon \mathbf{2}^X \leftarrow X, \quad x \mapsto \overline{\{x\}} . \tag{17}$$

We construct a natural isomorphism $\beta\colon \langle \mathbf{2}^{\mathcal{A} \times - +1}, \cup \rangle \Rightarrow \langle \mathbf{2} \times (\mathbf{2}^-)^{\mathcal{A}}, \sqcup \rangle$ in order to get a lifting. For every join semilattice $\langle X, \sqcup \rangle$ we take β_X to be the same function as in (8) of our first example given in Sect. 3.2 and claim that this is a join-preserving function: for two sets $Q_1, Q_2 \in \mathbf{2}^{\mathcal{A} \times X + 1}$ let $\langle o, s \rangle := \beta_X(Q_1 \cup Q_2)$ and $\langle o_i, s_i \rangle = \beta_X(Q_i)$ for $i \in \{1, 2\}$ then we have $o = \chi_{Q_1 \cup Q_2}(\checkmark) = \max\{\chi_{Q_1}(\checkmark), \chi_{Q_2}(\checkmark)\} = \max\{o_1, o_2\} = o_1 \sqcup o_2$ and for each $a \in A$ we have

$$s(a) = \{x \in X \mid \langle a, x \rangle \in Q_1 \cup Q_2\}$$
$$= \{x \in X \mid \langle a, x \rangle \in Q_1\} \cup \{x \in X \mid \langle a, x \rangle \in Q_2\} = s_1(a) \cup s_2(a)$$

which can be generalized to arbitrary unions.

We calculate $\alpha \colon \mathbf{2} \times \_^{\mathcal{A}} \Leftarrow \mathcal{A} \times \_ + \mathbf{1}$ where for every join semilattice $\langle X, \sqcup \rangle$ the function $\alpha_{\langle X, \sqcup \rangle}$ is given by $\alpha_{\langle X, \sqcup \rangle}(\checkmark) = \langle 0, (s_{\checkmark} \colon \mathcal{A} \to X, s_{\checkmark}(a) = \top) \rangle$ and

$$\alpha_{\langle X, \sqcup \rangle}(\langle a, x \rangle) = \left\langle 1, \left(s_{\langle a, x \rangle} \colon \mathcal{A} \to X, s_{\langle a, x \rangle}(a') = \begin{cases} x, & a' = a \\ \top, & a' \neq a \end{cases} \right) \right\rangle. \quad (18)$$

The lifted left adjoint $\overline{L} \colon \mathbf{DJA} \to \mathbf{CDA}$ maps a $\mathbf{DJA}$ $c \colon \langle X, \sqcup \rangle \to \langle \mathbf{2} \times X^{\mathcal{A}}, \sqcup \rangle$ to the $\mathbf{CDA}$ $\overline{L}c \colon X \leftarrow \mathcal{A} \times X + \mathbf{1}$ whose unique final state is the join of the non-final states of the original automaton, i.e. $\overline{L}c(\checkmark) = \bigsqcup \{x' \in X \mid \pi_1(c(x')) = 0\}$. For every action $a \in \mathcal{A}$ and every state $x \in X$ the unique a-predecessor of x is the join of all the states of the original automaton whose a-successor is less or equal to x, i.e. $\overline{L}c(\langle a, x \rangle) = \bigsqcup \{x' \in X \mid \pi_2(c(x'))(a) \sqsubseteq x\}$.

The lifted right adjoint $\overline{R} \colon \mathbf{CDA} \to \mathbf{DJA}$ maps a $\mathbf{CDA}$ $d \colon Y \leftarrow \mathcal{A} \times Y + \mathbf{1}$ (which can also be regarded as nondeterministic automaton) to its determinization $\overline{R}(d) \colon \langle \mathbf{2}^Y, \cup \rangle \to \langle \mathbf{2} \times (\mathbf{2}^Y)^{\mathcal{A}}, \sqcup \rangle$ via the usual powerset construction, i.e. for every $S \in \mathbf{2}^Y$ we have $\overline{R}(d)(S) = \beta_Y \circ d^{-1}[S]$. Since the reverse image of any function preserves arbitrary unions, this is indeed a $\mathbf{DJA}$.

Example 3. Take another look at Fig. 1. The $\mathbf{DJA}$ of Example 2 (in the middle) is transformed into a $\mathbf{CDA}$ with the same state set $X = \{\bot, x, y, u, \top\}$ (the diagram on the right). Its unique final state is $\bot$ (the join of the non-final states) and the unique a-predecessor of a state is the join of the previous a-predecessors, for instance the new a-predecessor of $\bot$ is y (the join of $\bot, u, y$). If we transfer this automaton into a $\mathbf{DJA}$ with state set $\mathbf{2}^X$ via the right adjoint the unit $\eta_{\langle X, \sqcup \rangle}$ maps every state to the complement of its upward-closure. For instance state x is mapped to $\{\bot, u, y\}$.

5   Linear Weighted Automata

In the previous sections we looked at automata over $\mathbf{Set}$, $\mathbf{Rel}$ and $\mathbf{JSL}$. Now we will look at automata over the category $\mathbf{Vect}$ of linear maps between vector spaces called *linear weighted automata*. There are two flavors: *input linear weighted automata* (category: $\mathbf{WAut}_i$) and *output linear weighted automata* ($\mathbf{WAut}_o$). We will use Theorem 1 to obtain an adjunction between $\mathbf{WAut}_i$ and $\mathbf{WAut}_o$. We have taken this example from [3, Section 4] and extended it from finite to arbitrary vector spaces. For this section, let $\mathcal{A}$ be an arbitrary set (not necessarily finite).

5.1   Vector Spaces

Let us recall some facts about the category $\mathbf{Vect}$ of vector spaces. To begin $\mathbf{Vect}$ has all products and coproducts. Let V and W be vector spaces. The product of V and W is simply the cartesian product $V \times W$ with coordinatewise addition and scalar multiplication. The coproduct of V and W is $V \times W$ as well; the coprojections $V \xrightarrow{\kappa_0} V \times W \xleftarrow{\kappa_1} W$ are given by $\kappa_0(v) = (v, 0)$ and $\kappa_1(w) = (0, w)$ for $v \in V$ and $w \in W$.

Similarly, given a set B the vector space V^B of functions from B to V is the B-fold product of V. However, the coproduct of infinitely many vector spaces is a bit more interesting. Let B be a set and V a vector space. The B-fold coproduct of V is the following subspace of V^B.

$$B \cdot V := \{ f \in V^B \mid \operatorname{supp} f \text{ is finite} \}$$

Here $\operatorname{supp} f := \{ b \in B \mid f(b) \neq 0 \}$ denotes the *support* of $f \in V^B$. Let $b \in B$ be given. The b-th coprojection $\kappa_b \colon V \to B \cdot V$ is given by, for every $v \in V$: $\kappa_b(v)(b) = v$ and $\kappa_b(v)(b') = 0$ for all $b' \in B$ with $b' \neq b$.

Note that the B-fold product and the B-fold coproduct coincide, i.e. $B \cdot V = V^B$, if and only if $V = \{0\}$ or B is finite. In fact, we even have $B \cdot \mathbb{R} \cong \mathbb{R}^B$ if and only if B is finite.

5.2 Input Linear Weighted Automata

An input linear weighted automaton is a linear map of the form $c \colon \mathcal{A} \cdot V + \mathbb{R} \to V$. That is, it is an algebra on **Vect** of type $\mathcal{A} \cdot (\_) + \mathbb{R}$. Accordingly we define the category **WAut**$_i$ of input linear weighted automata to be $\mathbf{Alg}\,(\mathcal{A} \cdot (\_) + \mathbb{R})$.

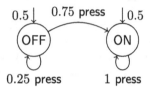

For an example of an input linear weighted automaton we draw on our experiences with turning on electrical devices such as printers and projectors. Let us say that the state of such device is either "ON" or "OFF" or a superposition of both. We can represent the state space as $\mathbb{R}^2$ where $\mathsf{ON} := \langle 1, 0 \rangle$ and $\mathsf{OFF} := \langle 0, 1 \rangle$. When we approach the device it is fair to say it is as likely that we find it running as it is that we find it turned off. So the initial state is

$$I := \langle 0.5, 0.5 \rangle = 0.5 \cdot \mathsf{ON} + 0.5 \cdot \mathsf{OFF}.$$

When the device is turned on and we press the "power on" button, surely nothing will happen, but when the device is turned off and the button is pressed the device will only turn on with probability, say, 0.75 (see diagram above).

Formally, pressing the button is represented by a linear map $P \colon \mathbb{R}^2 \to \mathbb{R}^2$ with $P(\mathsf{ON}) = \langle 1, 0 \rangle$ and $P(\mathsf{OFF}) = \langle 0.75, 0.25 \rangle$. There is precisely one such P:

$$P(\langle x, y \rangle) = \begin{pmatrix} 1 & 0.75 \\ 0 & 0.25 \end{pmatrix} \begin{pmatrix} x \\ y \end{pmatrix} \qquad \text{for } x, y \in \mathbb{R}.$$

Together, the initial state I and P form an input linear weighted automaton $c \colon \mathcal{A} \cdot \mathbb{R}^2 + \mathbb{R} \to \mathbb{R}^2$ with $\mathcal{A} := \{\mathsf{press}\}$. Indeed, c is given by, for $x, y, \mu \in \mathbb{R}$,

$$c(\langle (\mathsf{press}, \langle x, y \rangle), \mu \rangle) = \begin{pmatrix} 1 & 0.75 \\ 0 & 0.25 \end{pmatrix} \begin{pmatrix} x \\ y \end{pmatrix} + \mu \begin{pmatrix} 0.5 \\ 0.5 \end{pmatrix}.$$

5.3 Output Linear Weighted Automata

An output linear weighted automaton is a linear map of the form $d\colon V \to \mathbb{R} \times V^{\mathcal{A}}$. That is, it is a coalgebra on **Vect** of type $\mathbb{R} \times (\_)^{\mathcal{A}}$. The category of output linear weighted automata **WAut**$_o$ is the category **Coalg** $\left(\mathbb{R} \times (\_)^{\mathcal{A}}\right)$ of coalgebras for $\mathbb{R} \times (\_)^{\mathcal{A}}$.

We get an example of an output linear weighted automaton (see diagram) if we reverse all arrows of the input linear weighted automaton of Sect. 5.2. Formally, let $d\colon \mathbb{R}^2 \to \mathbb{R} \times (\mathbb{R}^2)^{\mathcal{A}}$ with $\mathcal{A} = \{\mathsf{press}\}$ be given by, for $x, y \in \mathbb{R}$,

$$d(\langle x, y\rangle) = \left((0.5\ 0.5) \begin{pmatrix} x \\ y \end{pmatrix}, \ \lambda a^{\mathcal{A}}. \begin{pmatrix} 1 & 0 \\ 0.75 & 0.25 \end{pmatrix} \begin{pmatrix} x \\ y \end{pmatrix} \right).$$

We could say that the output linear weighted automaton d is the *dual* (or *transpose*) of the input linear weighted automaton c. We will generalize this construction to arbitrary input linear weighted automata using the dual of a vector space (see below) and Theorem 1.

5.4 Dual of a Vector Space

Let V be a vector space. The dual of V is the following subspace of $\mathbb{R}^V$.

$$V^* := \left\{ \varphi \in \mathbb{R}^V \mid \varphi \text{ is linear} \right\}$$

Let us determine the dual of $\mathbb{R}^n$ for some natural number n. Recall that for every linear map $\varphi\colon \mathbb{R}^n \to \mathbb{R}$ there is a unique $u \in \mathbb{R}^n$ with, for all $x \in \mathbb{R}^n$,

$$\varphi(x) = u_1 x_1 + \cdots + u_n x_n \equiv u \cdot x.$$

So we get a bijection $\Phi\colon \mathbb{R}^n \to (\mathbb{R}^n)^*$ given by, for $x, u \in \mathbb{R}^n$

$$\Phi(u)(x) = u \cdot x. \tag{19}$$

It is not hard to see that Φ is linear, so $(\mathbb{R}^n)^*$ is isomorphic to $\mathbb{R}^n$. Consequently, $V^* \cong V$ for any finite dimensional vector space V.

For an infinite dimensional vector space V the situation is different. Let B be a basis for V (so $B \cdot \mathbb{R} \cong V$ and B is infinite). Define $\Psi\colon (B \cdot \mathbb{R})^* \to \mathbb{R}^B$ by, for $\varphi \in (B \cdot \mathbb{R})^*$ and $b \in B$, $\Psi(\varphi)(b) = \varphi(b)$. Then it is not hard to see that Ψ is an isomorphism. Since B is infinite we know that $B \cdot \mathbb{R} \ncong \mathbb{R}^B$, so $B \cdot \mathbb{R} \ncong (B \cdot \mathbb{R})^*$. Hence $V^* \ncong V$ for any infinite dimensional vector space V.

To apply Theorem 1 to the dual vector space construction we need to recognize the assignment $V \mapsto V^*$ as part of an adjunction. To begin, note that $V \mapsto V^*$ extends to a functor $(\_)^*\colon \mathbf{Vect} \to \mathbf{Vect}^{\mathrm{op}}$ as follows. Given a linear map $f\colon V \to W$ define $f^*\colon W^* \to V^*$ by, for $\varphi \in W^*$, $f^*(\varphi) = \varphi \circ f$.

This also gives us a functor $(\_)^*\colon \mathbf{Vect}^{\mathrm{op}} \to \mathbf{Vect}$. Now, given a vector space V define $\iota_V\colon V \to V^{**}$ by, for $v \in V$ and $\varphi \in V^*$, $\iota_V(v)(\varphi) = \varphi(v)$. Then we have an adjunction $\langle (\_)^* \dashv (\_)^*, \iota, \iota \rangle : \mathbf{Vect} \to \mathbf{Vect}^{\mathrm{op}}$.

If $V \equiv \mathbb{R}^n$ for some $n \in \mathbb{N}$, then ι_V is an isomorphism. So if we restrict the adjunction to the category **FVect** of linear maps between finite dimensional vector spaces we get a duality $\langle (\_)^* \dashv (\_)^*, \iota, \iota \rangle : \mathbf{FVect} \to \mathbf{FVect}^{\mathrm{op}}$.

5.5   Dual of a Linear Weighted Automaton

We will now lift the adjunction between **Vect** and **Vect**^{op}. The result will be:

$$
\begin{array}{ccc}
\mathbf{WAut}_o = \mathbf{Coalg}\left(\mathbb{R}\times(\_)^{\mathcal{A}}\right) & \underset{\underset{(\_)^o}{\xleftarrow{\hspace{1cm}}}}{\overset{(\_)^i}{\xdashrightarrow{\hspace{1cm}}}} \bot & \mathbf{Alg}\left(\mathcal{A}\cdot\_+\mathbb{R}\right) = \mathbf{WAut}_i \\
\downarrow & & \downarrow \\
\mathbb{R}\times(\_)^{\mathcal{A}} \;\hookrightarrow\; \mathbf{Vect} & \underset{\underset{R=(\_)^*}{\xrightarrow{\hspace{2cm}}}}{\overset{L=(\_)^*}{\xleftarrow{\hspace{2cm}}}} \bot & \mathbf{Vect}^{\mathrm{op}} \;\circlearrowright\; \mathcal{A}\cdot\_+\mathbb{R}
\end{array}
$$

For a vector space V define $\beta_V \colon (\mathcal{A}\cdot V+\mathbb{R})^* \to \mathbb{R}\times(V^*)^{\mathcal{A}}$ by, for $f \in (\mathcal{A}\cdot V+\mathbb{R})^*$,

$$
\beta_V(f) \;=\; \Big\langle f\big(\kappa_1(1)\big),\, \lambda a^{\mathcal{A}}\lambda v^V.\, f\big(\kappa_0(\kappa_a(v))\big)\Big\rangle.
$$

Then β_V is invertible and we get a natural iso $\beta \colon (\mathcal{A}\cdot(\_)+\mathbb{R})^* \to \mathbb{R}\times((\_)^*)^{\mathcal{A}}$. Hence we get an adjunction lifting by Theorem 1 as depicted above.

Let $c \colon \mathcal{A}\cdot V+\mathbb{R} \to V$ be an input linear weighted automaton. Then

$$
c^o \colon V^* \to \mathbb{R}\times(V^*)^{\mathcal{A}} \qquad \text{and} \qquad c^o \;=\; \beta_V \circ c^*.
$$

If we take c to be as depicted in the diagram in Sect. 5.2 then c^o will be as depicted in the diagram in Sect. 5.3 (if we identify $(\mathbb{R}^2)^*$ with $\mathbb{R}^2$ via the isomorphism in Eq. 19).

Let $d \colon V \to \mathbb{R}\times V^{\mathcal{A}}$ be an output linear weighted automaton. Then

$$
d^i \colon \mathcal{A}\cdot V^* +\mathbb{R} \to V^* \qquad \text{and} \qquad d^i \;=\; d^* \circ \alpha_V,
$$

where $\alpha_V \colon \mathcal{A}\cdot V^* +\mathbb{R} \to (\mathbb{R}\times V^{\mathcal{A}})^*$ and for $h \in \mathcal{A}\cdot V^*$, $\mu,\nu \in \mathbb{R}$, $f \in V^{\mathcal{A}}$,

$$
\alpha_V(\langle h,\nu\rangle)(\langle \mu,f\rangle) = \mu\cdot\nu + \sum_{a\in\mathcal{A}} h(a)\big(f(a)\big). \tag{20}
$$

If we take d as in the diagram of Sect. 5.3, then d^i will be as depicted in the diagram of Sect. 5.2. In particular, we see that $(d^i)^o \cong c$. More generally, since we have a duality $\langle(\_)^* \dashv (\_)^*, \iota, \iota\rangle \colon \mathbf{FVect} \to \mathbf{FVect}^{\mathrm{op}}$ we get a duality

$$
\langle(\_)^i \dashv (\_)^o, \iota, \iota\rangle \colon \mathbf{FWAut}_o \to \mathbf{FWAut}_i
$$

where $\mathbf{FWAut}_o$ and $\mathbf{FWAut}_i$ are variants of $\mathbf{WAut}_o$ and $\mathbf{WAut}_i$, respectively, for finite dimensional vector spaces.

6   Checking Behavioral Equivalences

Finally, we will show how the results on adjunctions can be used to check behavioral equivalences. Since our coalgebras do not necessarily live in **Set**, where we could address elements of a carrier set, we use the following alternative definition, where we specify whether two arrows are behaviorally equivalent. This is reminiscent of equipping a coalgebra with start states, similar to the initial states of an automaton.

Definition 4 (Behavioral Equivalence). *Let* **C** *be a category,* $F: \mathbf{C} \to \mathbf{C}$ *be an endofunctor such that a final F-coalgebra* $\omega: \Omega \to F\Omega$ *exists. Furthermore let* $c_1: X_1 \to FX_1$ *and* $c_2: X_2 \to FX_2$ *be two F-coalgebras and U be a* **C**-*object.*

We say that two **C**-*arrows* $x_1: U \to X_1$, $x_2: U \to X_2$ *are* behaviorally equivalent *(in symbols* $x_1 \sim_F^{c_1,c_2} x_2$*), whenever the diagram on the left commutes where* $!_1: c_1 \to \omega$ *and* $!_2: c_2 \to \omega$ *are the unique coalgebra homomorphisms into the final coalgebra.*

$$
\begin{array}{ccc}
X_1 & \xleftarrow{\ x_1\ } U \xrightarrow{\ x_2\ } & X_2 \\
c_1 \downarrow & & \downarrow c_2 \\
FX_1 & \overset{!_1}{\searrow} \ \Omega \ \overset{!_2}{\swarrow} & FX_2 \\
& \overset{\omega}{\searrow} \ \downarrow \ \overset{}{\swarrow} F!_2 & \\
F!_1 & F\Omega &
\end{array}
$$

In **Set** the choice for U will typically be a singleton set and then the problem reduces to asking whether two given states are behaviorally equivalent.

Now assume that we have an adjunction $\langle L \dashv R, \eta, \varepsilon \rangle : \mathbf{C} \to \mathbf{D}$ that is lifted to coalgebras as specified in Definition 3. Now, since $\overline{R}$ is a right adjoint, it preserves limits, specifically it preserves the final coalgebra. Let us take another look at the first diagram in Sect. 3: It can easily be determined that $\mathcal{A}^*$, the set of all finite words, is the carrier of the final coalgebra in **Set**^{op}. Via the right adjoint, this translates to the carrier set $\mathcal{A}^*$ in **Rel**, where the arrow $!_1$ into the final coalgebra is a relation, relating each state with the words that are accepted by it (in [8] this final coalgebra is discussed in detail). The final coalgebra can also be transferred into **JSL** and **Set**, where it has carrier set $2^{\mathcal{A}^*}$.

Hence, the adjunctions allow to construct final coalgebras and to transfer results about final semantics from other categories. Furthermore, it is possible to check behavioral equivalence in a different category, by translating queries via the adjunction.

Proposition 1. *Let* **C** *be a category,* $F: \mathbf{C} \to \mathbf{C}$ *and* $G: \mathbf{D} \to \mathbf{D}$ *be endofunctors,* $\langle L \dashv R, \eta, \varepsilon \rangle : \mathbf{C} \to \mathbf{D}$ *be an adjunction together with a lifting in the sense of Definition 3, i.e. an adjunction* $\langle \overline{L} \dashv \overline{R}, \eta, \varepsilon \rangle : \mathbf{Coalg}(F) \to \mathbf{Coalg}(G)$. *Furthermore assume that a final G-coalgebra exists and that R is faithful.*

Let $d_1: Y_1 \to GY_1$, $d_2: Y_2 \to GY_2$ *be two G-coalgebras and let* $y_1: U \to Y_1$, $y_2: U \to Y_2$ *be two arrows in* **D**. *Then the following equivalence holds.*

$$
y_1 \sim_G^{d_1,d_2} y_2 \qquad \Longleftrightarrow \qquad Ry_1 \sim_F^{\overline{R}d_1, \overline{R}d_2} Ry_2
$$

In all our examples the right adjoint R is faithful. If this is the case and the final coalgebra exists, this allows us to check behavioral equivalence in a different category, where this might be easier or more straightforward. The classical example is of course the lifted adjunction between **Set** and **Rel**. In order to check language equivalence for nondeterministic automata, the standard technique is to determinize them via the right adjoint into **Set**. Then, language equivalence can be checked on the powerset automaton.

7   Conclusion, Related and Future Work

We have shown how to lift adjunctions between categories to adjunctions on coalgebras. Furthermore we gave several examples for such adjunction liftings and showed how they can be used to transfer behavioral equivalence checks from one category to another.

Several open questions remain, both concerning the examples and the general technique. Our main example involving deterministic and nondeterministic automata is strongly related to the powerset monad, since one adjunction is based on the Kleisli and another on the Eilenberg-Moore construction for this monad. In this specific case we obtain two more adjunctions by considering **Set**^{op}, however this will not work for any monad. Still, it would be interesting to investigate which monads allow such a rich structure of adjunctions, in what way these adjunctions can be lifted to adjunctions to coalgebras and whether this results in well-known constructions. There exists also the comparison functor between the Kleisli and the Eilenberg-Moore category, which however is not necessarily part of an adjunction. It would be interesting to find out which behavioral information can be transported over the comparison functor.

In [1] the authors used the adjunction between **Rel** and **Set**^{op} in order to characterize a factorization structure that is employed for a minimization algorithm. Hence, an obvious question is whether other adjunctions can be used for such algorithmic purposes, for instance for minimizing a coalgebra in one category, but using the structure of another category. It also seems plausible that up-to techniques can be explained in this way, for instance by checking language equivalence for nondeterministic automata in **Rel**, using the algebraic structure of **JSL** via the comparison functor (similar to [5]).

We also showed how to transfer equivalence checking queries through a right adjoint. Can they also be transferred in the other direction, via a left adjoint?

Finally, the conditions in Theorem 1 (Lifting an Adjunction to Coalgebras) are sufficient for the lifting to exist. However, it is unclear whether they are also necessary.

Related Work. The adjunction between **Rel** and **Set**^{op}, transforming nondeterministic automata into codeterministic automata has already been considered in [1] and similarly in [8]. The paper [17] is concerned with the adjunction between **Set** and **JSL** and uses it to determinize automata, but different from the approach in this paper. More concretely, in [17] a *nondeterministic automaton* specified by a function $X \to 2 \times (2^X)^A$ in **Set** is translated into a join-preserving function $2^X \to 2 \times (2^X)^A$, which does not give an adjunction on coalgebras. Another closely related paper is [11] which is also concerned with the Kleisli and Eilenberg-Moore constructions for the powerset monad and uses the comparison functor in order to determinize automata. Furthermore our example in Sect. 5 is based on a duality of categories [3].

Hence, many ideas that are summarized in this paper are not completely new, but have been stated in various forms. However, we think that it is insightful to present this theory strictly from the point of view of adjunction lifting and to clearly spell out what it means to preserve and reflect behavioral equivalences by adjoints. Furthermore, to our knowledge, the adjunction between join semi-lattices and $\mathbf{Set}^{op}$, that gives rise to a quite surprising and unusual construction, has never been studied in this setting.

Acknowledgements. We would like to thank Marcello Bonsangue, Alexandra Silva and Filippo Bonchi for raising and discussing the problem with us. Furthermore we would like to acknowledge Ana Sokolova for interesting discussions on this topic.

References

1. Adámek, J., Bonchi, F., Hülsbusch, M., König, B., Milius, S., Silva, A.: A coalgebraic perspective on minimization and determinization. In: Birkedal, L. (ed.) FOSSACS 2012. LNCS, vol. 7213, pp. 58–73. Springer, Heidelberg (2012)
2. Awodey, S.: Category Theory. Clarendon Press, Oxford (2006)
3. Bezhanishvili, N., Kupke, C., Panangaden, P.: Minimization via duality. In: Ong, L., de Queiroz, R. (eds.) WoLLIC 2012. LNCS, vol. 7456, pp. 191–205. Springer, Heidelberg (2012)
4. Bonchi, F., Bonsangue, M.M., Rutten, J.J.M.M., Silva, A.: Brzozowski's algorithm (co)algebraically. In: Constable, R.L., Silva, A. (eds.) Kozen Festschrift. LNCS, vol. 7230, pp. 12–23. Springer, Heidelberg (2012)
5. Bonchi, F., Pous, D.: Checking NFA equivalence with bisimulations up to congruence. In: Proceedings of POPL '13. pp. 457–468. ACM (2013)
6. Brzozowski, J., Tamm, H.: Theory of átomata. In: Mauri, G., Leporati, A. (eds.) DLT 2011. LNCS, vol. 6795, pp. 105–116. Springer, Heidelberg (2011)
7. Fiore, M., Turi, D.: Semantics of name and value passing. In: Proceedings of LICS '01, pp. 93–104. IEEE (2001)
8. Hasuo, I., Jacobs, B., Sokolova, A.: Generic trace semantics via coinduction. Log. Methods Comput. Sci. **3**(4:11), 1–36 (2007)
9. Hermida, C., Jacobs, B.: Structural induction and coinduction in a fibrational setting. Inf. Comput. **145**(IC982725), 107–152 (1998)
10. Jacobs, B., Rutten, J.: A tutorial on (co)algebras and (co)induction. Bull. Eur. Assoc. Theor. Comput. Sci. **62**, 222–259 (1997)
11. Jacobs, B., Silva, A., Sokolova, A.: Trace semantics via determinization. In: Pattinson, D., Schröder, L. (eds.) CMCS 2012. LNCS, vol. 7399, pp. 109–129. Springer, Heidelberg (2012)
12. Kelly, G., Street, R.: Review of the elements of 2-categories. In: Kelly, G.M. (ed.) Category Seminar, Lecture Notes in Mathematics, vol. 420, pp. 75–103. Springer, Heidelberg (1974). (http://dx.doi.org/10.1007/BFb0063101)
13. Kerstan, H., König, B.: Coalgebraic trace semantics for continuous probabilistic transition systems. Log. Methods Comput. Sci. **9**(4:16)(834) (2013). http://arxiv.org/abs/1310.7417v3

14. Mac Lane, S.: Categories for the Working Mathematician, 2nd edn. Springer, New York (1998)
15. Panangaden, P.: Labelled Markov Processes. Imperial College Press, London (2009)
16. Rutten, J.: Universal coalgebra: a theory of systems. Theor. Comput. Sci. **249**, 3–80 (2000)
17. Silva, A., Bonchi, F., Bonsangue, M.M., Rutten, J.J.M.M.: Generalizing determinization from automata to coalgebras. Log. Methods Comput. Sci. **9**(1:09), 1–27 (2013)
18. Sokolova, A.: Probabilistic systems coalgebraically: a survey. Theor. Comput. Sci. **412**(38), 5095–5110 (2011). (CMCS Tenth Anniversary Meeting)

Canonical Nondeterministic Automata

Robert S.R. Myers[1], Jiří Adámek[1], Stefan Milius[2], and Henning Urbat[1(✉)]

[1] Institut für Theoretische Informatik,
Technische Universität Braunschweig, Braunschweig, Germany
urbat@iti.cs.tu-bs.de
[2] Lehrstuhl für Theoretische Informatik,
Friedrich-Alexander-Universität Erlangen-Nürnberg, Erlangen, Germany

Abstract. For each regular language L we describe a family of canonical nondeterministic acceptors (nfas). Their construction follows a uniform recipe: build the minimal dfa for L in a locally finite variety $\mathcal{V}$, and apply an equivalence between the finite $\mathcal{V}$-algebras and a category of finite structured sets and relations. By instantiating this to different varieties we recover three well-studied canonical nfas (the átomaton, the jiromaton and the minimal xor automaton) and obtain a new canonical nfa called the distromaton. We prove that each of these nfas is minimal relative to a suitable measure, and give conditions for state-minimality. Our approach is coalgebraic, exhibiting additional structure and universal properties.

1 Introduction

One of the core topics in classical automata theory is the construction of state-minimal acceptors for a given regular language. It is well-known that the difficulty of this task depends on whether one has deterministic or nondeterministic acceptors in mind. First, every regular language $L \subseteq \Sigma^*$ is accepted by a unique minimal *deterministic* finite automaton (dfa): its states Q_L are the derivatives of L, i.e.,

$$Q_L = \{w^{-1}L : w \in \Sigma^*\} \quad \text{where } w^{-1}L = \{v \in \Sigma^* : wv \in L\},$$

the transitions are $K \xrightarrow{a} a^{-1}K$ for $K \in Q_L$ and $a \in \Sigma$, the initial state is L, and a state is final iff it contains the empty word. This construction is due to Brzozowski [9], and is the basis for efficient dfa minimization algorithms. For *nondeterministic* finite automata (nfas) the situation is significantly more complex: a regular language may have many non-isomorphic state-minimal nfas, and generally there is no way to identify a "canonical" one among them. However, several authors have recently proposed nondeterministic acceptors that *are* in some sense canonical (though not necessarily state-minimal), e.g. the *átomaton* of Brzozowski and Tamm [8], the *jiromaton*[1] of Denis, Lemay and Terlutte [10],

---

[1] In [10] the authors called their acceptor "canonical residual finite state automaton". We propose the shorter "jiromaton" because this is analogous to the átomaton terminology.

© IFIP International Federation for Information Processing 2014
M.M. Bonsangue (Ed.): CMCS 2014, LNCS 8446, pp. 189–210, 2014.
DOI: 10.1007/978-3-662-44124-4_11

and the *minimal xor automaton* of Vuillemin and Gama [17]. In each case, the respective nfa is formed by closing the set Q_L of derivatives under certain algebraic operations and taking a minimal set of generators as states. Specifically,

1. the states of the átomaton are the atoms of the boolean algebra generated by Q_L, obtained by closing Q_L under finite union, finite intersection and complement;
2. the states of the jiromaton are the join-irreducibles of the join-semilattice generated by Q_L, obtained by closing Q_L under finite union;
3. the states of the minimal xor automaton form a basis for the $\mathbb{Z}_2$-vector space generated by Q_L, obtained by closing Q_L under symmetric difference.

In this paper we demonstrate that all these canonical nfas arise from a coalgebraic construction. For this purpose we first consider *deterministic* automata interpreted in a locally finite variety $\mathcal{V}$, where *locally finite* means that finitely generated algebras are finite. A *deterministic $\mathcal{V}$-automaton* is a coalgebra for the functor $T_\Sigma = 2 \times \mathsf{Id}^\Sigma$ on $\mathcal{V}$, for a fixed two-element algebra 2. In Sect. 2 we describe a Brzozowski-like construction that yields, for every regular language, the minimal deterministic finite $\mathcal{V}$-automaton accepting it. Next, for certain varieties $\mathcal{V}$ of interest, we derive an equivalence between the full subcategory $\mathcal{V}_f$ of finite algebras and a suitable category $\overline{\mathcal{V}}$ of finite structured sets, whose morphisms are relations preserving the structure. In each case, the objects of $\overline{\mathcal{V}}$ are "small" representations of their counterparts in $\mathcal{V}_f$, based on specific generators of algebras in $\mathcal{V}_f$. The equivalence $\mathcal{V}_f \cong \overline{\mathcal{V}}$ then induces an equivalence between deterministic finite $\mathcal{V}$-automata and coalgebras in $\overline{\mathcal{V}}$ which are *nondeterministic* automata.

This suggests a two-step procedure for constructing a canonical nfa for a given regular language L: (i) form L's minimal deterministic $\mathcal{V}$-automaton, and (ii) use the equivalence of $\mathcal{V}_f$ and $\overline{\mathcal{V}}$ to obtain an equivalent nfa. Applying this to different varieties $\mathcal{V}$ yields the three canonical nfas mentioned above. For the átomaton one takes $\mathcal{V} = \mathsf{BA}$ (boolean algebras). Then the minimal deterministic BA-automaton for L arises from the minimal dfa by closing its states Q_L under boolean operations. The category $\overline{\mathcal{V}} = \overline{\mathsf{BA}}$ is based on Stone duality: $\overline{\mathsf{BA}}$ is the dual of the category of finite sets, so it has a objects all finite sets, as morphisms all converse-functional relations, and the equivalence functor $\mathsf{BA}_f \to \overline{\mathsf{BA}}$ maps each finite boolean algebra to the set of its atoms. This equivalence applied to the minimal deterministic BA-automaton for L gives precisely L's átomaton. Similarly, by taking $\mathcal{V} =$ join-semilattices and $\mathcal{V} =$ vector spaces over $\mathbb{Z}_2$ and describing a suitable equivalence $\mathcal{V}_f \cong \overline{\mathcal{V}}$, we recover the jiromaton and the minimal xor automaton, respectively. Finally, for $\mathcal{V} =$ distributive lattices we get a new canonical nfa called the *distromaton*, which bears a close resemblance to the universal automaton [14].

Example 1.1. Consider the language $L = (a + b)^* b(a + b)^n$ where $n \in \omega$. Its minimal dfa has $\geq 2^n$ states and its (A) átomaton, (X) minimal xor automaton, (J) jiromaton and (D) distromaton are the nfas with $\leq n + 3$ states depicted below (see Sect. 3.3).

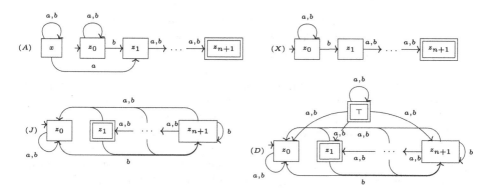

The minimal xor automaton accepts L by $\mathbb{Z}_2$-weighted acceptance, which is the usual acceptance in this case. It is a state-minimal nfa, as is the jiromaton. The state-minimality of the latter follows from a general result (Theorem 4.4).

Generally, the sizes of the four canonical nfas and the minimal dfa are related as follows:

(a) all the four canonical nfas can have exponentially fewer states than the minimal dfa;

(b) the minimal xor automaton and jiromaton have no more states than the minimal dfa;

(c) the átomaton and distromaton have the same number of states, although their structure can be very different.

In Sect. 4 we characterize the átomaton, jiromaton, minimal xor automaton and distromaton by a minimality property. This provides an explanation of the canonicity of these acceptors that is missing in the original papers. We then use this additional structure to identify conditions on regular languages that guarantee the *state-minimality* of the canonical nfas. That is, there exists a natural class of languages where *canonical* state-minimal nfas exist and can be computed relatively easily.

Related work. Our paper unifies the constructions of canonical nfas given in [8,10,17] from a coalgebraic perspective. Previously, several authors have studied coalgebraic methods for constructing minimal and canonical representatives of machines, including Adámek, Bonchi, Hülsbusch, König, Milius and Silva [1], Adámek, Milius, Moss and Sousa [2] and Bezhanishvili, Kupke and Panangaden [4]. Only the first of these three papers, however, treats the case of nondeterministic automata explicitly – in particular, there the átomaton is recovered as an instance of projecting coalgebras in a Kleisli category into a reflective subcategory. This approach is methodologically rather different from the present paper where a categorical equivalence (rather than a reflection) is the basis for the construction of nfas.

In [8] the authors propose a surprisingly simple algorithm for constructing the átomaton of a language L: take the minimal dfa for L's reversed language, and reverse this dfa. These steps form a fragment of a classical dfa minimization algorithm due to Brzozowski. Recently Bonchi, Bonsangue, Rutten and Silva [6] gave a (co-)algebraic explanation of this procedure, based on the classical duality between observability and reachability of dfas. We provide another explanation in Sect. 3.3.

A coalgebraic treatment of linear weighted automata (of which xor automata considered here are a special case) appears in [5]; this paper also provides a procedures for computing the minimal linear weighted automaton.

Finally, our work is somewhat related to work on coalgebraic trace semantics [11]. However, while that work considers coalgebras whose carrier is a the free algebra of a variety we consider coalgebras whose carriers are arbitrary algebras from the given variety; this means we consider coalgebras over an Eilenberg-Moore category (cf. [7,12]).

2   Deterministic Automata

We start with recalling the concept of a finite automaton. Throughout this paper let us fix a finite input alphabet Σ.

Definition 2.1. *(a) A nondeterministic finite automaton (nfa) is a triple $N = (Z, R_a, F)$ consisting of a finite set Z of states, transition relations $R_a \subseteq Z \times Z$ for each $a \in \Sigma$ and final states $F \subseteq Z$. Morphisms of nfas are the usual bisimulations, i.e., relations that preserve and reflect transitions and final states. If N is equipped with initial states $I \subseteq Z$ we write $N = (Z, R_a, F, I)$. In this case, N accepts a language $\mathcal{L}_N(I) \subseteq \Sigma^*$ in the usual way.*
(b) A deterministic finite automaton (dfa) is an nfa with a single initial state whose transition relations are functions.

Although the goal of our paper is constructing canonical nondeterministic automata, we first consider deterministic ones from a coalgebraic perspective. Given an endofunctor $T : \mathcal{V} \to \mathcal{V}$ of a category $\mathcal{V}$, a T-coalgebra (Q, γ) consists of a $\mathcal{V}$-object Q and a $\mathcal{V}$-morphism $\gamma : Q \to TQ$. A coalgebra homomorphism into another coalgebra $\gamma' : Q' \to TQ'$ is a $\mathcal{V}$-morphism $h : Q \to Q'$ such that $Th \circ \gamma = \gamma' \circ h$. This defines a category $\mathsf{Coalg}(T)$. If it exists, its terminal object νT is called the *final T-coalgebra*.

Assumption 2.2. From now on $\mathcal{V}$ is a locally finite variety with a specified two-element algebra $2 = \{0, 1\}$. That is, $\mathcal{V}$ is the category of algebras for some finitary signature and equations, its morphisms being the usual algebra homomorphisms. That $\mathcal{V}$ is *locally finite* means its finitely generated algebras are finite, equivalently its finitely generated *free* algebras are finite.

Example 2.3. (a) The category Set_* of pointed sets is a locally finite variety, given by the signature with a constant 0 and no equations. Let $2 \in \mathsf{Set}_*$ have point 0.
(b) The category BA of boolean algebras is a locally finite variety: a boolean algebra on n generators has at most 2^{2^n} elements. 2 is the 2-chain $0 < 1$.

(c) The category $\mathsf{Vect}(\mathbb{Z}_2)$ of vector spaces over the binary field $\mathbb{Z}_2$ is a locally finite variety. Here $2 = \mathbb{Z}_2$ as a one-dimensional vector space.

(d) The category JSL of (join-)semilattices with a least element 0 is locally finite: the finite powerset $\mathcal{P}_f X$ is the free semilattice on X, so a semilattice on n generators has at most 2^n elements. 2 is the 2-chain $0 < 1$.

(e) The category DL of distributive lattices with a least and largest element $\bot$ and $\top$ is locally finite. Again, 2 is the 2-chain $0 < 1$.

Definition 2.4. *If Q is a join-semilattice then $q \in Q$ is join-irreducible if (i) $q \neq 0$ and (ii) $q = r \vee r'$ implies $q = r$ or $q = r'$. The set of join-irreducibles is written $J(Q) \subseteq Q$.*

Definition 2.5. *A T-coalgebra (Q', γ') is a subcoalgebra of (Q, γ) if there exists an injective coalgebra homomorphism $m : (Q', \gamma') \rightarrowtail (Q, \gamma)$, and a quotient coalgebra of (Q, γ) if there exists a surjective coalgebra homomorphism $e : (Q, \gamma) \twoheadrightarrow (Q', \gamma')$.*

Definition 2.6. *A deterministic $\mathcal{V}$-automaton is a coalgebra for the functor*

$$T_\Sigma : \mathcal{V} \to \mathcal{V}, \quad T_\Sigma = 2 \times \mathsf{Id}^\Sigma = 2 \times \mathsf{Id} \times \cdots \times \mathsf{Id}.$$

Remark 2.7. Hence, by the universal property of the product, a deterministic $\mathcal{V}$-automaton $Q \to 2 \times Q^\Sigma$ is given by an algebra Q of states, a $\mathcal{V}$-morphism $\gamma_\epsilon : Q \to 2$ defining final states via $\gamma_\epsilon^{-1}(\{1\})$ and, for each $a \in \Sigma$, a $\mathcal{V}$-morphism $\gamma_a : Q \to Q$ representing the a-transitions. In particular, deterministic Set-automata are precisely the classical (possibly infinite) deterministic automata without initial states, shortly *da*'s.

Example 2.8. (a) A deterministic Set_*-automaton is a da whose carrier is a pointed set and whose point is a non-final sink state; these are the partial automata of [16].

(b) A deterministic BA-automaton is a da with a boolean algebra structure on the states Q such that (i) the final states form an ultrafilter, (ii) $q \xrightarrow{a} q'$ and $r \xrightarrow{a} r'$ implies $q \vee r \xrightarrow{a} q' \vee r'$ and $\neg q \xrightarrow{a} \neg q'$, and (iii) $\bot$ is a non-final sink state.

(c) A deterministic $\mathsf{Vect}(\mathbb{Z}_2)$-automaton is a da with a $\mathbb{Z}_2$-vector space structure on the states Q such that (i) the final states $F \subseteq Q$ satisfy $0 \notin F$ and also $q + r \in F$ iff either $q \in F$ or $r \in F$ but not both, (ii) $q \xrightarrow{a} q'$ and $r \xrightarrow{a} r'$ implies $q + q \xrightarrow{a} r + r'$, and (iii) 0 is a non-final sink state.

(d) A deterministic JSL-automaton is a da with a join-semilattice structure on the states Q such that (i) the final states form a prime filter, (ii) $q \xrightarrow{a} q'$ and $r \xrightarrow{a} r'$ implies $q + r \xrightarrow{a} q' + r'$, and (iii) 0 is a non-final sink state. Recall that a *prime filter* is an upwards closed $F \subseteq Q$ where $0 \notin F$ and $q + q' \in F$ iff $q \in F$ or $q' \in F$.

(e) A deterministic DL-automaton is a da with a distributive lattice structure on the states Q such that (i) the final states form an prime filter, (ii) $q \xrightarrow{a} q'$ and $r \xrightarrow{a} r'$ implies $q \vee r \xrightarrow{a} q' \vee r'$ and $q \wedge r \xrightarrow{a} q' \wedge r'$, and (iii) $\bot$ is a non-final sink state and $\top$ is a final one.

Remark 2.9. For finitary endofunctors T, Milius [15] introduced the concept of a locally finitely presentable coalgebra: it is a filtered colimit of coalgebras carried by finitely presentable objects. In the present context the finitely presentable objects are precisely the finite algebras in $\mathcal{V}$, so we speak about *locally finite coalgebras.* A T_Σ-coalgebra is locally finite iff from each state only finitely many states are reachable by transitions.

Remark 2.10. 1. The final T_Σ-coalgebra in Set is $\nu T_\Sigma = \mathcal{P}\Sigma^*$, the set of formal languages over Σ, with transitions $L \xrightarrow{a} a^{-1}L$ for $a \in \Sigma$ and final states precisely those languages containing ϵ. Importantly, νT_Σ arises as the ω^{op}-limit of T_Σ's terminal sequence $(T_\Sigma^n 1)_{n<\omega}$, see [3]. Since for any variety $\mathcal{V}$ the forgetful functor from $\mathcal{V}$ to Set creates limits, the final T_Σ-coalgebra νT_Σ in $\mathcal{V}$ exists and lifts the one in Set, so νT_Σ has underlying set $\mathcal{P}\Sigma^*$ and the transitions and final states are as above.
   2. The *final locally finite T_Σ-coalgebra* is denoted by ρT_Σ. In $\mathcal{V} =$ Set this is the sub-da of $\nu T_\Sigma = \mathcal{P}\Sigma^*$ given by the set of all regular languages over Σ. This generalizes to any locally finite variety $\mathcal{V}$: ρT_Σ is a subcoalgebra of νT_Σ and its underlying set is the set of regular languages.

Example 2.11. (a) In Set$_*$ the carrier of the final coalgebra νT_Σ has the constant $\emptyset$, which ρT_Σ inherits.
(b) In BA, νT_Σ has the usual set-theoretic boolean algebra structure. The principal filter $\uparrow\epsilon$ is an ultrafilter and the transition maps $L \mapsto a^{-1}L$ are boolean morphisms.
(c) In Vect($\mathbb{Z}_2$) the vector space structure on νT_Σ and ρT_Σ is given by symmetric difference and $\emptyset$ is the zero vector.
(d) In JSL the join-semilattice structure on νT_Σ is union and $\emptyset$. The final states form a one-generated upset $\uparrow\epsilon$ which is a prime filter because the language $\{\epsilon\}$ is join-irreducible in νT_Σ. The transitions maps are join-semilattice morphisms.
(e) In DL we have the usual set-theoretic lattice structure on νT_Σ. The final states form a prime filter and the transition maps are lattice morphisms.

Notation 2.12. Let (Q, γ) be a locally finite T_Σ-coalgebra. The unique coalgebra homomorphism into ρT_Σ is written:

$$\mathcal{L}_\gamma : Q \to \rho T_\Sigma.$$

The function $\mathcal{L}_\gamma$ sends $q \in Q$ to the regular language $\mathcal{L}_\gamma(q) \subseteq \Sigma^*$ the state q accepts.

Definition 2.13. *Let $V \in \mathcal{V}$ denote the free algebra on one generator g. Then a pointed T_Σ-coalgebra (Q, γ, q_0) is a T_Σ-coalgebra (Q, γ) with a morphism $q_0 : V \to Q$. The latter may be viewed as the initial state $q_0(g) \in Q$. The language accepted by (Q, γ, q_0) is $\mathcal{L}_\gamma(q_0)$. We say that (Q, γ, q_0) is*

1. reachable *if it is generated by q_0, i.e., no proper subcoalgebra contains q_0;*
2. simple *if it has no proper quotients, i.e., for every quotient coalgebra e : $(Q, \gamma) \twoheadrightarrow (Q', \gamma')$ the map e is bijective;*
3. minimal *if it is reachable and simple.*

Lemma 2.14. *(Q, γ, q_0) is reachable iff the algebra Q is generated by those $q \in Q$ reachable from q_0 by transitions. It is simple iff $\mathcal{L}_\gamma$ is injective.*

Brozozowski's construction of the minimal dfa for a regular language (see Introduction) generalizes to deterministic $\mathcal{V}$-automata as follows:

Construction 2.15. For any regular language $L \subseteq \Sigma^*$ let $A_\mathcal{V}^L$ be the pointed T_Σ-coalgebra (Q_L, γ, L) where:

1. Q_L is the subalgebra of $\nu T_\Sigma = \mathcal{P}\Sigma^*$ generated by all derivatives $w^{-1}L$ ($w \in \Sigma^*$).
2. The transitions are $K \xrightarrow{a} a^{-1}K$ for $a \in \Sigma$ and $K \in Q_L$.
3. $K \in Q_L$ is final iff $\epsilon \in K$.

Lemma 2.16. *For every regular language $L \subseteq \Sigma^*$, $A_\mathcal{V}^L$ is a well-defined finite pointed T_Σ-coalgebra.*

Proof. L is regular so it has only finitely many distinct derivatives $w^{-1}L$. Hence Q_L is a finite algebra because $\mathcal{V}$ is a locally finite variety. It remains to show that $\gamma_a : Q_L \to Q_L$ and $\gamma_\epsilon : Q_L \to 2$ as specified in points 2. and 3. are well-defined $\mathcal{V}$-morphisms. Recall the final locally finite T_Σ-coalgebra $(\rho T_\Sigma, \gamma_\rho)$. Then

$$\gamma_\epsilon = Q_L \hookrightarrow \rho T_\Sigma \xrightarrow{(\gamma_\rho)_\epsilon} 2$$

is a $\mathcal{V}$-morphism since ρT_Σ is a lifting of the da of regular languages, see Remark 2.10. Furthermore $(\gamma_\rho)_a : \rho T_\Sigma \to \rho T_\Sigma$ is defined $(\gamma_\rho)_a(K) = a^{-1}K$ i.e. the derivative $a^{-1}(-)$ preserves the algebraic operations. Thus Q_L is closed under derivatives, so γ_a is a well-defined algebra morphism. $\square$

Example 2.17. (a) In $\mathsf{Set}_\star$, we have $Q_L = \{\emptyset\} \cup \{w^{-1}L : w \in \Sigma^*\}$.
(b) In BA, Q_L is the closure of $\{\emptyset\} \cup \{w^{-1}L : w \in \Sigma^*\}$ under union and complement.
(c) In $\mathsf{Vect}(\mathbb{Z}_2)$, Q_L is the closure of $\{w^{-1}L : w \in \Sigma^*\}$ under symmetric difference.
(d) In JSL, Q_L is the closure of $\{\emptyset\} \cup \{w^{-1}L : w \in \Sigma^*\}$ under union.
(e) In DL, Q_L is the closure of $\{\emptyset, \Sigma^*\} \cup \{w^{-1}L : w \in \Sigma^*\}$ under union and intersection.

Remark 2.18. The category $\mathsf{Coalg}(T_\Sigma)$ of T_Σ-coalgebras has a factorization system (surjective homomorphism, injective homomorphism) lifting the usual factorization system (surjective, injective) $=$ (regular epi, mono) in $\mathcal{V}$.

Construction 2.19. (see [2]). These factorizations give a two-step minimization of any finite pointed T_Σ-coalgebra (Q, γ, q_0):

1. Construct the reachable subcoalgebra $(R, \delta) \hookrightarrow (Q, \gamma)$ generated by q_0.
2. Factorize the unique T_Σ-coalgebra homomorphism $\mathcal{L}_\delta : (R, \delta) \to (\rho T_\Sigma, \gamma_\rho)$ as:

$$(R, \delta) \xrightarrow{s} (R', \delta') \xrightarrow{m} (\rho T_\Sigma, \gamma_\rho)$$

Then $(R', \delta', s(q_0))$ is minimal.

Theorem 2.20. *Let $L \subseteq \Sigma^*$ be a regular language. Then $A_\mathcal{V}^L$ is (up to isomorphism) the unique minimal pointed $\mathcal{V}$-automaton accepting L. It arises from any pointed finite $\mathcal{V}$-automaton (Q, γ, q_0) accepting L by Construction 2.19.*

Proof. Viewed as a da, $A_\mathcal{V}^L$ is a subautomaton of the da ρT_Σ of regular languages. Then the state L accepts L. It is reachable because every state is a $\mathcal{V}$-algebraic combination of those states reachable from L by transitions i.e. L's derivatives. It is simple because different states accept different languages, so it is minimal.

Now let (Q, γ, q_0) be any pointed T_Σ-coalgebra accepting L and (R, δ, q_0) its reachable subautomaton, so every $q' \in R$ arises as a $\mathcal{V}$-algebraic combination of those states reachable from q_0 by transitions. Now $\mathcal{L}_\delta : R \to \rho T_\Sigma$ is an automata morphism, so the languages of states reachable from q_0 are precisely the derivatives of L. Since $\mathcal{L}_\delta$ is an algebra morphism its image is Q_L.   □

3   From Deterministic to Nondeterministic Automata

We now know that each regular language L has *many* canonical deterministic acceptors: one for each locally finite variety $\mathcal{V}$ containing a two-element algebra 2. However this canonical acceptor $A_\mathcal{V}^L$ is generally larger than the minimal dfa in Set because one has to close under the $\mathcal{V}$-algebraic operations on the regular languages. In this section we will show how these larger *deterministic* machines induce smaller *nondeterministic* ones. Let us outline our approach:

1. We restrict attention to finite da's in $\mathcal{V}$, i.e., T_Σ-coalgebras with finite carrier.
2. For each of our varieties $\mathcal{V}$ of interest, we describe an equivalence G of categories between the finite algebras $\mathcal{V}_f$ and another category $\overline{\mathcal{V}}$ where (i) $\overline{\mathcal{V}}$'s objects are "small" representations of their counterparts in $\mathcal{V}_f$, and (ii) $\overline{\mathcal{V}}$'s morphisms are relations, not functions (see Lemmas 3.4, 3.8 and 3.10).
3. From G we derive equivalences $\mathbb{G}$ and $\mathbb{G}_*$ between (pointed) deterministic finite $\mathcal{V}$-automata and (pointed) coalgebras in $\overline{\mathcal{V}}$ which are *nondeterministic* finite automata, see Lemma 3.17.
4. Applying this equivalence to the minimal deterministic $\mathcal{V}$-automaton $A_\mathcal{V}^L$ gives a canonical nondeterministic acceptor for L. This is illustrated in Sect. 3.3.

3.1   The Equivalence Between $\mathcal{V}_f$ and $\overline{\mathcal{V}}$

For each of our varieties $\mathcal{V}$ of interest there is a well-known description of the dual category of $\mathcal{V}_f$: we have Stone duality ($\mathsf{BA}_f \cong \mathsf{Set}_f^{op}$), Priestley duality ($\mathsf{DL}_f \cong \mathsf{Poset}_f^{op}$), where Poset_f is the category of finite posets and monotone functions, and the self-dualities $\mathsf{JSL}_f \cong \mathsf{JSL}_f^{op}$ and $\mathsf{Vect}_f(\mathbb{Z}_2) \cong \mathsf{Vect}_f(\mathbb{Z}_2)^{op}$. We now describe each of these dually equivalent categories as a category $\overline{\mathcal{V}}$ of finite structured sets and relations. The idea is to represent the finite algebras in $\mathcal{V}$ in terms of a minimal set of generators.

Example 3.1. (a) For any $Q \in \mathsf{Set}_\star$ the subset $Q \setminus \{0\}$ generates Q; that means that we can always drop one element.

(b) Any finite boolean algebra $Q \in \mathsf{BA}_f$ is generated by its atoms $\mathsf{At}(Q)$, these being the join-irreducible elements.

(c) Any finite join-semilattice $Q \in \mathsf{JSL}_f$ is generated by its join-irreducibles $J(Q)$.

(d) A finite dimensional vector space $Q \in \mathsf{Vect}_f(\mathbb{Z}_2)$ is generated by any basis $B \subseteq Q$, although there is no canonical choice of a basis.

(e) Any finite distributive lattice $Q \in \mathsf{DL}_f$ is generated by its join-irreducibles $J(Q)$.

In the case of $\mathsf{Set}_{\star f}$, BA_f and $\mathsf{Vect}_f(\mathbb{Z}_2)$ we can replace each algebra by a set of generators and each algebra morphism by a relation between these generators.

Definition 3.2. *Let* $\overline{\mathsf{Set}_\star}$ *be the category* Par_f *of finite sets and partial functions.* $\overline{\mathsf{BA}}$ *is obtained from the category* Rel_f *of finite sets and relations by restricting to relations whose converse is a function. Finally* $\overline{\mathsf{Vect}(\mathbb{Z}_2)}$ *has the same objects and morphisms as* Rel_f *although now the composition of* $R_1 \subseteq X \times Y$ *and* $R_2 \subseteq Y \times Z$ *is defined by*

$$R_2 \bullet R_1 := \{(x, z) : |\{y : (x, y) \in R_1, (y, z) \in R_2\}| \text{ is odd}\}.$$

Notation 3.3. Given a basis GQ of a vector space Q, for each basis vector $z \in GQ$ denote by $\pi_z : Q \to \{0, 1\}$ the projection onto the z-coordinate.

Lemma 3.4. *The following functors* G *are equivalences of categories where* $f : Q \to Q'$ *is any* $\mathcal{V}_f$-*morphism:*

1. $G : \mathsf{Set}_{\star f} \to \mathsf{Par}_f$ *defined by*

$$GQ = Q \setminus \{0\} \quad Gf(z) = \begin{cases} f(z) & if f(z) \neq 0, \\ undefined & otherwise. \end{cases}$$

2. $G : \mathsf{BA}_f \to \overline{\mathsf{BA}}$ *where* $GQ = \mathsf{At}(Q)$ *is the set of atoms and* $Gf = \{(z, z') \in \mathsf{At}(Q) \times \mathsf{At}(Q') : z' \leq_{Q'} f(z)\}$.

3. $G : \mathsf{Vect}_f(\mathbb{Z}_2) \to \overline{\mathsf{Vect}(\mathbb{Z}_2)}$ *where* GQ *chooses a basis and* $Gf = \{(z, z') \in GQ \times GQ' : \pi_{z'} \circ f(z) = 1\}$.

Finite join-semilattices are represented using closure spaces:

Definition 3.5. *For any set* X *a* closure operator *(shortly, a* closure*) on* X *is a function* $\mathbf{cl}_X : \mathcal{P}X \to \mathcal{P}X$ *such that for all* $S, S' \subseteq X$:

$$\frac{S \subseteq S'}{\mathbf{cl}_X(S) \subseteq \mathbf{cl}_X(S')}, \quad \mathbf{cl}_X(S) \supseteq S, \quad \mathbf{cl}_X \circ \mathbf{cl}_X = \mathbf{cl}_X.$$

A closure space $X = (X, \mathbf{cl}_X)$ *is a set with a closure defined on it. It is* finite *if* X *is finite,* strict *if* $\mathbf{cl}_X(\emptyset) = \emptyset$, separable *if* $x \neq x'$ *implies* $\mathbf{cl}_X(x) \neq \mathbf{cl}_X(x')$, *and* topological *if* $\mathbf{cl}_X(A \cup B) = \mathbf{cl}_X(A) \cup \mathbf{cl}_X(B)$ *for all* $A, B \subseteq X$. *A subset* $S \subseteq X$ *is* closed *if* $\mathbf{cl}_X(S) = S$ *and* open *if its complement is closed.*

Finite posets are well-known to be equivalent to finite T_0 topological spaces, which amount to finite separable topological closures. For finite join-semilattices we instead use *finite strict closures* i.e. we do not require separability or preservation of unions.

Example 3.6. Each finite join-semilattice Q has an associated finite strict closure space $GQ = (J(Q), \mathbf{cl}_{J(Q)})$ where $J(Q) \subseteq Q$ is the set of join-irreducibles and

$$\mathbf{cl}_{J(Q)}(S) = \{j \in J(Q) : j \leq \sum_{s \in S} s\} \quad \text{for any } S \subseteq J(Q).$$

For example the closure space associated to the free join-semilattice $\mathcal{P}n$ is $(n, \mathsf{id}_{\mathcal{P}n})$, identifying $J(\mathcal{P}n)$ with n.

Definition 3.7. *The category* $\overline{\mathsf{JSL}}$ *has as objects all finite strict closure spaces as morphisms all continuous relations. Here a relation* $R \subseteq X \times Y$ *between two finite strict closure spaces* X *and* Y *is called* continuous *if, for all* $x \in X$ *and* $S \subseteq X$,

1. $R[x] \subseteq Y$ *is closed, and*
2. *if* $x \in \mathbf{cl}_X(S)$ *then* $R[x] \subseteq \mathbf{cl}_Y(R[S])$.

The composition of $R_1 \subseteq X \times Y$ *and* $R_2 \subseteq Y \times Z$ *is defined by*

$$R_2 \bullet R_1 := \{(x, z) \in X \times Z : z \in \mathbf{cl}_Z(R_2 \circ R_1[x])\},$$

and the identity morphism on X *is* $\mathsf{id}_X = \{(x, x') \in X \times X : x' \in \mathbf{cl}_X(\{x\})\}$.

The following equivalence was derived from a similar one due to Moshier [13].

Lemma 3.8. *The functor* $G : \mathsf{JSL}_f \to \overline{\mathsf{JSL}}$, *defined on objects* Q *as in Example 3.6 and for morphisms* $f : Q \to Q'$ *by*

$$Gf = \{(j, j') \in J(Q) \times J(Q') : j' \leq_{Q'} f(j)\},$$

is an equivalence of categories.

Proof. (Sketch) We describe the opposite equivalence $H : \overline{\mathsf{JSL}} \to \mathsf{JSL}_f$ and also the unit and counit. Given $X = (X, \mathbf{cl}_X)$ then $HX = \{S \subseteq X : \mathbf{cl}_X(S) = S\} \subseteq \mathcal{P}X$ is the join-semilattice of closed subsets where $0_{HX} = \emptyset$ and $S +_{HX} S' = \mathbf{cl}_X(S \cup S')$. Given a continuous relation $R \subseteq X \times Y$ then $HR = \lambda S.\mathbf{cl}_Y(R[S]) : HX \to HY$ is the corresponding algebra morphism. The unit $\eta : \mathsf{Id} \Rightarrow HG$ is defined

$$\eta_Q = \lambda q \in Q.\{j \in J(Q) : j \leq_Q q\},$$

and for $X = (X, \mathbf{cl}_X)$ the counit $\epsilon : GH \Rightarrow \mathsf{Id}$ is defined:

$$\epsilon_X = \{(K, x) \in J(HX) \times X : K \in J(HX), x \in K\}.$$

It is well-typed because $J(HX) \subseteq HX \subseteq \mathcal{P}X$.   □

Definition 3.9. $\overline{\mathsf{DL}}$ *has finite posets as objects and as morphisms those relations* $R \subseteq P \times Q$ *such that:*

1. *Each $R[p] \subseteq Q$ is downclosed,*
2. *If $p \leq_P p'$ then $R[p] \subseteq R[p']$,*
3. *R preserves all intersections of downclosed subsets.*

id_P *is the relation $\{(p, p') \in P \times P : p' \leq_P p\}$ and composition is relational composition.*

Lemma 3.10. *The functor $G : \mathsf{DL}_f \to \overline{\mathsf{DL}}$ where $GQ = J(Q)$ (considered as a subposet of Q) and for morphisms $f : Q \to Q'$*

$$Gf = \{(z, z') \in J(Q) \times J(Q') : z' \leq_{Q'} f(z)\}$$

is an equivalence of categories.

Proof. G is restriction of the equivalence $\mathsf{JSL}_f \cong \overline{\mathsf{JSL}}$ described above. The closure spaces associated to distributive lattices are precisely the separable topological ones, so we can replace them by finite posets. This gives the first two conditions on morphisms, where closed means downwards closed. However semilattice morphisms between distributive lattices need not preserve meets. This is captured by the third condition.   □

3.2 From Determinism to Nondeterminism

We first restrict the endofunctor T_Σ of Definition 2.6 to finite algebras:

$$T_\Sigma = 2 \times \mathsf{Id}^\Sigma : \mathcal{V}_f \to \mathcal{V}_f$$

Then for each of our five equivalences $G : \mathcal{V}_f \to \overline{\mathcal{V}}$ described in the previous section we have a corresponding functor

$$\overline{T}_\Sigma = \mathbb{1} \times \mathsf{Id}^\Sigma : \overline{\mathcal{V}} \to \overline{\mathcal{V}}$$

where $\mathbb{1} = G2 \in \overline{\mathcal{V}}$. In each case $\mathbb{1}$ has carrier $\{1\}$.

$\mathcal{V}$	$\mathbb{1}$
Set$_\star$	$2 \setminus \{0\} = \{1\}$
BA	$At(2) = \{1\}$ the unique atom
Vect($\mathbb{Z}_2$)	$\{1\}$ unique basis of $2 = \mathbb{Z}_2$
JSL	$(J(2), \mathbf{cl})$ where $J(2) = \{1\}$, $\mathbf{cl} = \mathrm{id}_{\mathcal{P}\{1\}}$
DL	$J(2) = \{1\}$ a discrete poset

Lemma 3.11. *There is an equivalence* $\mathbb{G} : \mathsf{Coalg}(T_\Sigma) \to \mathsf{Coalg}(\overline{T}_\Sigma)$ *defined by*

$$\mathbb{G}(Q, \gamma) = (GQ, \gamma') \text{ on objects} \quad \text{and} \quad \mathbb{G}f = Gf \text{ on morphisms,}$$

where $\gamma' : GQ \to \mathbb{1} \times (GQ)^\Sigma$ *is the* $\overline{\mathcal{V}}$-*morphism uniquely determined by the morphisms* $G\gamma_\epsilon : GQ \to \mathbb{1}$ *and* $G\gamma_a : GQ \to GQ$ *for each* $a \in \Sigma$.

Given a $\overline{T}_\Sigma$-coalgebra $\delta : Z \to \overline{T}_\Sigma Z = \mathbb{1} \times Z^\Sigma$ we write its component maps as $\delta_\epsilon : Z \to \mathbb{1}$ and $\delta_a : Z \to Z$ for $a \in \Sigma$. Notice that these are *relations* rather than functions, so $\overline{T}_\Sigma$-coalgebras are *nondeterministic* automata.

Example 3.12. (a) When $\mathcal{V} = \mathsf{Set}_\star$ a $\overline{T}_\Sigma$-coalgebra $\delta : X \to \overline{T}_\Sigma X$ consists of:
1. A finite set X.
2. A partial function $\delta_\epsilon : X \to \{1\}$ whose domain defines the final states.
3. A partial function $\delta_a : X \to X$ for each $a \in \Sigma$, defining the transitions.
Hence $\overline{T}_\Sigma$-coalgebras are *partial dfas*. The equivalence $\mathbb{G}$ assigns to each deterministic $\mathsf{Set}_\star$-automaton (Q, γ) the partial dfa $(Q \setminus \{0_Q\}, \delta)$ whose final states are the given ones and $q \xrightarrow{a} q'$ iff $\gamma_a(q) = q' \neq 0_Q$.
(b) When $\mathcal{V} = \mathsf{BA}$ a $\overline{T}_\Sigma$-coalgebra $\delta : X \to \overline{T}_\Sigma X$ consists of:
1. A finite set X.
2. A converse-functional relation $\delta_\epsilon \subseteq X \times \{1\}$ whose domain defines a single final state.
3. Converse-functional relations $\delta_a \subseteq X \times X$ for $a \in \Sigma$.
Hence $\overline{T}_\Sigma$-coalgebras are *reverse-deterministic nfas*, i.e., reversing all transitions yields a dfa. The equivalence $\mathbb{G}$ assigns to each deterministic BA-automaton (Q, γ) an nfa $(\mathsf{At}(Q), \delta)$ whose states are $Q's$ atoms. Moreover, its single final state is the unique atom generating the ultrafilter $\gamma_\epsilon^{-1}(\{1\})$ and $z \xrightarrow{a} z'$ iff $z' \leq_Q \gamma_a(z)$.
(c) If $\mathcal{V} = \mathsf{Vect}(\mathbb{Z}_2)$ then a $\overline{T}_\Sigma$-coalgebra $\delta : X \to \overline{T}_\Sigma X$ consists of:
1. A finite set X.
2. An arbitrary relation $\delta_\epsilon \subseteq X \times \{1\}$, amounting to an arbitrary set of final states by taking the domain.
3. Arbitrary relations $\delta_a \subseteq X \times X$ for each $a \in \Sigma$.
Hence $\overline{T}_\Sigma$-coalgebras are classical nfas. The equivalence $\mathbb{G}$ assigns to a deterministic $\mathsf{Vect}(\mathbb{Z}_2)$-automaton (Q, γ) the nfa (Z, δ) for some chosen basis $Z \subseteq Q$. The final states are $Z \cap \gamma_\epsilon^{-1}(\{1\})$ and $z \xrightarrow{a} z'$ iff $\pi_{z'} \circ \gamma_a(z) = 1$, cf. Notation 3.3.
(d) If $\mathcal{V} = \mathsf{JSL}$ then a $\overline{T}_\Sigma$-coalgebra $\delta : Z \to \overline{T}_\Sigma Z$ consists of:

1. A finite strict closure space $Z = (Z, \mathbf{cl}_Z)$.
2. A continuous relation $\delta_\epsilon \subseteq Z \times \{1\}$, equivalently δ_ϵ's domain $F \subseteq Z$ is an open set of final states.
3. Continuous relations $\delta_a \subseteq Z \times Z$.

We call $\overline{T}_\Sigma$-coalgebras *nondeterministic closure automata*. The equivalence $\mathbb{G}$ assigns to each deterministic JSL-automaton (Q, γ) the nondeterministic closure automaton $((J(Q), \mathbf{cl}_Q), \delta)$ whose states are Q's join-irreducibles. The open set of final states is $J(Q) \cap \gamma_\epsilon^{-1}(\{1\})$ and $z \xrightarrow{a} z'$ iff $z' \leq_Q \gamma_a(z)$. Note that every nfa can be turned into a nondeterministic closure automaton by endowing the states with the identity closure, so classical nfas form a proper subclass.

(e) If $\mathcal{V} = \mathsf{DL}$ then a $\overline{T}_\Sigma$-coalgebra $\delta : P \to \overline{T}_\Sigma P$ consists of:
1. A finite poset P.
2. A non-empty relation $\delta_\epsilon \subseteq P \times \{1\}$ whose domain is a filter (i.e., a down-directed upset), these being the final states.
3. Transition relations $\delta_a \subseteq P \times P$ such that:
    (i) $\delta_a[p]$ is downclosed for each $p \in P$.
    (ii) $p \leq_P q$ implies $\delta_a[p] \subseteq \delta_a[q]$.
    (iii) $\delta_a[\bigcap_I A_i] = \bigcap_I \delta_a[A_i]$ for downclosed A_i.

Note that reverse-deterministic nfas are the special case where P is discrete. An important non-discrete example is the *universal automaton* [14], we recall it after Corollary 3.21.

The equivalence $\mathbb{G}$ assigns to each deterministic DL-automaton (Q, γ) the $\overline{T}_\Sigma$-coalgebra $(J(Q), \delta)$ where $J(Q)$ is a subposet of Q. The final states form the upwards closed set $J(Q) \cap \gamma_\epsilon^{-1}(1)$ and $z \xrightarrow{a} z'$ iff $z' \leq_Q \gamma_a(z)$.

Remark 3.13. A morphism $f : (Z, \delta) \to (Z', \delta')$ of $\overline{T}_\Sigma$-coalgebras is, by definition, a $\mathcal{V}$-morphism $f : Z \to Z'$ satisfying $\overline{T}_\Sigma f \circ \delta = \delta' \circ f$, or equivalently:

$$\delta_\epsilon = \delta'_\epsilon \circ f, \qquad \delta'_a \circ f = f \circ \delta_a \quad (a \in \Sigma).$$

For $\mathcal{V} = \mathsf{Set}_*$, BA and DL, these morphisms are those relations (from $\overline{\mathcal{V}}$) which (i) reflect and preserve transitions and (ii) have $z \in Z$ final iff some $z' \in f[z]$ is final. The cases $\mathcal{V} = \mathsf{JSL}, \mathsf{Vect}(\mathbb{Z}_2)$ are different because composition in $\overline{\mathcal{V}}$ is not relational.

3.3   Canonical Nondeterministic Automata

So far we have seen equivalences between deterministic and nondeterministic automata without initial states. Next, for each of our five running examples $\mathcal{V} = \mathsf{Set}_*$, BA, $\mathsf{Vect}(\mathbb{Z}_2)$, JSL, DL we will extend $\mathbb{G} : \mathsf{Coalg}(T_\Sigma) \to \mathsf{Coalg}(\overline{T}_\Sigma)$ to an equivalence of pointed coalgebras.

Definition 3.14. $\mathsf{Coalg}_*(T_\Sigma)$ *is the category whose objects are the pointed T_Σ-coalgebras and whose morphisms* $f : (Q, \gamma, q_0) \to (Q', \gamma', q'_0)$ *are those T_Σ-coalgebra homomorphisms* $f : (Q, \gamma) \to (Q', \gamma')$ *preserving initial states, i.e.,* $f \circ q_0 = q'_0$.

Using the equivalence $G : \mathcal{V}_f \to \overline{\mathcal{V}}$, a pointed $\overline{T}_\Sigma$-coalgebra is a $\overline{T}_\Sigma$-coalgebra (Z, δ) equipped with a $\overline{\mathcal{V}}$-morphism $i : GV \to Z$. And pointed $\overline{T}_\Sigma$-coalgebra homomorphisms are those $\overline{T}_\Sigma$-coalgebra homomorphisms f from (Z, δ) to (Z', δ') such that $f \circ i = i'$. Just as a morphism $q_0 : V \to Q$ corresponds to an initial state $q_0(g)$, it turns out that a morphism $i : GV \to Z$ corresponds to a *set* of initial states $I = i[g] \subseteq Z$, as one would expect for nfas.

Example 3.15. For each $\mathcal{V}$ we describe the possible sets of initial states $I \subseteq Z$ for a $\overline{T}_\Sigma$-coalgebra (Z, δ).

(a) If $\mathcal{V} = \mathsf{Set}_\star$ then $V = \{0, g\}$ and $GV = \{g\}$. Partial functions $i : \{g\} \to Z$ are determined by their codomain $I = i[g]$. Then I is either empty or any singleton subset.

(b) If $\mathcal{V} = \mathsf{BA}$ then $V = \{\bot, g, \neg g, \top\}$ and $GV = \{g, \neg g\}$. Given $i \subseteq \{g, \neg g\} \times Z$ then $i[g], i[\neg g]$ partition Z so i is determined by $I = i[g]$. Then I is any subset of Z.

(c) If $\mathcal{V} = \mathsf{Vect}(\mathbb{Z}_2)$ then $V = \{0, g\}$ and $GV = \{g\}$, so the arbitrary relation $i \subseteq \{g\} \times Z$ is determined by its codomain $I = i[g]$. Then I is any subset of Z.

(d) If $\mathcal{V} = \mathsf{JSL}$ then $V = \{0, g\}$ and $GV = \{g\}$ with closure $\mathrm{id}_{\mathcal{P}\{g\}}$. The relation $i \subseteq \{g\} \times Z$ is determined by $I = i[g]$. By continuity $I \subseteq Z$ is any closed subset.

(e) If $\mathcal{V} = \mathsf{DL}$ then $V = \{\bot, g, \top\}$ is a 3-chain and $GV = \{g, \top\}$ a 2-chain. Given $i \subseteq \{g, \top\} \times Z$ then $i[g] \subseteq i[\top]$ and $i[\{g, \top\}] = Z$ implies $i[\top] = Z$, so i is determined by $I = i[g]$. Then I is any downclosed subset of Z.

By reinterpreting point preservation relative to I we can finally define the category of pointed $\overline{T}_\Sigma$-coalgebras.

Definition 3.16. *For each of our five running examples,* $\mathsf{Coalg}_\star(\overline{T}_\Sigma)$*'s objects are triples* (Z, δ, I) *where* (Z, δ) *is a* $\overline{T}_\Sigma$*-coalgebra and* $I \subseteq Z$ *is restricted as in Example 3.15. The pointed* $\overline{T}_\Sigma$*-coalgebra homomorphisms* $f : (Z, \delta, I) \to (Z', \delta', I')$ *are* $\overline{T}_\Sigma$*-coalgebra homomorphisms* $f : (Z, \delta) \to (Z', \delta')$ *such that:*

1. *If* $\mathcal{V} = \mathsf{Set}_\star$, BA *or* DL *then* $I' = f[I]$.
2. *If* $\mathcal{V} = \mathsf{JSL}$ *then* I' *is the closure of* $f[I]$.
3. *If* $\mathcal{V} = \mathsf{Vect}(\mathbb{Z}_2)$ *then* $I' = \{z' \in Z' : |I \cap \check{f}[z']| \text{ is odd}\}$.

where $\check{f} \subseteq Z' \times Z$ *is the converse relation.*

Lemma 3.17. *There is an equivalence of pointed coalgebras* $\mathbb{G}_\star : \mathsf{Coalg}_\star(T_\Sigma) \to \mathsf{Coalg}_\star(\overline{T}_\Sigma)$ *defined by*

$$\mathbb{G}_\star(Q, \gamma, q_0) = (\mathbb{G}(Q, \gamma), I) \qquad \mathbb{G}_\star f = \mathbb{G} f$$

where $I = Gq_0[g] \subseteq GQ$.

Let us spell out the equivalence $\mathbb{G}_\star$ for each of our varieties $\mathcal{V}$. For the rest of this section fix a T_Σ-coalgebra $A = (Q, \gamma, q_0)$ and a regular language $L \subseteq \Sigma^*$.

We give an explicit description of the nfa G_*A and, in particular, of the canonical nfa for L obtained by applying $\mathbb{G}_*$ to $A_{\mathcal{V}}^L$ from Construction 2.15.

(a) **The Minimal Partial Dfa.** If $\mathcal{V} = \mathsf{Set}_*$ then $\mathbb{G}_*A$ is the partial dfa $(Q \setminus \{0_Q\}, \delta, I)$ that arises from A by deleting the state 0_Q along with all in- and outgoing transitions. Hence the initial states are $I = \{q_0\}$ if $q_0 \neq 0_Q$ and $I = \emptyset$ if $q_0 = 0_Q$. Clearly $\mathbb{G}_*A$ (viewed as an nfa) accepts A's language.

In particular, $\mathbb{G}_*(A_{\mathsf{Set}_*}^L)$ is the *minimal partial dfa* of L. It has states

$$\mathcal{Q}_L = \{w^{-1}L : w \in \Sigma^*\} \setminus \{\emptyset\},$$

transitions $K \xrightarrow{a} a^{-1}K$ whenever $a^{-1}K \neq \emptyset$, and a state is final iff it contains ϵ. The initial states are $\{L\}$ if $L \neq \emptyset$ and $\emptyset$ otherwise. Hence the minimal partial dfa is the trim part of L's minimal dfa (obtained by deleting its sink state, if it exists).

(b) **The Átomaton.** If $\mathcal{V} = \mathsf{BA}$ then $\mathbb{G}_*A$ is the nfa $(\mathsf{At}(Q), \delta, I)$ with initial states $I = \{q \in \mathsf{At}(Q) : q \leq_Q q_0\}$. It accepts A's language. In particular, $\mathbb{G}_*(A_{\mathsf{BA}}^L)$ is called the *átomaton* of L, see [8]. Its states

$$\mathcal{Q}_L = \mathsf{At}(\langle \{w^{-1}L : w \in \Sigma^*\} \rangle_{\nu T_\Sigma})$$

are the atoms of the finite boolean subalgebra of $\mathcal{P}\Sigma^*$ generated by L's derivatives. An atom K is an initial state if $K \subseteq L$, the final states are the atoms containing ϵ, and one has transitions $K \xrightarrow{a} K'$ whenever $K' \subseteq a^{-1}K$. Explicitly constructing $\mathcal{Q}_L$ can be difficult. Fortunately, a simpler method is known [8]:

1. Construct the minimal dfa for L's reversed language.
2. Construct its reversed nfa i.e. flip initial/final states and reverse all transitions.

The átomaton is isomorphic to the resulting nfa as we now explain coalgebraically. Let $T_\Sigma' = 2 \times \mathsf{Id}^\Sigma : \mathsf{Set}_f \to \mathsf{Set}_f$. Then the usual reversal of finite pointed deterministic automata defines a dual equivalence:

$$H : (\mathsf{Coalg}_*(T_\Sigma'))^{op} \to \mathsf{Coalg}_*(\overline{T}_\Sigma)$$
$$Hf^{op} = \{(z', z) : z \in f^{-1}(\{z'\})\} \subseteq Z' \times Z,$$

Since reachability (no proper subobjects) and simplicity (no proper quotients) are *dual* concepts (see Definition 2.13), a T_Σ'-coalgebra is minimal iff its image under H is minimal, implying the above description.

Example 3.18. 1. The átomaton for $L = (a+b)^*b(a+b)^n$ in Example 1.1 arises by constructing the minimal dfa for the reversed language $\mathsf{rev}(L)$ and taking the reverse nfa. Its atoms are $\{(a+b)^*a(a+b)^n, L\} \cup \{(a+b)^j : 0 \leq j \leq n\}$.
   2. The átomaton can have exponentially many more states than the minimal dfa, e.g. for $L = (a+b)^nb(a+b)^*$ it has $\geq 2^n$ states.

(c) **The Minimal Xor Automaton.** If $\mathcal{V} = \mathsf{Vect}(\mathbb{Z}_2)$ then $\mathbb{G}_*A$ is the nfa (Z, δ, I) where $Z \subseteq Q$ is a basis and $I = \{z \in Z : \pi_z(q_0) = 1\}$, see Notation 3.3.

It accepts A's language by $\mathbb{Z}_2$-*weighted* nondeterministic acceptance: a word $w \in \Sigma^*$ is accepted iff its number of accepting paths is odd (this is different than the usual acceptance condition of standard nondeterministic automata).

The nfa $\mathbb{G}_*(A^L_{\mathsf{Vect}(\mathbb{Z}_2)})$ is called the *minimal xor automaton* of L, see [17]. Note that its construction depends on the choice of a basis, so the minimal xor automaton is only determined up to isomorphism in the category of pointed $\overline{T}_\Sigma$-coalgebras. We provide a new way to construct it:

1. Construct L's átomaton (Z, R_a, F, I) and determine the collection $C \subseteq \mathcal{P}Z$ of all subsets of Z which are reachable from I.
2. Find any minimal $\mathcal{Q} \subseteq \mathcal{P}Z$ whose closure under set-theoretic symmetric difference equals C's closure.
3. Build the nfa $(\mathcal{Q}, R'_a, \mathcal{Q} \cap F, I)$ where $R'_a(y, y')$ iff $\pi_{y'}(R_a[y]) = 1$ and $I = \{y \in \mathcal{Q} : \pi_y(I) = 1\}$.

Briefly, closure under boolean operations implies closure under symmetric difference. Then $A^L_{\mathsf{Vect}(\mathbb{Z}_2)} \subseteq A^L_{\mathsf{BA}}$ as da's, leading to the above algorithm. Since the basis $\mathcal{Q}$ has $|\mathcal{Q}| \leq |C| = |\{w^{-1}L : w \in \Sigma^*\}|$ it follows that *the minimal xor automaton is never larger than the minimal dfa* of L, see [17].

Example 3.19. Take the átomaton of Example 1.1, with states $Z = \{x\} \cup \{z_i : 0 \leq i \leq n+1\}$ and reachable subsets $C = \{S \subseteq Z : x \notin S, z_0 \in S\}$. One can verify that (i) the closure of $\mathcal{Q} = \{\{z_i\} : 0 \leq i \leq n+1\}$ under symmetric difference is the closure of C and (ii) $\mathcal{Q}$ is minimal. The induced nfa is the minimal xor automaton of Example 1.1. Alternatively $\mathcal{Q} = \{\{z_0, z_i\} : 0 \leq i \leq n+1\} \subseteq C$ yields a different nfa.

(d) **The Jiromaton.** If $\mathcal{V} = \mathsf{JSL}$ then $\mathbb{G}_*A$ is the nondeterministic closure automaton $(J(Q), \delta, I)$ with initial states $I = \{z \in J(Q) : z \leq_Q q_0\}$ where $J(Q)$ is the closure space of Example 3.6. The underlying nfa (forgetting the closure) accepts A's language. In particular, $\mathbb{G}_*(A^L_{\mathsf{JSL}})$'s underlying nfa is called the *jiromaton* of L, see [10]. Its states

$$\mathcal{Q}_L = J(\langle\{w^{-1}L : w \in \Sigma^*\}\rangle_{\nu T_\Sigma})$$

are the join-irreducibles of the finite join-subsemilattice of $\mathcal{P}\Sigma^*$ generated by L's derivatives. Since the latter form the minimal generating set, $\mathcal{Q}_L$ consists of those L-derivatives not arising as unions of other derivatives – the *prime* derivatives. Therefore, the jiromaton has no more states than the minimal dfa. Its structure is analogous to the átomaton: $K \in \mathcal{Q}_L$ is initial iff $K \subseteq L$, final iff $\epsilon \in K$ and $K \xrightarrow{a} K'$ iff $K' \subseteq a^{-1}K$.

An algorithm to construct the jiromaton from any nfa accepting L is given in [10].

Example 3.20. In the jiromaton of Example 1.1, the state z_0 accepts L and state z_i accepts $L + (a+b)^{i-1}$ for each $i > 0$. These are the prime derivatives of L. The closure is defined $\mathbf{cl}_Z(\emptyset) = \emptyset$, $\mathbf{cl}_Z(S) = \{z_0\} \cup \{S\}$ for $S \neq \emptyset$. It is topological: the closed sets are the downsets of the poset where $z_0 \leq_Z z_i$ for all $0 \leq i \leq n+1$.

(e) **The Distromaton.** If $\mathcal{V} = \mathsf{DL}$ then $\mathbb{G}_* A = (J(Q), \delta, I)$ with initial states $I = \{z \in J(Q) : z \leq_Q q_0\}$. Forgetting $J(Q)$'s poset structure, the underlying nfa accepts A's language. We call $\mathbb{G}_*(A_{\mathsf{DL}}^L)$ the *distromaton* of L. Its states

$$\mathcal{Q}_L = J(\langle\{w^{-1}L : w \in \Sigma^*\}\rangle_{\nu T_\Sigma})$$

are the join-irreducibles of the sublattice of $\mathcal{P}\Sigma^*$ generated by L's derivatives. One can close under intersections and then unions (which cannot add or remove join-irreducibles) so $\mathcal{Q}_L$ consists of finite intersections $\bigcap_i w_i^{-1}L$ not arising as finite unions of other such intersections. The structure is again analogous to the átomaton and the jiromaton: $K \in \mathcal{Q}_L$ is initial iff $K \subseteq L$, final iff $\epsilon \in K$ and $K \xrightarrow{a} K'$ iff $K' \subseteq a^{-1}K$. There is another way to construct the distromaton, analogous to the construction of the átomaton:

1. Take the minimal pointed dfa $(Z, \xrightarrow{a}, z_0, F)$ for the reversed language $\mathsf{rev}(L)$ where Z is ordered by language-inclusion.
2. Build the pointed $\overline{T}_\Sigma$-coalgebra (Z^{op}, δ, F) with final states $\downarrow_Z z_0$ and $z' \in \delta_a[z]$ iff $z' \xrightarrow{a} y \geq_Z z$.

The initial states F are downclosed in Z^{op} and the final states are upclosed in Z^{op}, as required. The proof that this is isomorphic to the distromaton is analogous to our earlier argument regarding the átomaton. Briefly, let $T'_\Sigma = 2 \times \mathsf{Id}^\Sigma : \mathsf{Poset}_f \to \mathsf{Poset}_f$ where 2 is the two-chain. Then there is a dual equivalence

$$H : (\mathsf{Coalg}_*(T'_\Sigma))^{op} \to \mathsf{Coalg}_*(\overline{T}_\Sigma),$$

which 'reverses' finite pointed deterministic automata equipped with a compatible ordering. The minimal T'_Σ-coalgebra for L is the usual minimal dfa, now equipped with the language-inclusion ordering. Its image under H is again minimal, yielding the above description of the distromaton.

Corollary 3.21. *L's átomaton and distromaton have the same number of states, namely, the number of states of the minimal dfa for the reversed language $\mathsf{rev}(L)$.*

Example 3.22. The distromaton in Example 1.1 has order $z_0 \leq_Z z_i$ and $z_i \leq_Z \top$ for all $0 \leq i \leq n+1$. We have the state $\top$ because Σ^* is not the union of non-empty intersections of L's derivatives, see Example 3.20. It arises from the jiromaton by adding a final sink state, see Corollary 4.6.

We finally mention the well-studied *universal automaton* for L [14]. It is the nfa with states

$$\mathcal{Q} = \{\bigcap_{w \in I} w^{-1}L : I \subseteq_\omega \Sigma^*\}$$

ordered by inclusion, where K is final iff $\epsilon \in K$ and $K \xrightarrow{a} K'$ iff $K' \subseteq a^{-1}K$. The distromaton is never larger and often much smaller because one restricts to the join-irreducible intersections. However the universal automaton has its own advantages: in a sense every state-minimal nfa lies inside it.

4   State Minimality and Universal Properties

This final section is split into three parts.

1. We prove L's jiromaton is minimal amongst all nondeterministic acceptors of L relative to a suitable measure (Sect. 4.1).
2. We give a sufficient condition on L such that the jiromaton is state-minimal and the distromaton and átomaton have at most one more state (Sect. 4.2).
3. We characterize each of our canonical nfas amongst subclasses of nondeterministic acceptors (Sect. 4.3).

4.1   The Jiromaton is Minimal

There is a measure on finite nondeterministic automata such that L's jiromaton is smaller than any other nfa accepting L. For any nfa $N = (Q, R_a, F)$ and $I \subseteq Q$ let $\mathcal{L}_N(I) \subseteq \Sigma^*$ be the accepted language. Define the following measures:

$$|N| = |Q|, \quad \mathrm{acc}(N) = |\{\mathcal{L}_N(I) : I \subseteq Q\}|, \quad \mathrm{tr}(N) = \sum_{a \in \Sigma} |R_a|.$$

These are the number of states, the number of distinct languages accepted and the number of transitions. Let J_L be L's jiromaton without initial states. Recall that isomorphisms of nfas are bijective bisimulations (see Definition 2.1).

Theorem 4.1. *The jiromaton J_L is (up to isomorphism) the unique nfa accepting L such that for every nfa N accepting L:*

(1) $\mathrm{acc}(J_L) \leq \mathrm{acc}(N)$,
(2) If additionally $\mathrm{acc}(J_L) = \mathrm{acc}(N)$ then either:
   (a) $|J_L| < |N|$ or
   (b) $|J_L| = |N|$ and $\mathrm{tr}(N) \leq \mathrm{tr}(J_L)$.

Proof. Since J_L's individual states accept derivatives of L, it follows that J_L accepts precisely the unions of derivatives of L. Any nfa N accepting L accepts these languages, so $\mathrm{acc}(J_L) \leq \mathrm{acc}(N)$. Suppose $\mathrm{acc}(J_L) = \mathrm{acc}(N)$, so N accepts precisely the unions of L's derivatives. Then each prime derivative has a distinct state in N accepting it, as it cannot arise as the union of other derivatives, so $|J_L| \leq |N|$. Lastly if $\mathrm{acc}(J_L) = \mathrm{acc}(N)$ and $|J_L| = |N|$ then there is language preserving bijection between N's states and the set of prime derivatives P_L, so assume N's carrier is P_L. Given $K \xrightarrow{a} K'$ in N we must have $K' \subseteq a^{-1}K$, so there is a corresponding transition in J_L. Hence $\mathrm{tr}(N) \leq \mathrm{tr}(J_L)$ and (2) holds. Moreover, in case $\mathrm{tr}(N) = \mathrm{tr}(J_L)$ the previous argument shows that N and J_L are isomorphic. Thus the conditions (1) and (2) determine J_L up to isomorphism. □

4.2   Conditions for Canonical State-Minimality

In the following let d_L and n_L be the minimal number of states of a dfa (respectively nfa) accepting the regular language L. For any state-minimal nfa $N = (n_L, R_a, F)$ accepting L via $I \subseteq n_L$, one can construct a simple pointed T_Σ-coalgebra $(\mathcal{Q}, \gamma', L)$ whose equivalent nondeterministic closure automaton is another state-minimal acceptor of L. First view N as the T_Σ-coalgebra $(\mathcal{P}n_L, \gamma)$ via the subset construction. Factorizing the unique homomorphism $\mathcal{L}_\gamma$ we obtain $(\mathcal{Q}, \gamma')$ where $\mathcal{Q}$ is the semilattice of languages accepted by N. Then $(\mathcal{Q}, \gamma')$ is equivalent to a nondeterministic closure automaton accepting L. Since $\mathcal{P}n_L \twoheadrightarrow \mathcal{Q}$ implies $n_L = |J(\mathcal{P}n_L)| \geq |J(\mathcal{Q})|$, by forgetting the closure we obtain a state-minimal nfa accepting L.

Hence instead of working with state-minimal nfas we may work with simple T_Σ-coalgebras which are *supercoalgebras* of A_{JSL}^L. This follows because A_{JSL}^L's carrier is the semilattice S_L of unions of L's derivatives, which $\mathcal{Q}$ necessarily contains. We now provide a condition ensuring that $|J(S_L)|$ is the minimal size of an nfa accepting L and hence L's jiromaton is *state-minimal*.

Definition 4.2. *A regular language L is* intersection-closed *if every binary intersection of L's derivatives is a union of L's derivatives.*

Example 4.3. 1. $L = (a + b)^* b(a + b)^n$ where $n \in \omega$ is intersection-closed.
2. $\emptyset$, Σ^* and $\{w\}$ for $w \in \Sigma^*$ are intersection-closed.
3. Fix $n \in \omega$, $t \in \mathbb{R}$ and $k_i \in \mathbb{R}$ ($1 \leq i \leq n$). Then the language $L = \{w \in 2^n : \sum_i k_i w_i \geq t\}$ (modeling the behaviour of an artificial neuron) is intersection-closed.
4. Every linear subspace $L \subseteq \mathbb{Z}_2^n$ (viewed as a language over the alphabet $\{0, 1\}$) is intersection-closed.

Theorem 4.4. *If L is intersection-closed then its jiromaton is state-minimal.*

Proof. By assumption the carrier S_L of A_{JSL}^L is closed under both unions *and* non-empty intersections, so $D = S_L \cup \{\Sigma^*\}$ is a distributive lattice of languages. Let N be any state-minimal nfa accepting L via initial states I, and $S \subseteq \mathcal{P}\Sigma^*$ be the semilattice of languages accepted by N (by varying I). The nfa N must at least accept L's derivatives. Since S is closed under unions we have $S_L \subseteq S$. By the surjective morphism $\mathcal{P}n_L \twoheadrightarrow S$ it follows that $|N| \geq |J(S)|$, so it suffices to prove that $|J(S)| \geq |J(S_L)|$. Let $S_* = S \cup \{\Sigma^*\}$ be the semilattice obtained by adding a top element if necessary. We have a JSL_f-morphism $\iota : D \hookrightarrow S_*$. The meets in D are also meets in S_* so the same function defines a JSL_f-morphism $\iota : D^{op} \hookrightarrow S_*^{op}$. By the self-duality of JSL_f we obtain a surjective morphism $\iota' : S_* \twoheadrightarrow D$, hence $|J(S_*)| \geq |J(D)|$. If $D = S_L$ then $S_* = S$, so $|J(S)| \geq |J(S_L)|$ and we are done. Otherwise $\Sigma^* \notin S_L$ and we now prove $\Sigma^* \notin S$. By state minimality N is reachable, so each state q accepts a subset of some L-derivative. Then if $\Sigma^* \in S$ we deduce Σ^* is the union of L's derivatives, so $\Sigma^* \in S_L$ – a contradiction. Consequently $|J(D)| = 1 + |J(S_L)|$ and $|J(S_*)| = 1 + |J(S)|$ hence $|J(S)| \geq |J(S_L)|$ again. $\square$

Remark 4.5. The converse of this theorem is generally false: the language $L = \overline{\{aa\}}$ is not intersection-closed, but its jiromaton is state-minimal.

Corollary 4.6. *If L is intersection-closed then its átomaton and distromaton have at most one more state than the jiromaton.*

Proof. By the above proof the distromaton may only have an additional final sink state – otherwise it has the same transition structure. By Corollary 3.21 the átomaton has the same number of states.   □

By Corollary 3.21 we further deduce:

Corollary 4.7. *If $L \subseteq \Sigma^*$ is intersection-closed then any state-minimal nfa accepting L has (i) $d_{\mathsf{rev}(L)}$ states if Σ^* is a union of L's derivatives and (ii) $d_{\mathsf{rev}(L)} - 1$ otherwise.*

Theorem 4.8. *If $d_L = 2^{n_L}$ then the jiromaton of L is state-minimal.*

Proof. Let $N = (n_L, R_a, F)$ be a state-minimal nfa accepting L via $I \subseteq n_L$. View it as a pointed T_Σ-coalgebra $A = (\mathcal{P}n_L, \gamma, I)$ via the subset-construction. By assumption $d_L = |\mathcal{P}n_L|$, so this is a state-minimal dfa accepting L; in particular, it is a reachable pointed T_Σ-coalgebra. Then the surjective morphism $A \twoheadrightarrow A^L_{\mathsf{JSL}}$ implies that A^L_{JSL} has no more than n_L join-irreducibles, so the jiromaton is state-minimal.   □

4.3   Characterizing the Canonical Nfas

Although the canonical nfas are generally not state-minimal, they are state-minimal amongst certain subclasses of nfas.

Theorem 4.9. *The átomaton of a regular language L is state-minimal amongst all nfas accepting L whose accepted languages are closed under complement.*

Proof. Assume the weaker condition that an nfa N accepts every language in the boolean algebra $\mathcal{B} \subseteq_\omega \mathcal{P}\Sigma^*$ generated by L's derivatives. By an earlier argument, N induces a simple T_Σ-coalgebra $(\mathcal{Q}, \gamma)$ whose states are the languages N accepts and $|N| \geq |J(\mathcal{Q})|$. By assumption $\mathcal{Q} \supseteq \mathcal{B}$ (a distributive lattice), so $|J(\mathcal{Q})| \geq |J(\mathcal{B})|$ by the proof of Theorem 4.4. The join-irreducibles of a finite boolean algebra are its atoms, so N has no less states than the átomaton.   □

The next result is from [17]. It follows because quotients and subspaces of finite-dimensional vector spaces cannot have larger dimension.

Theorem 4.10. *([17]). Any canonical xor nfa for L is state-minimal amongst nfas accepting L via $\mathbb{Z}_2$-weighted acceptance.*

We give a mild generalization of a result in [10]. Recall that nfas accepting L also accept all unions of its derivatives. Then we can conclude from Theorem 4.1:

Corollary 4.11. *The jiromaton of a regular language L is state-minimal amongst nfas accepting precisely the unions of L's derivatives.*

Example 4.12. Let N be an nfa accepting L via initial states I. If every singleton set of states is reachable from I then N accepts precisely the unions of L's derivatives. Thus, it is no smaller than L's jiromaton.

Theorem 4.13. *The distromaton of a regular language L is state-minimal amongst all nfas accepting L whose accepted languages are closed under intersection.*

Proof. Reuse the proof of Theorem 4.9. Again we actually have a stronger result: the distromaton is state-minimal amongst all nfas which can accept every intersection of L's derivatives.   □

5   Conclusions and Future Work

It is often claimed in the literature that canonical nondeterministic automata do not exist, usually as a counterpoint to the minimal dfa. On the contrary we have shown that they *do* exist and moreover arise from the minimal dfa interpreted in a locally finite variety. In so doing we have unified previous work from three sources [8,10,17] and introduced a new canonical nondeterministic acceptor, the distromaton. We also identified a class of languages where canonical *state-minimal* nfas exist. These results depend heavily on a coalgebraic approach to automata theory, providing not only new structural insights and construction methods but also a new perspective on what a state-minimal acceptor actually is.

In this paper we introduced nondeterministic closure automata, viz. $\overline{T}_\Sigma$-coalgebras in the category of closure spaces, mainly as a tool for constructing the jiromaton. However, nondeterministic closure automata bear interesting structural properties themselves, which we did not discuss here in depth. We expect that a proper investigation of these machines will lead to further insights about nondeterminism, in particular additional and more general criteria for the (state-)minimality of nfas.

Another point we aim to investigate in more detail are the algorithmic aspects of the state-minimization problem for nfas. Although this problem is known to be PSPACE-complete in general, the canonicity of our nfas suggests that – at least for certain natural subclasses of nfas – efficient state-minimization procedures may be in reach. We leave the study of such complexity-related issues for future work.

References

1. Adámek, J., Bonchi, F., Hülsbusch, M., König, B., Milius, S., Silva, A.: A coalgebraic perspective on minimization and determinization. In: Birkedal, L. (ed.) FOSSACS 2012. LNCS, vol. 7213, pp. 58–73. Springer, Heidelberg (2012)
2. Adámek, J., Milius, S., Moss, L.S., Sousa, L.: Well-pointed coalgebras (extended abstract). In: Birkedal, L. (ed.) FOSSACS 2012. LNCS, vol. 7213, pp. 89–103. Springer, Heidelberg (2012)
3. Barr, M.: Terminal coalgebras in well-founded set theory. Theor. Comput. Sci. **114**(2), 299–315 (1993)
4. Bezhanishvili, N., Kupke, C., Panangaden, P.: Minimization via duality. In: Ong, L., de Queiroz, R. (eds.) WoLLIC 2012. LNCS, vol. 7456, pp. 191–205. Springer, Heidelberg (2012)
5. Bonchi, F., Bonsangue, M.M., Boreale, M., Rutten, J.J.M.M., Silva, A.: A coalgebraic perspective on linear weighted automata. Inform. Comput. **211**, 77–105 (2012)
6. Bonchi, F., Bonsangue, M.M., Rutten, J.J.M.M., Silva, A.: Brzozowski's algorithm (co)algebraically. In: Constable, R.L., Silva, A. (eds.) Logic and Program Semantics, Kozen Festschrift. LNCS, vol. 7230, pp. 12–23. Springer, Heidelberg (2012)
7. Bonsangue, M.M., Milius, S., Silva, A.: Sound and complete axiomatizations of coalgebraic language equivalence. ACM Trans. Comput. Log. **14**(1), 7:1–7:52 (2013)
8. Brzozowski, J., Tamm, H.: Theory of átomata. In: Mauri, G., Leporati, A. (eds.) DLT 2011. LNCS, vol. 6795, pp. 105–116. Springer, Heidelberg (2011)
9. Brzozowski, J.A.: Canonical regular expressions and minimal state graphs for definite events. Mathematical Theory of Automata. MRI Symposia Series, vol. 12, pp. 529–561. Polytechnic Press/Polytechnic Institute of Brooklyn, New York (1962)
10. Denis, F., Lemay, A., Terlutte, A.: Residual finite state automata. Fund. Inform. **XX**, 1–30 (2002)
11. Hasuo, I., Jacobs, B., Sokolova, A.: Generic trace semantics via coinduction. Log. Methods Comput. Sci. **3**(4:11), 1–36 (2007)
12. Jacobs, B., Silva, A., Sokolova, A.: Trace semantics via determinization. In: Pattinson, D., Schröder, L. (eds.) CMCS 2012. LNCS, vol. 7399, pp. 109–129. Springer, Heidelberg (2012)
13. Jipsen, P.: Categories of algebraic contexts equivalent to idempotent semirings and domain semirings. In: Kahl, W., Griffin, T.G. (eds.) RAMICS 2012. LNCS, vol. 7560, pp. 195–206. Springer, Heidelberg (2012)
14. Lombardy, S., Sakarovitch, J.: The universal automaton. In: Flum, J., Grädel, E., Wilke, T. (eds.) Logic and Automata. Texts in Logic and Games, vol. 2, pp. 457–504. Amsterdam University Press, Amsterdam (2008)
15. Milius, S.: A sound and complete calculus for finite stream circuits. In: Proceedings of 25th Annual Symposium on Logic in Computer Science (LICS'10), pp. 449–458. IEEE Computer Society (2010)
16. Silva, A., Bonchi, F., Bonsangue, M.M., Rutten, J.J.M.M.: Generalizing determinization from automata to coalgebras. Log. Methods Comput. Sci **9**(1:9), 23 (2013)
17. Vuillemin, J., Gama, N.: Efficient equivalence and minimization for non deterministic Xor automata. Research report, LIENS (May 2010). http://hal.inria.fr/inria-00487031

Towards Systematic Construction of Temporal Logics for Dynamical Systems via Coalgebra

Baltasar Trancón y Widemann[(✉)]

Ilmenau University of Technology, Ilmenau, Germany
`baltasar.trancon@tu-ilmenau.de`

Abstract. Temporal logics are an obvious high-level descriptive companion formalism to dynamical systems which model behavior as deterministic evolution of state over time. A wide variety of distinct temporal logics applicable to dynamical systems exists, and each candidate has its own pragmatic justification. Here, a systematic approach to the construction of temporal logics for dynamical systems is proposed: Firstly, it is noted that dynamical systems can be seen as coalgebras in various ways. Secondly, a straightforward standard construction of modal logics out of coalgebras, namely Moss's coalgebraic logic, is applied. Lastly, the resulting systems are characterized with respect to the temporal properties they express.

1 Introduction

Dynamical systems are the classical constructive formalism for behaviour arising from the deterministic evolution of system state over time [1], dating back to the works of Newton and Laplace. Clearly *temporal logics*, with operators such as 'next', 'always', 'eventually' and 'for-at-least', constitute a companion descriptive formalism. However, the relation is not one-to-one: One the one hand, there is a unifying theory underlying the various perspectives on dynamical systems as monoid actions, which uniformly covers discrete and continuous, as well as hybrid systems [6]. But on the other hand, the diversity of temporal logics in literature is immense, see [13], and the choice for a particular system is often justified by ad-hoc pragmatic arguments. The present article explores a potentially systematic approach to the construction of temporal logics for dynamical systems, via the relatively recent mathematical field of *universal coalgebra* which has been shown to be intimately connected to both dynamical systems [11] and modal logics [5]. A different approach, also based on coalgebras and the Stone duality, has been suggested [3] for constructing modal logics of *transition systems*, a close relative of dynamical systems from theoretical computer science.

The method outlined in the remainder of this article, while theoretically simple, touches on many different fields of mathematics: order theory, category theory, algebra, coalgebra, classical modal logics à la Kripke, and coalgebraic logics à la Moss [8]. Thus a significant fraction of this paper is dedicated to reviewing the relevant definitions and propositions from the respective standard literature.

© IFIP International Federation for Information Processing 2014
M.M. Bonsangue (Ed.): CMCS 2014, LNCS 8446, pp. 211–224, 2014.
DOI: 10.1007/978-3-662-44124-4_12

This review makes up the Sects. 2 and 3. The expert reader is encouraged to skip ahead: Sect. 4 ties up all the loose ends and gives a novel contribution. There a selection of obvious coalgebraic perspectives on dynamical systems is explored, and the respective logics entailed by applying Moss's construction are characterized.

2   Review: Classical Ingredients

This section reviews some basic definitions and propositions.

2.1   Order Relations

We assume that the reader is familiar with basic order-theoretic properties of binary relations, namely with *reflexive, transitive, symmetric* relations, and with *preorders, partial orders* and *equivalences*. We give two additional related definitions that are not quite as universal:

Definition 1 (Non-Branching & Linear Relations). *Let X be a set. A binary relation $R \subseteq X^2$ is called*

– ***non-branching*** *if and only if $x \mathrel{R} y$ and $x \mathrel{R} z$ imply $y \mathrel{R} z$ or $z \mathrel{R} y$, and*
– ***linear*** *if and only if $x \mathrel{R} y$ or $y \mathrel{R} x$,*

respectively, for all $x, y, z \in X$. Clearly, every linear relation is non-branching.

2.2   Monoids

We assume that the reader is familiar with the notions of a *monoid* $\mathbb{M} = (M, 0, +)$, and of monoid *generators*. We recall that every monoid induces an ordering relation:

Definition 2 (Monoid Order). *Let $\mathbb{M} = (M, 0, +)$ be a monoid. For any elements $a, b \in M$, we write $a \leq_M b$ if and only if there is some $c \in M$ such that $a + c = b$. We say that $a \leq_M b$ **via** c. It follows directly from the monoid axioms that $\leq_M$ is reflexive and transitive, and hence a preorder. By extension, $\mathbb{M}$ itself is called **symmetric/non-branching/linear** if and only if $\leq_M$ is symmetric/non-branching/linear, respectively.*

Note that being symmetric in this sense is different from being Abelian. In fact, symmetry characterizes a famous subclass of monoids, namely the groups:

Lemma 1 (Groups). *A monoid $\mathbb{M}$ is a group if and only if it is symmetric. Every symmetric monoid is trivially linear, with the degenerate order $(\leq_M) = M^2$, the full relation.*

The proof of this simple but non-standard proposition is left as an exercise to the reader.

2.3 Dynamical Systems

Definition 3 (Dynamical System). *Let* $\mathbb{T} = (T, 0, +)$ *be a monoid called* **time** *(durations). A* **dynamical system** *is an enriched structure* $\mathbb{S} = (\mathbb{T}, S, \Phi)$ *with*

– *a set* S *called* **state space***, and*
– *a map* $\Phi : S \times T \to S$ *called* **dynamics***,*

such that

$$\Phi(s, 0) = s \qquad\qquad \Phi\big(\Phi(s, t), u\big) = \Phi(s, t + u)$$

In other words, Φ *is a* right monoid action *of* $\mathbb{T}$ *on* S. $\mathbb{S}$ *is called*

– **linear-time** *if and only if* $\mathbb{T}$ *is linear, otherwise* **nonlinear-time***, and*
– **invertible** *if and only if* $\mathbb{T}$ *is symmetric.*

Corollary 1. *There are no invertible nonlinear-time dynamical systems.*

Dynamical systems are a fundamental model class of many natural and social sciences. In comparison with their younger counterparts in computer science, automata and transition systems, dynamical systems are typically

– behaviourally weaker; deterministic, non-pointed (without distinguished initial states) and total (without spontaneous termination), but
– structurally stronger; with additional features of time (density, completeness) and state space (topology, metric, differential geometry, measures).

Automata-like constructions can be emulated by dynamical systems; see Example 1 below.

Definition 4 (Step, Trajectory & Orbit). *From the dynamics map* Φ *we may derive three forms of secondary maps:*

$$\begin{array}{lll}
\Phi^t : S \to S & \Phi_s : T \to S & \Phi^\circ : S \to \mathcal{P}S \\
\Phi^t(s) = \Phi(s, t) & \Phi_s(t) = \Phi(s, t) & \Phi^\circ(s) = \mathrm{Img}(\Phi_s) \\
& & = \{\Phi(s, t) \mid t \in T\}
\end{array}$$

– Φ^t *is called the* **step** *of* **duration** t*, or just the* t-step.
– Φ_s *is called the* **trajectory** *of* **initial** *state* s.
– $\Phi^\circ(s)$ *is called the* **orbit** *of state* s.

Lemma 2 (Homomorphic Steps). *The dynamical systems with time* $\mathbb{T}$ *are precisely those systems* $(\mathbb{T}, S, \Phi)$ *such that the step construction is a monoid homomorphism from* $\mathbb{T}$ *into the monoid of maps of type* $S \to S$ *with right composition.*

$$\Phi^0 = \mathrm{id}_S \qquad\qquad \Phi^{t+u} = \Phi^u \circ \Phi^t$$

Corollary 2 (Generating Steps). *If $G \subseteq T$ generates the monoid $\mathbb{T}$, then Φ is determined uniquely by the collection of steps $(\Phi^t)_{t \in G}$.*

Example 1 (Instances of Time)

- The time monoid $(\mathbb{N}, 0, +)$ yields standard non-invertible, linear-time, discrete-time dynamical systems. The step Φ^1 is generating. Trajectories are (one-sided) infinite sequences.
- The time monoid $(\mathbb{Z}, 0, +)$ yields standard invertible, hence linear-time, discrete-time dynamical systems. The step Φ^1 is generating and must be invertible. Trajectories are two-sided infinite sequences.
- The time monoid $(\mathbb{R}_+, 0, +)$ yields standard non-invertible, linear-time, continuous-time dynamical systems. No finite step generator collection exists. Trajectories are one-sided parametric curves.
- The time monoid $(\mathbb{R}, 0, +)$ yields standard invertible, hence linear-time, continuous-time dynamical systems. No finite step generator exists; classically definitions are given as solutions to ordinary differential equations. Trajectories are two-sided parametric curves.
- The free "time" monoid $(\Sigma^*, \varepsilon, \cdot)$ over some finite alphabet Σ yields total semiautomata, or deterministic finitely-labelled transition systems. The steps $\{\Phi^a \mid a \in \Sigma\}$ (columns of the transition table) are generating. Trajectories are big-step transition functions of total automata, mapping input words to final states. □

2.4   Propositional Modal Logics

We assume that the reader is familiar with the syntax and semantics of classical propositional logics and their presentation solely in terms of the connectives $\neg$ and $\rightarrow$. For the modal extensions, see [2] or some other textbook.

Definition 5 (Syntax of Propositional Modal Logics). *The modal extension of classical propositional logics adds two unary connectives $\square$ and $\lozenge$, taking $\square$ as primitive and defining*

$$\lozenge A = \neg \square \neg A$$

Definition 6 (Semantics of Propositional Modal Logics). *A **normal** modal extension of classical propositional logics adds at least the deduction rule of **necessitation** or **generalization**, and the axiom of **distribution**:*

$$A \vdash \square A \qquad\qquad \square(A \rightarrow B) \rightarrow (\square A \rightarrow \square B)$$

Example 2. Important normal modal logics are obtained by adding certain axioms:

- $\square A \rightarrow A$ added to the minimal system results in the logic T.
- $\square A \rightarrow \square\square A$ added to T results in the logic $S4$.
- $\square(\square A \rightarrow B) \vee \square(\square B \rightarrow A)$ added to $S4$ results in the logic $S4.3$.
- $\lozenge A \rightarrow \square\lozenge A$ added to $S4$ or $S4.3$ results in the logic $S5$. □

2.5 Kripke Semantics

Definition 7 (Kripke Frame). *A Kripke frame is a structure (W, R) with a set W of **worlds** and a binary relation R on W called **accessibility**.*

Definition 8 (Kripke Model). *Let (W, R) be a Kripke frame. A Kripke model (of propositional modal logic) is an extended structure $(W, R, \Vdash)$, where $\Vdash$ is a relation between W and the language Form of logical formulas, such that:*

$$w \Vdash \neg A \quad \Longleftrightarrow \quad w \nVdash A$$
$$w \Vdash A \to B \quad \Longleftrightarrow \quad w \Vdash A \text{ implies } w \Vdash B$$
$$w \Vdash \Box A \quad \Longleftrightarrow \quad v \Vdash A \text{ whenever } w \, R \, v$$

*We say that w **satisfies** A in $(W, R, \Vdash)$ if and only if $w \Vdash A$.*

Lemma 3. *The satisfaction relation $\Vdash$ of a Kripke model is determined uniquely by the satisfaction of atomic propositions.*

Definition 9 (Validity). *A formula A is called **valid** in*

- *a Kripke model $(W, R, \Vdash)$ if and only if it is satisfied in all worlds $w \in W$,*
- *a Kripke frame (W, R) if and only if it is valid in all Kripke models $(W, R, \Vdash)$,*
- *a class C of Kripke frames if and only if it is valid in all members of C.*

Definition 10 (Soundness/Completeness). *A propositional modal logic L is called, with respect to a class C of Kripke frames,*

- ***sound** if and only if provability in L implies validity in C, and*
- ***complete** if and only if validity in C implies provability in L.*

Theorem 1. *The modal logics $S4/S4.3/S5$ are sound and complete for the classes of Kripke frames (W, R) where R is an arbitrary/non-branching/symmetric pre-order, respectively.*

Definition 11 (Finite Frame Property). *A propositional modal logic L is said to have the **finite frame property**, if and only if it is complete for a class of finite Kripke frames.*

Theorem 2. *The modal logics $S4/S4.3/S5$ have the finite frame property, for the finite-frame subclasses of the respective classes given in Theorem 1.*

3 Review: Non-classical Ingredients

This section briefly reviews some definitions and propositions from categorial coalgebra and coalgebraic logics. See [11] and [7,8], respectively, for full details.

3.1 Category Theory

We assume that the reader is familiar with basic endofunctors on the category **Set**, in particular the identical functor Id, the constant functor $\mathrm{Const}(C)$ for any object C, the covariant Hom-functors and the covariant powerset functor $\mathcal{P}$. All functors considered in the following are tacitly **Set**-endofunctors.

Definition 12 (Monotonic, Standard & Finitary Functors). *As usual, a functor is called*

- **monotonic** *if and only if it preserves inclusions,*
- **standard** *if and only if it preserves inclusions and weak pullbacks, and*
- **finitary** *if and only if it is determined completely by its action on finite sets.*

*A standard, infinitary functor F has a **finitary restriction** F_ω defined by*

$$F_\omega X = \bigcup \{FY \mid Y \subseteq X \wedge Y \ finite\} \qquad F_\omega(h : X \to Y) = Fh|_{F_\omega X}$$

Relation liftings for functors are conventionally defined in terms of *span* diagrams, but a pointwise notation will be more convenient for the following discussions:

Definition 13 (Relation Lifting). *Let F be a functor. Every relation $R \subseteq X \times Y$ has a lifting $F[R] \subseteq FX \times FY$ defined as the set of pairs $(\hat{x}, \hat{y})$ for which there is some $\hat{r} \in FR$ such that $(F\pi_1)(\hat{r}) = \hat{x}$ and $(F\pi_2)(\hat{r}) = \hat{y}$.*

Example 3. The liftings for some basic functors are as follows:

- The identical functor lift a relation to itself: $x\ \mathrm{Id}[R]\ y$ if and only if $x\ R\ y$.
- A constant functor lifts to the identical relation: $c\ \mathrm{Const}(C)[R]\ d$ if and only if $c = d$.
- $Y\ \mathcal{P}[R]\ Z$ if and only if for all $y \in Y$ there is a $z \in Z$, and vice versa, such that $y\ R\ z$.
- $f\ \mathrm{Hom}(C, -)[R]\ g$ if and only if $f(c)\ R\ g(c)$ for all $c \in C$.   □

3.2 Universal Coalgebra

We assume that the reader is familiar with the notion of F-coalgebras for a functor F, their associated homomorphisms, F-bisimulations and F-bisimilarity.

Definition 14 (Parallel Coalgebra Composition). *Coalgebras with the same carrier can be combined in parallel: Let (X, f) be an F-coalgebra and (X, g) be a G-coalgebra. Then $(X, \langle f, g\rangle)$ is an $(F \times G)$-coalgebra, where*

$$\langle f, g\rangle(x) = \big(f(x), g(x)\big)$$

3.3   Moss's Coalgebraic Logic

The idea of Moss's coalgebraic logic [8] is to replace Kripe frames by F-coalgebras for some functor F, and to derive a universal and natural modality from F itself:

Definition 15 (Moss's Coalgebraic Logic, Abstractly). *Fix a standard functor F. Extend the syntax of propositional logic by a pseudo-unary connective ∇ that, unlike the classical modalities like $\square$, applies not to a single formula $A \in Form$ but to an expression of type either $\widehat{A} \in F(Form)$ or $\widehat{A} \in F_\omega(Form)$. For infinitary F where the choice makes a difference, the cases are called **infinitary** and **finitary** F-coalgebraic logics, respectively. A Moss model is a structure $(X, f, \Vdash)$ where (X, f) is an F-coalgebra and $\Vdash$ is a relation between coalgebra states and formulas, such that*

$$x \Vdash \neg A \iff x \nVdash A \qquad x \Vdash A \to B \iff x \Vdash A \text{ implies } x \Vdash B$$

as for Kripke models, but

$$x \Vdash \nabla \widehat{A} \iff f(x) \, F[\Vdash] \, \widehat{A}$$

Moss's coalgebraic logic as presented here specifies satisfaction only up to atomic propositions, in analogy to Kripke frames. In Moss's original presentation, the specification is unique, in analogy to Kripke models:

Definition 16 (Moss's Coalgebraic Logic, Concretely). *Let (X, f) be an F-coalgebra. Let $s : X \to \mathcal{P}(Prop)$ be the map that assigns to each state $x \in X$ the desired set of valid atomic propositions. Then (X, s) is a $\mathrm{Const}(\mathcal{P}(Prop))$-coalgebra. For the parallel composite coalgebra $(X, g = \langle f, s \rangle)$, a unique Moss model is specified by the additional clause*

$$x \Vdash A \iff A \in s(x) \qquad (A \in Prop)$$

The following two propositions state that traditional Kripke frames are essentially equivalent to the special case $F = \mathcal{P}$:

Lemma 4. *$\mathcal{P}$-coalgebras (X, f) are in one-to-one correspondence to relations R on X by putting $x \, R \, y$ if and only if $y \in f(x)$.*

Theorem 3. *The Kripke modalities $\square, \Diamond$ and the Moss modality ∇ for finitary $\mathcal{P}$-coalgebraic logics are equivalent. For infinitary $\mathcal{P}$-coalgebraic logics, they are also equivalent in the presence of infinitary conjunction and disjunction; otherwise ∇ is generally more expressive.*

$$w \Vdash_K \square A \iff w \Vdash_M \nabla\{A\} \vee \nabla\emptyset \qquad w \Vdash_K \Diamond A \iff w \Vdash_M \nabla\{A, \top\}$$

$$w \Vdash_M \nabla\widehat{A} \iff w \Vdash_K \square(\bigvee \widehat{A}) \wedge \bigwedge \Diamond\widehat{A} \quad \text{where} \quad \Diamond\widehat{A} = \{\Diamond B \mid B \in \widehat{A}\}$$

where $\Vdash_K / \Vdash_M$ denote satisfaction à la Kripke/Moss, respectively.

In general, the infinitary version of the operator ∇ is better matched with a logic where conjunction and disjunction are also infinitary. While an uncommon topic classically, infinitary logics are an important topic in modal logic because of their connection to bisimulations; they always satisfy the Henessy–Milner property:

Theorem 4 (Expressivity). *In any fully $(\wedge, \vee, \nabla)$-infinitary F-coalgebraic logic, two states $s, t \in S$ satisfy the same set of formulas if and only if they are bisimilar.*

Generally finitary logics are nicer to work with, and of course more likely to be decidable. See Example 5 below for popular temporal operators that are *not* finitary in this framework.

4   Constructions

This section gives novel theoretical results by investigating the ramifications of the following recipe:

1. identify some generic F-coalgebraic view on dynamical systems;
2. use Moss's construction to obtain logics with ∇_F modality, depending on the functor F;
3. relate ∇_F to established temporal logic operators.

Note that all of the following constructions have the state space S of a fixed dynamical system as the carrier of some coalgebra for various standard functors. Hence the associated logical languages can coexist naturally in a single system, by the parallel composition given in Definition 14.

4.1   Step Logics

Definition 17 (Step Coalgebra). *Let $\mathbb{S} = (\mathbb{T}, S, \Phi)$ be a dynamical system. For any element $t \in T$, the Id-coalgebra (S, Φ^t) is called the t-**step coalgebra** of $\mathbb{S}$.*

Definition 18 (Multistep Coalgebra). *Let $\mathbb{S} = (\mathbb{T}, S, \Phi)$ be a dynamical system. For any subset $U \subseteq T$, the $\mathrm{Hom}(U, -)$-coalgebra $(S, s \mapsto \Phi_s \circ \mathrm{in})$, given the inclusion map $\mathrm{in} : U \to T$, is called the U-**multistep coalgebra** of $\mathbb{S}$.*

Lemma 5. *The ∇ modality of step coalgebras amounts to*

– *for the t-step:*
$$s \Vdash \nabla A \iff \Phi(s, t) \Vdash A$$

– *for the U-multistep:*
$$s \Vdash \nabla \widehat{A} \iff \Phi(s, t) \Vdash \widehat{A}(t) \ for \ all \ t \in U$$

The functors for t-steps and finite U-multisteps are finitary; hence no additional distinction between finitary and infinitary logics arises.

Definition 19 (Step Modality). *We define the temporal modality $\bigcirc$, with the intuitive meaning* next, *in terms of ∇.*

$$\bigcirc A = \nabla A \qquad\qquad \bigcirc_t A = \nabla u \mapsto \begin{cases} A & (t = u) \\ \top & (t \neq u) \end{cases}$$

Example 4. (Multi-)Step coalgebras are of particular interest for finite generators, since they specify the dynamics uniquely and concisely. The following are generating in the sense of Example 1:

- For time $(\mathbb{N}, 0, +)$, the 1-step coalgebra maps every state to its successor. The resulting temporal logic has $\bigcirc$ as the *next* operator of traditional unidirectional discrete-time temporal logic.
- For time $(\mathbb{Z}, 0, +)$, the (± 1)-multistep coalgebra maps every state to its successor/predecessor, respectively. The resulting temporal logic has $\bigcirc_{\pm 1}$ as the *next/previously* operators of traditional bidirectional discrete-time temporal logic, respectively.
- For "time" $(\Sigma^*, \varepsilon, \cdot)$, the Σ-multistep coalgebra maps every automaton state to its response function (row of the transition table). The resulting logic has $(\bigcirc_a)_{a \in \Sigma}$ as the generating cases of Pratt's *necessity* operators $[a]$ in dynamic logic [10], where they are extended to the free Kleene algebra over Σ.

Interesting infinite, non-generating examples include:

- For time $(\mathbb{R}, 0, +)$ and $\delta > 0$, let U denote the open interval $(-\delta, \delta)$. The U-multistep coalgebra maps every state to its temporal δ-neighbourhood. □

Step generators benefit from Moss's Expressivity Theorem 4:

Corollary 3. *The modality ∇ and the family of modalities $(\bigcirc_t)_{t \in U}$ for generating U are straightforwardly equivalent if U is finite, and equivalent in the presence of infinitary conjunction otherwise.*

$$x \Vdash \nabla \widehat{A} \iff x \Vdash \bigwedge_{t \in U} \bigcirc_t \widehat{A}(t)$$

The following construction is the multistep limit case $U = T$.

4.2 Trajectory Logics

Definition 20 (Trajectory Coalgebra). *Let $\mathbb{S} = (\mathbb{T}, S, \Phi)$ be a dynamical system. The $\mathrm{Hom}(T, -)$-coalgebra $(S, s \mapsto \Phi_s)$ is called the **trajectory coalgebra** of $\mathbb{S}$.*

Lemma 6. *The ∇ modality of trajectory coalgebras amounts to*

$$s \Vdash \nabla \widehat{A} \iff \Phi(s, t) \Vdash \widehat{A}(t) \text{ for all } t \in T$$

The ∇ trajectory modality is a surprisingly powerful logical operator, with the severe disadvantage that there is no canonical syntactic representation. The following examples are but a small subset of useful special cases.

Example 5. Arguments of the ∇ trajectory modality are maps of type $T \to$ *Form*. Various intensional notations for such maps, or time-dependent formulas, give rise to well-known temporal operators. Note that all following examples work for finitary ∇.

– Consider discrete time $(\mathbb{N}, 0, +)$ or $(\mathbb{Z}, 0, +)$. Define a *zip* operator

$$A \leftrightharpoons B = \nabla t \mapsto \begin{cases} A & t \text{ even} \\ B & t \text{ odd} \end{cases}$$

Then a dynamic system is bipartite, with characteristic formula A, if and only if $(A \leftrightharpoons \neg A) \vee (\neg A \leftrightharpoons A)$ is valid in the Moss model associated with its trajectories.

– Consider automaton time $(\Sigma^*, \varepsilon, \cdot)$. Define a *consumption* operator

$$eat(L, A, B) = \nabla t \mapsto \begin{cases} A & t \in L \\ B & t \notin L \end{cases}$$

for languages $L \subseteq \Sigma^*$ and formulas A, B. Now let A be a formula characterizing accepting states. Then an automaton, as a dynamical system, accepts at least/exactly the language $L \subseteq \Sigma^*$ if and only if $eat(L, A, \top)/eat(L, A, \neg A)$, respectively, is valid for its initial state(s) in the Moss model associated with its trajectories.

– Consider time with a linear antisymmetric order $<$. Define a *change* operator

$$chg(t, A, B, C) = \nabla u \mapsto \begin{cases} A & u < t \\ B & u = t \\ C & u > t \end{cases}$$

for time duration t and formulas A, B, C. Then minimum/maximum-duration operators can be defined directly, in two variants differing in the inclusion of boundary cases:

$$\min t.\, A = chg(t, A, \top, \top) \qquad \max t.\, A = chg(t, \top, \top, \neg A)$$
$$\min' t.\, A = chg(t, A, A, \top) \qquad \max' t.\, A = chg(t, \top, \neg A, \neg A)$$

Imprecise operators such as *until* can be expressed as infinitary disjunctions:

$$A \mathbf{U} B = \bigvee_{t \in T} chg(t, A, B, \top)$$

4.3   Orbit Logics

The following construction shifts the coalgebraic focus from trajectories to orbits which are images of trajectories, hence abstracting from durations. The result is a family of qualitative temporal logics that can be expressed naturally in the classical modal operators, uniformly for all kinds of time structure:

Definition 21 (Orbit Coalgebra). *Let* $\mathbb{S} = (\mathbb{T}, S, \Phi)$ *be a dynamical system. The $\mathcal{P}$-coalgebra (S, Φ°) is called the **orbit coalgebra** of $\mathbb{S}$. We say that in $\mathbb{S}$, y is **reachable** from x, written $x \leadsto_\mathbb{S} y$, if and only if $y \in \Phi^\circ(x)$. More precisely, we say $x \leadsto_\mathbb{S} y$ **via** t, if and only if $y = \Phi^t(x)$.*

Clearly, $x \leadsto_\mathbb{S} y$ if and only if there is some witness $t \in T$ such that $x \leadsto_\mathbb{S} y$ via t.

Lemma 7. *For dynamical systems $\mathbb{S}$, the reachability relation $\leadsto_\mathbb{S}$ is*

1. *always a preorder,*
2. *additionally non-branching, but* not *generally linear, if $\mathbb{S}$ is linear-time, and*
3. *additionally symmetric if $\mathbb{S}$ is invertible.*

Proof

1. Reflexivity and transitivity follow directly from the monoid axioms: $x \leadsto_\mathbb{S} x$ via 0, and if $x \leadsto_\mathbb{S} y$ via t and $y \leadsto_\mathbb{S} z$ via u, then $x \leadsto_\mathbb{S} z$ via $t + u$.
2. Assume that $x \leadsto_\mathbb{S} y$ via t and $x \leadsto_\mathbb{S} z$ via u. By linearity of $\mathbb{T}$ assume, without loss of generality, that $t \leq_\mathbb{T} u$ via v. Then $y \leadsto_\mathbb{S} z$ via v.
3. For symmetric $\mathbb{T}$, if $x \leadsto_\mathbb{S} y$ via t, then $y \leadsto_\mathbb{S} x$ via $-t$.   □

The caveat in case 2 of this proposition is necessary:

Example 6 (Nonlinear Linear-Time Dynamical System). Set $T = \{0\}$, giving rise to the singleton monoid which is trivially linear. This fixes Φ completely as $\Phi(s, t) = \Phi(s, 0) = s$, giving rise to a "still-life" structure of time. Then neither $x \leadsto_\mathbb{S} y$ nor $y \leadsto_\mathbb{S} x$ for $x \neq y$.

Definition 22 (Orbital Frame). *A Kripke frame is called **orbital** if and only if it corresponds, in the sense of Lemma 4, to the orbit coalgebra of some dynamical system. An orbital frame is called linear-time/invertible if and only if it corresponds to the orbit coalgebra of some linear-time/invertible dynamical system, respectively.*

Using this definition, Lemma 7 extends to Kripke frames:

Lemma 8. *For any orbital Kripke frame $\mathbb{F} = (W, R)$, the relation R is*

1. *always a preorder,*
2. *additionally non-branching if $\mathbb{F}$ is linear-time,*
3. *additionally symmetric if $\mathbb{F}$ is invertible.*

This statement has a partial, finitary converse, which is by far the most technical result of the present paper:

Lemma 9. *A finite Kripke frame* (W, R) *is*

1. *always orbital if R is a preorder,*
2. *additionally linear-time if R is non-branching,*
3. *additionally invertible if R is symmetric.*

Proof. Construct a dynamical system $\mathbb{S} = (\mathbb{T}, S, \Phi)$ with $(\leadsto_\mathbb{S}) = R$. In any case, clearly $S = W$. Proceed in reverse order and increasing flexibility of cases. For the latter two, consider the partition of W into *strongly connected components* (sccs) of the preorder R: nonempty, maximal subsets $C \subseteq W$ such that $x \mathrel{R} y$ for all $x, y \in C$. We write $x \sim y$ if and only if x, y are in the same scc, that is $x \mathrel{R} y$ and $y \mathrel{R} x$.

3. Set $\mathbb{T} = (\mathbb{Z}, 0, +)$. By symmetry of R there are no related pairs across sccs. For each scc C choose an arbitrary cyclic permutation. Set Φ^1 to their union. Then
   - $x \leadsto_\mathbb{S} y$ via some $i < k$, where k is the size of the scc containing both, if $x \mathrel{R} y$, and
   - otherwise $x \not\leadsto_\mathbb{S} y$.
2. Set $\mathbb{T} = (\mathbb{N}, 0, +)$. We say that y is a *successor* of x, writing $x \ll y$, if and only if $x \mathrel{R} y$ but not $y \mathrel{R} x$. Clearly, $x \mathrel{R} y$ if and only if either $x \sim y$ or $x \ll y$, exclusively. We say that x is *transient* if it has successors. Since W is finite and R is non-branching, every transient x has a unique scc of least successors, from which we may choose one, say x', and all elements reachable from x are successors. Set $\Phi^1(x) = x'$. For non-transient x, all elements reachable from x are in the same scc. Proceed as above. Then
   - $x \leadsto_\mathbb{S} y$ via some $i < k$, where k is the number of successors of x, if $x \ll y$,
   - $x \leadsto_\mathbb{S} y$ via some $i < k$, where k is the size of the scc containing both, if $x \sim y$, and
   - otherwise $x \not\leadsto_\mathbb{S} y$.
1. There are in general no least successors, and there may be non-successors reachable from transient elements. A more basic construction is needed: Set $\mathbb{T} = (\mathbb{N}^*, \varepsilon, \cdot)$, the free monoid over $\mathbb{N}$. For each $x \in W$ choose some infinite sequence $y = (y_0, y_1, \dots) \in W^\omega$ such that $x \mathrel{R} z$ if and only if $z = y_i$ for some i. This is always possible by invocation of countable choice, since the set $\{z \mid x \mathrel{R} z\}$ is finite and nonempty. For the generating steps $\{\Phi^n \mid n \in \mathbb{N}\}$, set $\Phi^n(x) = y_n$. Then
   - $x \leadsto_\mathbb{S} y$ via 1, if $x \mathrel{R} y$, and
   - otherwise $x \not\leadsto_\mathbb{S} y$. $\qquad\square$

Theorem 5. *The modal logics S4/S4.3/S5 are sound and complete for arbitrary/linear-time/invertible orbital frames, respectively.*

Proof. S4/S4.3/S5 are sound for the class of Kripke frames (W, R) where R is an arbitrary/non-branching/symmetric preorder, respectively. By Lemma 8, they are also sound for the subclasses of arbitrary/linear-time/invertible orbital frames, respectively.

Conversely, $S4/4.3/S5$ are complete for the class of Kripke frames (W, R) where R is an arbitrary/non-branching/symmetric preorder, respectively, and have the finite frame property. By Lemma 9, they are also complete for the subclasses of arbitrary/linear-time/invertible orbital frames, respectively.   □

Example 7. The operators $\square$ and $\Diamond$ are well-suited to express "long-term" behavioral properties of dynamical systems. For instance, let A be the characteristic formula of a subset $U \subseteq S$ of the state space. Then U is a stationary solution of a dynamical system if and only if $A \rightarrow \square A$ is valid in the Moss model associated with its orbits.   □

5   Conclusion and Outlook

Many operators discussed in the temporal logic literature can be subsumed under a common framework by viewing them as instances of Moss's modality ∇, for some coalgebraic presentation of the underlying dynamical system models. As a rule of thumb,

- step coalgebras go with discrete time,
- trajectory coalgebras go with quantitative operators for either discrete or dense time, and
- orbit coalgebras go with arbitrary time and qualitative operators, in particular the classical modal operators and the framework of normal modal logics.

The examples given in this article are of course only a small selection to prove the viability of the approach. There is considerable potential for generalization. The trajectory modality is an extremely expressive tool, and it is likely that many other temporal operators can be shown to coincide with particular intensional notations for it. Besides, coalgebraic perspectives on dynamical systems other than the three detailed above could be considered. An interesting open problem and direction for future research is the integration of measure-theoretic temporal operators, for instance in duration calculus [4], into the framework.

Coalgebraic modal logics from predicate liftings [9] are an alternative to Moss's logic, with somewhat different properties. A study of the predicate liftings arising from the functorial perspective on dynamical systems is expected provide some interesting additional insights.

Recently, some progress has been made in the use of non-standard analysis to clarify the relationship of discrete-time and continuous-time systems, in a way that is compatible with a coalgebraic perspective [12]. That approach might also be helpful in bridging the gap between our trajectory and orbit logics.

References

1. Birkhoff, G.D.: Dynamical Systems. American Mathematical Society, New York (1927)
2. Blackburn, P., van Benthem, J., van Wolter, F. (eds.): Handbook of Modal Logic. Elsevier, Amsterdam (2006). ISBN: 9780080466668. http://cgi.csc.liv.ac.uk/frank/MLHandbook/
3. Bonsangue, M.M., Kurz, A.: Duality for logics of transition systems. In: Sassone, V. (ed.) FOSSACS 2005. LNCS, vol. 3441, pp. 455–469. Springer, Heidelberg (2005). doi:10.1007/978-3-540-31982-5-29. ISBN: 3-540-25388-2
4. Chaochen, Z., Hoare, C.A.R., Ravn, A.P.: A calculus of durations. Inf. Process. Lett. **40**(5), 269–276 (1991)
5. C. Cirstea, Kurz, A., Pattinson, D., Schröder, L., Venema, Y.: Modal logics are coalgebraic. In: Gelenbe, E., Abramsky, S., Sassone, V. (eds.) BCS International Academic Conference, pp. 128–140. British Computer Society (2008)
6. Jacobs, B.: Object-oriented hybrid systems of coalgebras plus monoid actions. Theor. Comput. Sci. **239**(1), 41–95 (2000). doi:10.1016/S0304-3975(99)00213-3. ISSN: 0304-3975
7. Kupke, C., Pattinson, D.: Coalgebraic semantics of modal logics: an overview. Theor. Comput. Sci. **412**(38), 5070–5094 (2011). doi:10.1016/j.tcs.2011.04.023
8. Moss, L.S.: Coalgebraic logic. Ann. Pure Appl. Logic **96**(1–3), 277–317 (1999). doi:10.1016/S0168-0072(98)00042-6
9. Pattinson, D.: Expressivity results in the modal logic of coalgebras. Ph.D. thesis. Universitat Munchen (2001)
10. Pratt, V.: Semantical considerations on floyd-hoare logic. In: Proceedings of 17th Annual IEEE Symposium on Foundations of Computer Science, pp. 109–121. IEEE Computer Society (1976). doi:10.1109/SFCS.1976.27
11. Rutten, J.: Universal coalgebra: a theory of systems. Theor. Comput. Sci. **249**(1), 3–80 (2000). doi:10.1016/S0304-3975(00)00056-6
12. Suenaga, K., Sekine, H., Hasuo, I.: Hyperstream processing systems: nonstandard modeling of continuous-time signals. In: Giacobazzi, R., Cousot, R. (eds.) POPL, pp. 417–430. ACM (2013). doi:10.1145/2429069.2429120
13. Venema, Y.: Temporal logic. In: Goble, L. (ed.) The Blackwell Guide to Philosophical Logic, pp. 203–223. Blackwell, Oxford (2001). doi:10.1111/b.9780631206934. 2001.00013.x. Chapter 10. ISBN: 9780631206934

Algebraic–Coalgebraic Recursion Theory of History-Dependent Dynamical System Models

Baltasar Trancón y Widemann[1,2]($\boxtimes$) and Michael Hauhs[2]

[1] Programming Languages and Compilers,
Ilmenau University of Technology, Ilmenau, Germany
`baltasar.trancon@tu-ilmenau.de`
[2] Ecological Modelling, University of Bayreuth, Bayreuth, Germany
`michael.hauhs@uni-bayreuth.de`

Abstract. We investigate the common recursive structure of history-dependent dynamic models in science and engineering. We give formal semantics in terms of a hybrid algebraic–coalgebraic scheme, namely course-of-value iteration. This theoretical approach yields categories of observationally equivalent model representations with precise semantic relationships. Along the initial–final axis of these categories, history dependence can appear both literally and transformed into instantaneous state. The framework can be connected to philosophical and epistemological discourse on one side, and to algorithmic considerations for computational modeling on the other.

1 Introduction

Models of system dynamics are a cornerstone of science and engineering. They relate the future of a system to its present and/or past. An obvious qualitative distinction is whether observation of the present (*state*) alone suffices to predict or modify the future, or whether information about the past (*history*) is necessary. This question can be discussed on the philosophical level, or on the mathematical level, potentially leading to theoretical frameworks and tools for the working scientist and engineer.

In this paper, we explore the mathematical option, and present a formalization that puts the two model classes on equal footing. Specifically, they shall be demonstrated to form not a dichotomy, but a continuum along the initial–final axis of suitable categories of models, constructed from first principles of algebraic–coalgebraic recursion theory.

In philosophical terms, this framework gives precise semantic relationships between more and less history-dependent models, and thus renders the two most common objections against history-dependent modeling obsolete: arguments from naïve reductionism (invoking Laplace's Daemon) and arguments from parsimony (invoking Occam's Razor).

We begin with motivating examples that demonstrate the pervasive occurrence, and the vastly different relative reputability, of the two modeling approaches

© IFIP International Federation for Information Processing 2014
M.M. Bonsangue (Ed.): CMCS 2014, LNCS 8446, pp. 225–244, 2014.
DOI: 10.1007/978-3-662-44124-4_13

across various disciplines. The purpose of this digression is to demonstrate the broad applicability of our proposed framework, which is hard to see directly from the fairly modest formal results. Readers more narrowly interested in theoretical matters are encouraged to skip ahead to Sect. 2.

1.1 Basic Scientific Example: Simple Harmonic Oscillator

A simple harmonic oscillator is an ideal point mass m moving frictionlessly along a line and acted on by a restoring force proportional, with positive coefficient k, to its displacement x. An empirical study of (real approximations of) many such systems might reveal that, with good accuracy, *three* displacements observed at snapshots spaced equally in time, with a small delay δ, are related according to the following model formula:

$$x_{t+\delta} = \left(2 - \tfrac{k}{m}\delta^2\right)x_t - x_{t-\delta} \tag{1a}$$

On the other hand, with slightly less accuracy, *two* observations at snapshots with a small delay δ are related according to

$$\begin{pmatrix} x_{t+\delta} \\ v_{t+\delta} \end{pmatrix} = \begin{pmatrix} 1 & \delta \\ -\tfrac{k}{m}\delta & 1 \end{pmatrix} \begin{pmatrix} x_t \\ v_t \end{pmatrix} \tag{1b}$$

provided that the "virtual" observable $v = dx/dt$ is added to the data set.

From the fact that apparently each system has a period of $T = 2\pi/\omega$ where $\omega = \sqrt{k/m}$, one might get the intuition that the matrix entries in (1b) are actually linear approximations of trigonometric functions for $\delta \to 0$. Indeed, whereas the preceding models are only accurate for $\delta \ll T$, the following model gives the exact dynamics of the system:

$$\begin{pmatrix} x_{t+\delta} \\ v_{t+\delta} \end{pmatrix} = \begin{pmatrix} \cos(\omega\delta) & \omega^{-1}\sin(\omega\delta) \\ -\omega\sin(\omega\delta) & \cos(\omega\delta) \end{pmatrix} \begin{pmatrix} x_t \\ v_t \end{pmatrix} \tag{1c}$$

Elementary theoretical physics tells us that all three models given above contain a grain of the same truth, namely that they can be derived from the characteristic linear differential equation of the system:

$$\frac{d^2x}{dt^2} + \omega^2 x = 0 \tag{2}$$

This differential equation has a family of solutions of the form $x_t = A\sin(\omega t + \varphi)$ for arbitrary A and φ. The models are then obtained as follows:

– Model (1c) by substitution of $t + \delta$ for t and various trigonometric identities.
– Model (1b) by linear approximation:

$$x_{t+\delta} \approx x_t + \delta v_t \qquad\qquad v_{t+\delta} \approx v_t + \delta a_t \tag{3a}$$

  where $a = dv/dt = d^2x/dt^2 = -\omega^2 x$ according to (2).
– Model (1a) by slightly different linear approximation eliminating v:

$$x_{t+\delta} \approx x_t + \delta v_{t+\delta} \qquad v_{t+\delta} \approx v_t + \delta a_t \qquad v_t \approx \frac{x_t - x_{t-\delta}}{\delta} \tag{3b}$$

1.2   Complex Scientific Example: ARMA

History dependence in scientific modeling can also take more assertive forms, where explicit dependence on past values is not only taken at face value, but featured as the principal methodological design concept.

The Box–Jenkins approach [1] focuses on the *auto-regressive moving average* (ARMA) class of stochastic models. These can be thought of as filters that add statistical autocorrelation to a discrete input signal in a controlled way, by linear dependence on past output (auto-regressive) and input (moving average) values.

$$\underbrace{y_t = \phi_1 y_{t-1} + \cdots + \phi_p y_{t-p}}_{\text{AR}} + x_t + \underbrace{\theta_1 x_{t-1} + \cdots + \theta_q x_{t-q}}_{\text{MA}} \tag{4a}$$

The theory is phrased as transformation of bilaterally infinite stochastic processes, although deterministic variants, with at best pseudorandom input, are used for actual simulation. In practice, the models are used both *directly*, to simulate data with a prescribed autocorrelation structure, and *inversely*, to estimate the model coefficients that describe observed output data optimally, by minimizing the variance of the implied, unobserved random input.

Linear combinations of past values are formulated neatly in an operator calculus on sequences, namely as formal power series of the *backshift* operator B, defined as $(Bx)_t = x_{t-1}$, which gives the following compact formula:

$$\underbrace{(1 - \Phi)}_{\text{AR}} y = \underbrace{(1 + \Theta)}_{\text{MA}} x \quad \text{where} \quad \Phi = \sum_{k=1} \phi_k B^k \quad \text{and} \quad \Theta = \sum_{k=1} \theta_k B^k \tag{4b}$$

In the basic form, both power series are finite, and the summations have upper bounds. But on one hand, there are transformations of finitary pure AR into infinitary pure MA models, and of finitary pure MA into infinitary pure AR models. And on the other hand, many kinds of real-world modeling problems call for extensions, in particular the ARIMA class, where the output of the ARMA filter is subsequently *integrated*, either a whole number of times (ARIMA proper) or *fractionally* (FARIMA) [2]. Since integration is linear and invertible, the resulting model can be viewed not only as a composition of two sequence transforms, but also as an ARMA model where a corresponding *differencing* step is composed with the auto-regressive part, in terms of a formal backward differencing operator:

$$\underbrace{(1 - \Phi)}_{\text{AR}} \underbrace{(1 - B)^d}_{\text{I}} y = \underbrace{(1 + \Theta)}_{\text{MA}} x \tag{4c}$$

Differencing can be raised to arbitrary real powers d by Newton's generalized binomial theorem:

$$(1 - B)^d = \sum_{k=0}^{\infty} \binom{d}{k} (-1)^k B^k \quad \text{where} \quad \binom{d}{k} = \prod_{i=0}^{k-1} \frac{d - i}{1 + i}$$

This power series is easily seen to vanish only for positive integer d; hence the flagship models of FARIMA class, particularly effective for the structure of realistic time series data [5], depend on *actually infinite* history. Of course, simulations approximate by padding with zeroes, and luckily the consequent behavioral error tends to vanish in few steps.

1.3   Complex Engineering Example: TFM

The trace function method (TFM) [8] is a mathematically fundamental approach to behavioral description of software components. Components are taken to interact at their interface in discrete events. Data flows through the values of input and output variables, controlled by the environment and the component, respectively, at the time of an event.

The behavior of the component is represented in terms of sequences of events (*traces*). Traces come in two flavours: a *complete* trace has input and output values for all events; an *incomplete* trace omits output values for the latest event. Valid responses of the component are given as a collection of maps (optionally set-valued for nondeterminism) from incomplete traces to output values, one per output variable. The set of valid complete traces, the semantic object of behavior, is defined recursively: The empty trace is valid; a nonempty trace is valid if and only if its latest outputs can be reconstructed from its incomplete form, and the rest of the trace is valid.

As a toy example, a "stubborn" vending machine that offers a choice of either coffee (c) or tea (t) but, when asked for coffee the third time in a row, produces tea instead, can be described as a component with one input and one output variable, each of type $\{c, t\}$, and the output function

$$f(c, (c, c), (c, c), \ldots) = t \qquad \text{otherwise} \qquad f(x_n, (y_{n-1}, x_{n-1}), \ldots) = x_n \qquad (5)$$

where output precedes input, and the latest, incomplete event is leftmost.

TFM shines particularly for behavior that is far too complex and/or irregular to be discussed here in passing. See [8,13] for worked-out examples.

1.4   Discussion

The succession of three oscillator models, as given in Sect. 1.1, creates the impression of the growth of scientific knowledge along the following lines:

1. Empirical models establish candidates for causal relationships between past and future values of directly observable variables. Such models are necessarily approximations.
2. In theoretically informed models, history of directly observable variables is "explained away" by reference to auxiliary, indirectly observable variables. The rate of change of some variable often turns out to be a good candidate for an auxiliary variable; hence the usefulness of differential equations.
3. If done in the right way, approximative models can eventually be replaced by exact, albeit idealized, history-independent models: *dynamical systems.*

Classical physics has had its great successes in progressing from stage 1 (typically via stage 2) to stage 3 for many systems. On the other hand, most practically relevant models in self-styled "complex system" sciences seem to be stuck in stage 1. For instance, the Box–Jenkins framework, as given in Sect. 1.2, is not only resilient to replacement by more state-oriented models, but actually spreading virulently across disciplines: Though originally developed for economics, it is considered state of the art in environmental sciences as well [4,6]. The success of these models, their empirical and heuristical nature notwithstanding, is due to their pragmatic relevance as powerful forecasting tools [1].

In engineering applications, TFM even demonstrates a reversal of reputability of modeling with and without history dependence: A description in terms of traces of history is more abstract, and *precisely hence* more durable, reliable and valuable information than any more concrete, state-based "implementation". This view conforms to the software engineering doctrine that the "what" should be stated, not the "how"; here applied to the question of *memory*.

With the benefit of theoretical hindsight, the approximations (3a) and (3b) that justify the empirical relations (1a) and (1b), respectively, may seem strangely ad-hoc. But the situation in complex system sciences is often such that modeling and simulation precedes theory, or even may take its place indefinitely [7]. Hence, in absence of theoretical justification, we need a neutral approach to judge the epistemological quality of models. Since, as Rosen [9] observed, scientific modeling is essentially about the discovery of *recursive* functional relationships in data, thus elucidating how the future can be understood as entailed by the past, it seems only reasonable to turn to (categorial) recursion theory for answers.

2 Algebraic–Coalgebraic Recursion Theory

We recall some basic notions of categorial recursion theory. Since the key construct for the present work is from [15], we adopt that paper's notation with a few exceptions; in particular we write the more conventional $\iota_1, \ldots, \iota_n$ and $\pi_1, \ldots, \pi_n$ for n-ary coproduct injections and product projections, respectively. We also write $C^{a..b}$ for the set of streams (sequences) of Cs of lengths a to b, inclusively, where $a, b \in \mathbb{N} \cup \{\omega\}$. Note that the symbol ω henceforth refers to the limit ordinal, not the angular velocity as in Sect. 1.1.

We consider algebras and coalgebras for certain simple endofunctors on **Set**. Unless explicitly restricted to **Set**, results generalize implicitly to other distributive categories, and endofunctors that satisfy the small print. We take the liberty to call any morphisms in this category "maps". We assume that, for a given functor F of interest, an initial F-algebra $(\mu F, \text{in}_F : F(\mu F) \to \mu F)$ and a final F-coalgebra $(\nu F, \text{out}_F : \nu F \to F(\nu F))$ exist. By Lambek's Lemma, in_F and its dual out_F are isomorphisms.

The following three definitions and their properties are a minimal excerpt of [15] for our present needs. Note that we have added references to applications in scientific modeling from our own work [3]. For a broader introduction to (co)iterative definitions arising from (co)algebras, see [10].

2.1   Iteration

The simplest recursion scheme that can be described by these structures is *iteration*: given a F-algebra $(C, \varphi : F(C) \to C)$, there is a unique F-algebra homomorphism $(\!|\varphi|\!)_F : (\mu F, \mathrm{in}_F) \to (C, \varphi)$. That $(\!|\_|\!)_F$ is in fact a recursion operator can be seen from the slightly transformed universal property:

$$(\!|\varphi|\!)_F = \varphi \circ F\big((\!|\varphi|\!)_F\big) \circ \mathrm{in}_F^{-1}$$

Uses of iteration in mathematically structured functional programming are ubiquitous. We have argued in [3] that iteration over functors of the form $\mathcal{A}(X) = A \times X + B$ is a systematic approach to the recursion scheme of typical *structure-oriented* scientific modeling questions, in particular *prediction* and *identification of initial conditions*.

2.2   Coiteration

The dual of iteration, (still?) distinctly less popular in computation theory, is *coiteration*: given a F-coalgebra $(C, \varphi : C \to F(C))$, there is a unique F-coalgebra homomorphism $[\!(\varphi)\!]_F : (C, \varphi) \to (\nu F, \mathrm{out}_F)$. That $[\!(\_)\!]_F$ is in fact a recursion operator can be seen from the slightly transformed universal property:

$$[\!(\varphi)\!]_F = \mathrm{out}_F^{-1} \circ F\big([\!(\varphi)\!]_F\big) \circ \varphi$$

Since coiteration produces infinite data, it is hardly ever considered outside the realms of process theory and lazy functional programming. We have argued in [3], based on an earlier discussion in [10], that coiteration over functors of the form $\mathcal{A}(X)$, dual to the above, is a systematic approach to the recursion scheme of typical *behavior-oriented* scientific modeling questions, in particular *symbolic dynamics* and the study of *irreversibility*.

2.3   Course-of-Value Iteration

Course-of-value (cov) iteration is a recursion scheme where the function value for a structured argument may depend on the function values for subarguments at any nesting depth, as opposed to ordinary iteration where dependency is limited to the function values for immediate, maximal subarguments. The key ingredient for the additional power is to augment the functor F by taking the product with some object C, intuitively a coloring. Henceforth, we assume C to be a *nonempty* set, avoiding uninteresting pathological cases. Since the product with a fixed object is a pervasive construct in the following, we abbreviate the functor $C \times (-)$ to C:

$$C(X) = C \times X \qquad\qquad C(h) = \mathrm{id}_C \times h$$

Then we can consider final CF-coalgebras and, most importantly, operations of type $\varphi : F(\nu CF) \to C$. Note that νCF parses as $\nu(CF)$. We define the cov iteration of φ, written $\{\!|\varphi|\!\}_F : \mu F \to C$ as the following equation from [15]; the situation and type information is summarized in the diagram below:

$$\{\!|\varphi|\!\}_F = \underbrace{\pi_1 \circ \text{out}_{CF}}_{\text{top}_{CF}} \circ \underbrace{(\!|\text{out}_{CF}^{-1} \circ \langle \varphi, \text{id}_{F(\nu CF)} \rangle |\!)_F}_{\overline{\varphi}} \tag{6}$$

$$
\begin{array}{ccc}
F(\mu F) & \xrightarrow{\ \text{in}_F\ } & \mu F \\
{\scriptstyle F((\!|\overline{\varphi}|\!)_F)}\Big\downarrow & & \Big\downarrow{\scriptstyle (\!|\overline{\varphi}|\!)_F} \quad\searrow{\scriptstyle \{\!|\varphi|\!\}_F} \\
F(\nu CF) & \xrightarrow[\ \overline{\varphi}\ ]{} & \nu CF \xrightarrow[\ \text{top}_{CF}\ ]{} C
\end{array}
$$

That $\{\!|\_|\!\}_F$ is in fact a recursion operator can be seen from the, slightly convoluted, universal property

$$\{\!|\varphi|\!\}_F = \varphi \circ F([\langle \{\!|\varphi|\!\}_F, \text{in}_F^{-1} \rangle]_{CF}) \circ \text{in}_F^{-1}$$

which is a key result of [15]. This recursion scheme is not only defined in terms of ordinary iteration, complicating the domain of evaluation from C to νCF, but vice versa also contains ordinary iteration over the domain C as a degenerate case: It is easy to see that $\{\!|\varphi \circ F(\text{top}_{CF})|\!\}_F = (\!|\varphi|\!)_F$ for $\varphi : F(C) \to C$.

Note that in the data structure νCF of histories, the most recent results are on top.

2.4 Simple Examples

For the motivating examples of this paper and many others besides, consider the extremely simple functor $N(X) = 1 + X$. It has the following well-known initial algebra and final coalgebra:

$$\mu N = \mathbb{N} \qquad\qquad\qquad \text{in}_N = [0, succ]$$
$$\nu N = \mathbb{N} \cup \{\omega\} \qquad\qquad\qquad \text{out}_N = pred$$

where $pred$ is the partial predecessor function with $pred(\omega) = \omega$.

For cov iteration, we also need the C-colored final coalgebra, which consists of the nonempty finite and infinite streams over C together with the total $head$ and partial $tail$ operations.

$$\nu CN = C^{1..\omega} \qquad\qquad\qquad \text{out}_{CN} = \langle head, tail \rangle$$

The domain of cov operations $N(\nu CN)$ is the set $C^{0..\omega}$ of finite and infinite streams over C, now additionally including the empty stream (). Because of the intrinsic left bias of the $head/tail$ structure, history is encoded with the most recent element on the left.

Iteration and coiteration over N obey simple rules. Given a N-algebra (C, φ) with $\varphi = [z, s]$, we have:

$$(\!|\varphi|\!)_N(n) = s^n(z)$$

For instance, define powers of two as iteration of doubling, starting from one:

$$([one, double])_N(n) = 2^n$$

Given a N-coalgebra (C, φ) with $\varphi : C \to N(C)$ understood as the partial function $\varphi : C \nrightarrow C$, we have

$$[\varphi]_N(x) = \sup\{n \in \mathbb{N} \mid \varphi^n(x) \text{ defined}\}$$

where $\sup \mathbb{N} = \omega$ and iteration of partial functions is strict. For a useful example of coiteration over N, first define the *domain restriction* of a total function $f : X \to Y$ to a region $U \subseteq X$ as:

$$f|_U : X \nrightarrow Y \qquad\qquad (f|_U)(x) = \begin{cases} f(x) & x \in U \\ \text{undefined} & x \notin U \end{cases}$$

Then we can define the *iterated logarithm*, as it appears in the study of algorithmic complexity, concisely as the coiteration of the logarithm, domain-restricted to a certain open real interval:

$$[\log|_{(1,\infty)}]_N = \log^*$$

For examples in the realm of scientific modeling, consider a discrete-time dynamical system with state space S and single-step transition function $f : S \to S$. For a given initial state $x_0 \in S$, the operation $t = [x_0, f]$ gives rise to a N-algebra (S, t). Iteration produces precisely the *trajectory* of x_0, that is the infinite sequence of states obtained by repeated application of f:

$$([t])_N(n) = f^n(x_0)$$

Now consider a region $U \subseteq S$ and the partial step function $f|_U$ where the domain is restricted to U. This gives rise to a N-coalgebra $(U, f|_U)$. Coiteration classifies states in S according to *escape time*, that is the length of the longest prefix of the trajectory completely contained in U:

$$[f|_U]_N(x) = \sup\{n \in \mathbb{N} \mid f^n(x) \in U\}$$

In particular, U is a *stationary solution* of the system if and only if $[f|_U]_N(x) = \omega$ for all $x \in U$.

The paradigmatic example for a cov iteration is the Fibonacci sequence. Define an auxiliary operation $\varphi : \mathbb{N}^{0..\omega} \to \mathbb{N}$ such that

$$\varphi() = 0 \qquad\qquad \varphi(a) = a + 1 \qquad\qquad \varphi(a, b, \ldots) = a + b \qquad (7)$$

and conclude $fib = ([\varphi])_N$. Note that there is a generalization of the system from $\mathbb{N}$ to $\mathbb{Z}$ that has several advantages; see below.

The simple harmonic oscillator model (1a) has a straightforward reconstruction as a cov iteration, completely analogous to the Fibonacci sequence: Fix the model parameters ω and δ to obtain the loose specification

$$\varphi(a, b, \ldots) = (2 - (\omega\delta)^2)a - b$$

Now fix a reference time t and the first two observations $f() = x_t$ and $f(a) = x_{t+\delta}$. This specifies f uniquely. Then we have $x_{t+n\delta} = \{\![\varphi]\!\}_N(n)$. Models (1b) and (1c), being essentially independent of history, are more adequately reconstructed as ordinary iterations. We leave the generating maps as exercises to the reader.

ARMA and TFM models are more complicated due to having input. Their reconstruction as cov iterations over functor N is possible (with exponential carrier and some higher-order functional programming; see [13]), but not straightforward and out of scope here. A more direct approach is by the composite functor $M = DN$ for some input object D, which yields complete and incomplete traces:

$$\nu CM = (C \times D)^{1..\omega} \qquad\qquad M(\nu CM) = D \times (C \times D)^{0..\omega}$$

For any cov trace function φ, the syntactic trace space νCM splits into *valid* traces $\mathrm{Img}\{\![\varphi]\!\}_M$ and the complementary *invalid* traces. It is easy to see by induction that the recursive function $\{\![\varphi]\!\}_M$ is determined completely by its behavior on valid traces.

3   General Theory: State Systems

Our theoretical approach hinges on the observation that definition (6) as given in [15] is not at all the only ordinarily iterative representation of a given cov iterative function; in fact, the collection of such representations has a rich categorial structure, covering all possible degrees of nominal history-dependence.

In all of the following, consider an endofunctor F and object C fixed.

3.1   Abstract State Systems

Definition 1 (Abstract State System). *A pair $(S, \sigma : \nu CF \to S)$ is called an **abstract state system**, if and only if top_{CF} factors through σ, that is, there is a complementary map $\psi : S \to C$ that makes the following diagram commute:*

$$
\begin{array}{ccc}
\nu CF & \xrightarrow{\;\sigma\;} & S \\[2pt]
\mathrm{out}_{CF} \downarrow & {\scriptstyle \mathrm{top}_{CF}}\searrow & \vdots\, \psi \\[2pt]
CF(\nu CF) & \xrightarrow[\pi_1]{} & C
\end{array}
\qquad (8)
$$

*The object S is called **state space**, the maps σ/ψ are called **state abstraction/valuation**, respectively.*

For the rationale of not stating the valuation ψ explicitly in the definition of a state system, see Lemma 4 below and Corollary 14 at the end of this section.

Definition 2 (Abstract State System Homomorphism). *A homomorphism between abstract state systems (S_1, σ_1) and (S_2, σ_2) is defined in the obvious way as a map $h : S_1 \to S_2$ such that the following diagram commutes:*

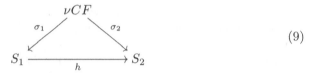

$$\text{(9)}$$

Abstract state systems and their homomorphisms give rise to a category **State**(F, C) (of coslices under νCF) with the obvious identity and composition.

Definition 3 (Epic Abstract State System). *An abstract state system (S, σ) is called **epic** if and only if σ is an epimorphism in the underlying category.*

We shall argue in the following that epic state systems are the "right" ones for various mathematical and epistemological reasons. Note that there are several subclasses of special epimorphisms, such as *regular*, *strong*, and *split* epimorphisms, all of which notoriously coincide in the base category **Set**, obscuring the subtle differences. Because there is no experience with relevant examples in other categories, we reserve judgement which exactly is the appropriate class for epic state systems in full generality. Most of the following definitions use only the characteristic property of ordinary epimorphisms e, namely

$$f \circ e = g \circ e \implies f = g$$

for all possible maps f, g but Lemma 10 also assumes that F preserves the class of epimorphisms in question, which is granted only for split epimorphisms in arbitrary categories, but generally in **Set** [14].

Lemma 4. *An epic state system determines the state valuation ψ uniquely. Homomorphisms between epic state system are unique epimorphisms.*

Hence the restriction of **State**(F, C) to epic systems gives a full, thin subcategory **EpicState**(F, C).

Lemma 5. *The categories (**Epic**)**State**(F, C) have (the same) initial objects, namely the state system $(\nu CF, \text{id}_{\nu CF})$, with the unique valuation top_{CF}, and the unique homomorphism to any abstract state system (S, σ) equalling σ.*

Lemma 6. *At least over **Set**, the category **State**(F, C) has generally no final objects.*

Proof. Consider the initial abstract state system $\mathcal{I} = (\nu CF, \text{id}_{\nu CF})$. Add a distinguished element to obtain the system $\mathcal{I}' = (\nu CF + 1, \iota_1)$ which clearly satisfies (8). Now consider any abstract state system $\mathcal{S} = (S, \sigma)$. The morphisms $\text{Hom}(\mathcal{I}', \mathcal{S})$ in **State**(F, C) are easily seen to correspond one-to-one to $\text{Hom}(1, S)$ in **Set** and hence to S itself, namely by

$$\text{Hom}(\mathcal{I}', \mathcal{S}) = \{[\sigma, i] \mid i \in \text{Hom}(1, S)\}$$

where the first component σ is fixed by Lemma 5. Hence if a final abstract state system existed, it would need to have a one-element state space, so (8) would imply $\text{top}_{CF}(x) = \text{top}_{CF}(x')$ for all $x, x' \in \nu CF$, which is generally false. □

Lemma 7. *At least over* **Set**, *the category* **EpicState**(F, C) *has a final object, namely the state system* (C, top_{CF}), *with the valuation* id_C, *and the unique homomorphism from any abstract state system* (S, σ) *equalling its valuation* ψ.

This last result, while not very useful in itself, gives hints for the construction of more richly structured final systems; see Sect. 4.2 below.

3.2   Concrete State Systems

Definition 8 (Concrete State System). *A triple* $(S, \sigma, \tau : F(S) \to S)$, *where* (S, σ) *is an abstract state system, is called a* **concrete state system** *for a cov operation* $\varphi : F(\nu CF) \to C$, *or* φ*-system for short, with* **state transition** τ, *if and only if the following (F-algebra homomorphism) diagram commutes:*

$$
\begin{array}{ccc}
F(\nu CF) & \xrightarrow{\ F(\sigma)\ } & F(S) \\
{\scriptstyle \overline{\varphi}}\Big\downarrow & & \Big\downarrow{\scriptstyle \tau} \\
\nu CF & \xrightarrow[\ \sigma\ ]{} & S
\end{array}
\tag{10}
$$

Definition 9 (Concrete State System Homomorphism). *A homomorphism between concrete state systems* (S_1, σ_1, τ_1) *and* (S_2, σ_2, τ_2) *is defined, in the obvious way, as a map* $h : S_1 \to S_2$ *that is both an abstract state homomorphism between* (S_1, σ_1) *and* (S_2, σ_2) *and a F-algebra homomorphism between* (S_1, τ_1) *and* (S_2, τ_2); *that is, the following diagram additionally commutes:*

$$
\begin{array}{ccc}
F(S_1) & \xrightarrow{\ F(h)\ } & F(S_2) \\
{\scriptstyle \tau_1}\Big\downarrow & & \Big\downarrow{\scriptstyle \tau_2} \\
S_1 & \xrightarrow[\ h\ ]{} & S_2
\end{array}
\tag{11}
$$

These definitions give rise to a (not necessarily full) subcategory **State**(φ) of **State**(F, C). The following lemma shows that the corresponding subcategory **EpicState**(φ) of **EpicState**(F, C) is full.

Lemma 10. *The additional consistency condition on concrete state system homomorphisms is essentially redundant for epic state systems; assuming F preserves enough epimorphisms, any map between state spaces of epic systems satisfying* (9) *automatically satisfies* (11).

Proof. Consider the diagram

$$
\begin{array}{ccccc}
& & \overset{F(\sigma_2)}{\overbrace{\qquad\qquad\qquad}} & & \\
F(\nu CF) & \underset{F(\sigma_1)}{\longrightarrow} & F(S_1) & \underset{F(h)}{\longrightarrow} & F(S_2) \\
{\scriptstyle \overline{\varphi}}\Big\downarrow & & {\scriptstyle \tau_1}\Big\downarrow & & \Big\downarrow{\scriptstyle \tau_2} \\
\nu CF & \underset{\sigma_1}{\longrightarrow} & S_1 & \underset{h}{\longrightarrow} & S_2 \\
& & \underset{\sigma_2}{\underbrace{\qquad\qquad\qquad}} & &
\end{array}
$$

where the inner left and outer quadrangle commute by (10), the triangles commute by (9), and the inner right quadrangle is the goal (11). A short diagram chase yields $h \circ \tau_1 \circ F(\sigma_1) = \tau_2 \circ F(h) \circ F(\sigma_1)$. Since σ_1 is an epimorphism, so is $F(\sigma_1)$ under mild assumptions as discussed above; conclude $h \circ \tau_1 = \tau_2 \circ F(h)$. $\qquad\square$

Lemma 11. *The initial abstract state system in* (**Epic**)**State**(F, C) *extends to an initial concrete state system in* (**Epic**)**State**(φ), *namely* $(\nu CF, \mathrm{id}_{\nu CF}, \overline{\varphi})$.

By contrast, the dual problem of final concrete state systems is relatively complex. Results are few and mostly negative, and many questions remain open.

Lemma 12. *The final epic abstract state system does not generally extend to concrete state system, let alone a final one.*

Proof. Assume there is a φ-system $(C, \mathrm{top}_{CF}, \tau)$. Then (10) implies that $\tau \circ F(\mathrm{top}_{CF}) = \mathrm{top}_{CF} \circ \overline{\varphi} = \varphi$. That is, φ factors through $F(\mathrm{top}_{CF})$. This is not only generally false, but defies the very purpose of cov iteration, covering only the degenerate case of ordinary iteration over C, as discussed in Sect. 2.3. $\qquad\square$

The constructive disproof of final objects for the abstract case **State**(F, C) in Lemma 7 does not carry over to the concrete case **State**(φ). A family of nontrivial and well-known examples for final concrete epic state systems will be given below in Sect. 4.2. No such example for the non-epic case is known, leaving the problem open.

3.3   Simulation by State Systems

Now we can clarify the notion of representation introduced at the start of this section, namely in terms of simulation.

Theorem 13 (Simulation). *Let* $S = (S, \sigma, \tau)$ *be a* φ-*system for a fixed cov operation* φ. *Let* ψ *be any suitable state valuation. Then the cov iteration of* φ *is simulated by ordinary iteration in terms of* S.

$$\{\!|\varphi|\!\}_F = \psi \circ (\!|\tau|\!)_F$$

Note that for the initial φ-system from Lemma 11 this equation reduces to (6).

Proof. Consider the following diagram in the category of F-algebras, which commutes by initiality:

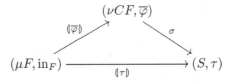

Hence back in the underlying category:

$$\psi \circ (\!|\tau|\!)_F = \psi \circ \sigma \circ (\!|\overline{\varphi}|\!)_F = \mathrm{top}_{CF} \circ (\!|\overline{\varphi}|\!)_F = \{\!|\varphi|\!\}_F \qquad\qquad \square$$

Corollary 14. *The choice of a particular state valuation ψ for a given cov operation φ is irrelevant for the purpose of simulation.*

This concludes the "universal" part of the theory. For particular choices of functor F and classes of operations φ, results are greatly more specific and practically useful.

4 Specific Theory: Autonomous Systems

4.1 Stream-Like State Spaces

For the special case $F = N$ already discussed in Sect. 2.4, a closer look reveals details hinting at a more concrete design strategy for state-based models (automata). Up to natural isomorphisms, the operations of a φ-system are of the following types:

$$\varphi : C^{0..\omega} \to C \qquad\qquad \sigma : C^{0..\omega} \to S$$
$$\overline{\varphi} : C^{a..b} \to C^{a'..b'} \qquad\qquad \tau : 1 + S \to S$$

The F-algebra operation $\overline{\varphi}$ acts predictably on stream lengths, always adding one element. Hence it can be soundly given many different types, with arbitrarily wide integer bounds $a' \le a+1$ and $b' \ge b+1$. Usage will be clear from the context.

The state transition is conveniently decomposed as $\tau = [\varepsilon, \triangleright]$ with $\varepsilon \in S$ and $\triangleright : S \to S$. Thus an algorithm that enumerates the values of $\{\!|\varphi|\!\}_N$ takes the form of a stream generator loop, as depicted in Fig. 1.

```
var x : S
x := ε
forever
   output ψ(x)
   x := ▷(x)
```

Fig. 1. Enumeration in concrete state systems over N, pseudocode

Lemma 15. *For a concrete φ-state system over N we have for all $n \ge 0$:*

$$\triangleright^n \circ \sigma = \sigma \circ \overline{\varphi}^n \qquad\qquad \{\!|\varphi|\!\}_N(n) = (\psi \circ \triangleright^n)(\varepsilon) = (\varphi \circ \overline{\varphi}^n)()$$

Concrete state systems are of particular interest, as theoretical objects rather than as algorithmic specifications, when the structure of their state space S is significantly simpler than $C^{1..\omega}$. For interpretation as physical systems, state spaces with simple product structure, which can be read as a small collection of independently observable variables, are preferred; see the oscillator example in Sect. 1.1. By contrast, interpretation as an automaton prefers a simple coproduct structure, read as a finite enumeration of alternative states.

4.2 Dependency Patterns

Definition 16 (Bounded/Regular Iteration). *A cov operation* $\varphi : C^{0..\omega} \rightarrow C$ *is called k-**bounded**, for k > 0, if and only if*

$$\varphi = \varphi \circ take(k)$$

*where take(k) takes the k first elements of its argument stream, or less if insufficient. It is called k-**regular** if and only if there is a surrogate history* $\mathfrak{h} \in C^k$ *such that*

$$\varphi = \varphi \circ take(k) \circ append(\mathfrak{h})$$

where append(s)(t) appends the stream s to t.

1-bounded cov iteration coincides with ordinary iteration. We write $\widehat{\varphi} : C^{0..k} \rightarrow C$ and $\widehat{\varphi} : C^k \rightarrow C$ for the domain restrictions of φ in bounded and regular systems, respectively. The auxiliary operation $take(k) : C^{a..b} \rightarrow C^{a'..b'}$ is polymorphic with $a' \leq \min(a, k)$ and $b' \geq \min(b, k)$.

Lemma 17. *Any k-bounded operation is (k + 1)-bounded. Any k-regular operation is k-bounded.*

The TFM example (5) is 3-bounded, as evident from the ellipsis (...) in the defining equation. The Fibonacci operation is a prototypic example of a 2-regular operation, see below. A cov operation may depend on point-wise finitely *many* elements without being k-bounded for any k (analogous to continuity versus uniform continuity); for instance, consider the well-known recursive definition of Bell numbers [11],

$$B_{n+1} = \sum_{k=0}^{n} \binom{n}{k} B_k$$

where each value depends on *all* of the preceding. Note that this example is *potentially infinitary*, as opposed to the actually infinitary FARIMA example.

The theory of φ-systems for bounded φ is particularly fruitful. In order to exploit it, we need to introduce some machinery for stream computation by bounded C-coiteration.

Definition 18 (Bounded Lookahead). *The family* $(\searrow^n)$ *of **lookahead** operators for all n ≥ 0, that take a pair of maps* $f : A \rightarrow A$ *and* $g : A \rightarrow B$ *to a map* $(g \searrow^n f) : A \rightarrow B^n$, *is defined inductively as:*

$$g \searrow^0 f = \langle \rangle \qquad\qquad g \searrow^{n+1} f = cons \circ \langle g \circ f^n, g \searrow^n f \rangle$$

That is, in closed form, $g \searrow^n f = \langle g \circ f^{n-1}, \ldots, g \circ f^0 \rangle$. Note the right-to-left "historical" enumeration order; the corresponding left-to-right enumeration would be given by $take(n) \circ [\langle g, f \rangle]_C$.

Lemma 19. *Lookahead operators obey the following laws:*

$$(g \searrow^n f) \circ f = take(n) \circ (g \searrow^{n+1} f)$$
$$(\forall m < n.\, g \circ f^m \circ e = k \circ i^m \circ h) \implies (g \searrow^n f) \circ e = (k \searrow^n i) \circ h$$

Corollary 20. *As special cases of the latter we obtain:*

$$(g \searrow^n f) \circ f = (g \circ f) \searrow^n f$$
$$(\forall m < n.\, f^m \circ e = k \circ i^m) \implies (g \searrow^n f) \circ e = (g \circ k) \searrow^n i$$

Definition 21 (FIFO System). *A first-in-first-out buffer of $k > 0$ elements of C gives rise to a canonical concrete state system candidate, the k-**fifo** system, for any k-bounded cov operation φ:*

$$S_{[k]} = C^k \qquad\qquad \sigma_{[k]} = head \searrow^k \overline{\varphi}$$
$$\varepsilon_{[k]} = (\widehat{\varphi} \searrow^k \overline{\varphi})() \qquad\qquad \triangleright_{[k]} = cons \circ \langle \widehat{\varphi}, take(k-1) \rangle$$

Note that $\overline{\varphi}$ is used polymorphically as a map on $C^{0..\omega}$ and $C^{1..\omega}$ in the definitions of $\epsilon_{[k]}$ and $\sigma_{[k]}$, respectively. These can be made precise as $\overline{\varphi} \circ \iota_2$ and $\iota_2 \circ \overline{\varphi}$, respectively, where $\iota_2 : C^{1..\omega} \to C^{0..\omega}$ is the natural inclusion.

Lemma 22. *The fifo operations have alternative forms:*

$$\sigma_{[k]} = take(k) \circ \overline{\varphi}^{k-1} \qquad \varepsilon_{[k]} = \overline{\varphi}^k() \qquad \tau_{[k]} = take(k) \circ \overline{\varphi}$$

This seems to suggest that a "lookbehind" abstraction $take(k)$ might be more straightforward than the "lookahead" $\sigma_{[k]}$. However, that would unnecessarily complicate the construction of homomorphisms; see Theorem 25 below.

Corollary 23.

$$\widehat{\varphi} \circ \sigma_{[k]} = head \circ \overline{\varphi}^k$$

Now we can verify our educated guess.

Theorem 24 (FIFO Simulation). *If φ is k-bounded, then the k-fifo system is a valid φ-system with valuation $\psi_{[k]} = \pi_k$.*

Proof. It is easy to see that (8) is satisfied:

$$\psi_{[k]} \circ \sigma_{[k]} = \pi_k \circ (head \searrow^k \overline{\varphi}) = head \circ \overline{\varphi}^0 = top_{CN}$$

In the following we use the simplified fifo operations of Lemma 22. A low-level proof in terms of lookahead laws is also possible, and will be required for the more advanced Theorem 25 below. To prove that (10) is satisfied, distinguish cases of F. On one hand:

$$(\sigma_{[k]} \circ \overline{\varphi})() = (take(k) \circ \overline{\varphi}^k)() = \varepsilon_{[k]}$$

On the other hand, exploiting polymorphism as discussed:

$$\begin{aligned}
\sigma_{[k]} \circ \overline{\varphi} \circ \iota_2 &= \sigma_{[k]} \circ \overline{\varphi} \\
&= take(k) \circ \overline{\varphi} \circ \overline{\varphi}^{k-1} \\
&= take(k) \circ cons \circ \langle \widehat{\varphi} \circ take(k), id_{C^{k..\omega}} \rangle \circ \overline{\varphi}^{k-1} \\
&= cons \circ \langle \widehat{\varphi} \circ take(k), take(k-1) \rangle \circ \overline{\varphi}^{k-1} \\
&= cons \circ \langle \widehat{\varphi}, take(k-1) \rangle \circ take(k) \circ \overline{\varphi}^{k-1} \\
&= \triangleright_{[k]} \circ \sigma_{[k]} \qquad\qquad\qquad\qquad\qquad\qquad \square
\end{aligned}$$

```
var x : C[1 .. k]
for i in k to 1
  x[i]  := φ̂(x[i + 1], ..., x[k])
forever
  output x[k]
  x := φ̂(x[1], ..., x[k]), x[1], ..., x[k − 1]
```

Fig. 2. Enumeration in fifo systems over N, pseudocode

The resulting, specialized enumeration algorithm is depicted in Fig. 2. In the case of the Fibonacci operation with $k = 2$ and $\hat{\varphi}$ defined as in (7), we obtain the familiar iterative algorithm by straightforward code specialization, depicted in Fig. 3.

```
var x : ℤ[1 .. 2]
x := 1, 0
forever
  output x[2]
  x := (x[1] + x[2]),  x[1]
```

Fig. 3. Enumeration in Fibonacci fifo system, pseudocode

The existence of fifo systems is merely a new formalization of a well-known algorithmic technique. The added value of the theory explored here is demonstrated by the following novel, surprisingly strong characterization of semantic adequacy.

Theorem 25 (FIFO Finality). *If a fifo φ-system is epic, then it is final in* **EpicState**(φ). *The unique homomorphism from any other φ-system $(S, \sigma, [\varepsilon, \triangleright])$ with valuation ψ is* $h = \psi \diagdown^k \triangleright$.

Note that there is no obvious *take*-based expression for h in analogy to Lemma 22.

Proof. By Lemma 4, it suffices to show that h is a homomorphism (†). We have

$$h \circ \sigma = (\psi \diagdown^k \triangleright) \circ \sigma$$

(Lemma 15, Corollary 20)
$$= (\psi \circ \sigma) \diagdown^k \overline{\varphi}$$

$$(8) \qquad = head \diagdown^k \overline{\varphi} = \sigma_{[k]}$$

thus verifying that h satisfies (9). Regarding (11) we have on one hand:

$$h(\varepsilon) = (\psi \diagdown^k \triangleright)(\varepsilon)$$

(Lemmas 15 and 19)
$$= (\varphi \diagdown^k \overline{\varphi})() = \varepsilon_{[k]}$$

By Lemma 15, observe that

$$\psi \circ \rhd^k \circ \sigma = \psi \circ \sigma \circ \overline{\varphi}^k$$

(8)
$$= head \circ \overline{\varphi}^k$$

(Corollary 23)
$$= \widehat{\varphi} \circ \sigma_{[k]} = \widehat{\varphi} \circ h \circ \sigma$$

With σ epi (†) conclude $\psi \circ \rhd^k = \widehat{\varphi} \circ h$. Then on the other hand:

$$h \circ \rhd = (\psi \setminus^k \rhd) \circ \rhd$$

(Lemma 19)
$$= take(k) \circ (\psi \setminus^{k+1} \rhd)$$

$$= take(k) \circ cons \circ \langle \psi \circ \rhd^k, (\psi \setminus^k \rhd) \rangle$$

$$= take(k) \circ cons \circ \langle \psi \circ \rhd^k, h \rangle$$

(above)
$$= take(k) \circ cons \circ \langle \widehat{\varphi} \circ h, h \rangle$$

$$= take(k) \circ cons \circ \langle \widehat{\varphi}, id_{C^k} \rangle \circ h$$

$$= cons \circ \langle \widehat{\varphi}, take(k-1) \rangle \circ h$$

$$= \rhd_{(k)} \circ h$$

Together conclude that h satisfies (11).   □

As a concrete example, we verify that the Fibonacci fifo system over $C = \mathbb{Z}$ is final: for every pair $(a, b) \in \mathbb{Z}^2$, we find that $\sigma_{[2]}(b, a - b, \ldots) = (a, b)$.

The fairly elegant proof makes use of the epimorphism property of σ in crucial ways, marked with (†); thus the restriction to **EpicState**(φ) is essential. There is no obvious strategy how to generalize the result to final objects in **State**(φ). Additionally, a "moral converse" of the theorem holds also.

Lemma 26. *Non-epic fifo φ-systems are not generally final, in* **State**(φ).

Proof. As a counterexample, consider the Fibonacci sequence over $C = \mathbb{N}$ instead of $\mathbb{Z}$. We can give a hierarchy of four distinct, nested valid φ-systems with state $S_0 \supset S_1 \supset S_2 \supset S_3$, namely

$$S_0 = \mathbb{Z}^2$$
$$S_1 = \{(a, b) \in S_1 \mid a \geq 0; a + b \geq 0\}$$
$$S_2 = \{(a, b) \in S_1 \mid a \geq 0; b \geq 0\} = \mathbb{N}^2$$
$$S_3 = \{(a, b) \in S_1 \mid a \geq b; b \geq 0\} = \text{Img } \sigma_{[2]}$$

arising from the fifo system over $\mathbb{Z}$ discussed above. For S_2 and S_3, the valuation $\psi_{[2]} = \pi_2$ is already $\mathbb{N}$-valued and determined uniquely. For S_0 and S_1, we can choose arbitrary valuations for elements outside S_2 without violating (8).

The system S_3 is epic, but the canonical $\mathbb{N}$-valued fifo system S_2 is not. Neither S_2 nor S_3 is final: Assume, for contradiction, a homomorphism $h : S_{0/1} \to S_{2/3}$. Now observe that the following point diagram, where the squares and the triangle are instantiations of (11) and (9), respectively, must commute.

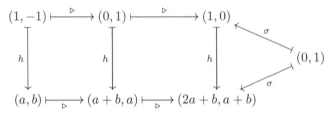

So $h(1, -1) = (a, b)$ implies $2a + b = 1$ and $a + b = 0$, which has no solution in $a, b \in \mathbb{N}$.   □

Lemma 27. *If the k-fifo φ-system is epic, then the corresponding $(k + 1)$-fifo φ-system is generally not.*

This straightforward observation clarifies that the parameter k of a final epic fifo φ-system is an objective measure of the order of history dependence in φ.

4.3   Regularization

Regularity is a stronger condition on cov operations than boundedness. We give some simple characterizations of regular operations.

Theorem 28 (Regularization). *If, for a k-bounded cov operation φ, the state transition $\triangleright_{[k]}$ is invertible, then there is an equivalent k-regular cov operation φ' such that $\{\!\{\varphi\}\!\}_N = \{\!\{\varphi'\}\!\}_N$.*

Proof. Choose $\varphi' = \varphi \circ take(k) \circ append(\mathfrak{h})$ with $\mathfrak{h} = \triangleright_{[k]}^{-k}(\varepsilon_{[k]})$. It suffices to show that the respective fifo systems coincide. Since $take(k) \circ append(\mathfrak{h})$ has no effect on the fifo state space C^k, the clause $\triangleright_{[k]} = \triangleright'_{[k]}$ is trivial. For the $\varepsilon_{[k]}$ clause, we have by Lemma 22:

$$
\begin{aligned}
\varepsilon'_{[k]} &= \overline{\varphi}'^k() \\
&= (take(k) \circ \overline{\varphi}^k)(\mathfrak{h}) \\
&= (take(k) \circ \overline{\varphi}^k \circ \triangleright_{[k]}^{-k})(\varepsilon_{[k]}) \\
\text{(Lemma 22)} \qquad &= (\triangleright_{[k]}^k \circ \triangleright_{[k]}^{-k})(\varepsilon_{[k]}) = \varepsilon_{[k]} \qquad\qquad \square
\end{aligned}
$$

For the concrete example of the Fibonacci operation, we find that $\triangleright_{[2]}$ is specified by the invertible matrix $\left(\begin{smallmatrix} 1 & 1 \\ 1 & 0 \end{smallmatrix}\right)$. The operation as defined above is its own regularization with $\mathfrak{h} = (1, -1)$. Note that there are equivalent definitions which are distinct from their regularization, for instance with the more commonly found (such as in [15]) second clause $\varphi(a) = 1$ instead of $a + 1$.

More algebraic structure on C gives more powerful results.

Corollary 29. *If $\widehat{\varphi} : C^k \to C$ is a linear form over a scalar field C, with coefficients $\alpha_1, \ldots, \alpha_k$, then $\triangleright_{[k]}$ is linear, and φ regularizes if and only if $\alpha_k \neq 0$.*

This result subsumes all finitary ARMA-like models.

5   Conclusion

The role of time in science has been hotly debated since antiquity, and quite remains so today [12]. The cov iteration approach to discrete-time history-dependent models goes a long way in retelling the meta-level discourse in a down-to-earth, formally precise and useful style. The categories $(\mathbf{Epic})\mathbf{State}(\varphi)$ of concrete systems span a solution space of hypothetical state-based realizations of a fixed black-box model with observable function $\{\![\varphi]\!\}_F$.

At one extreme, the initial system is a purely syntactic solution which takes history at face value. It is always given trivially, but yields little scientific insight, and is also deficient as an algorithmic specification, because it prescribes a *space leak*: iterated invocation of the history constructor out_F^{-1} makes state representations grow boundlessly, even if the computation could also easily be performed on bounded space, such as by a fifo system. The sheer size and coalgebraic structure of νCF as a datatype may also be distressing for the working modeler unfamiliar with coalgebraic techniques.

At the other extreme, a final system, if such a thing exists, is a purely semantic, fully abstract solution, which makes only the empirically necessary distinctions, and is hence the "holy grail" of model semantics and epistemology. Final systems can reconstruct and justify established algorithmic design techniques of computational modeling.

Since the quotient structure of a final system may be complex and hard to find and work with, the practice of modeling often deals with intermediate forms, which have a more compact and regular state space than the initial rendering, yet tractable transition and valuation rules. In particular, product-shaped state spaces welcome the interpretation of their projections as independent, objectively real variables, both analytically in science and synthetically in engineering.

Epic state systems are of particular interest. From the formal perspective, they come with vastly more benign properties. From the epistemological perspective, they can be justified by an invocation of Occam's Razor: states are not to be multiplied without necessity (here for simulation). On the other hand, epic homomorphisms, that is unique surjective structure-preserving maps $h : S_1 \to S_2$, can be understood as precise semantic relationships between alternatives of hypothetical state, by turning state space S_1 into a possibly redundant but complete system of representatives for the more abstract but elusive state space S_2. This view renders philosophical arguments against the structure of S_1 as a datatype obsolete. It is foreseeable that advanced aspects of model analysis can be integrated into the picture as equational specifications of such homomorphisms. Furthermore, we point out other (co)limit constructions than initial/final objects in the thin category of epic state systems as an interesting open problem.

The recursion scheme of the elementary functor N is sufficient for many textbook examples, as well as for straightforward modeling of discrete-time systems without input. It comes with an extensive theory of canonical, even final systems, with well-understood algorithmic properties. It remains to be seen whether behaviorally more complex applications are better dealt with by reduction to this base case, or by extension of the theory to more advanced functors.

Bounded and regular operations are of particular algorithmic interest, because they do away with space leaks and base cases, respectively. We expect that algorithmic treatment of more complicated patterns of history dependence is also possible, and can lead to formally derived implementations of TFM. On the other hand, dependency on actually infinite histories is a feature of the cov operation only, not of the generated recursive function. This should serve as a warning for the philosophical discourse to properly distinguish the theoretical role of the former from the empirical role of the latter, such as in inverse FARIMA model estimation.

And finally, we are happy to have given yet another meaning to the phrase "universal coalgebra: a theory of systems."

References

1. Box, G.E., Jenkins, G.M.: Time Series Analysis: Forecasting and Control. Holden-Day, San Francisco (1970)
2. Granger, C.W.J., Joyeux, R.: An introduction to long-memory time series models and fractional differencing. J. Time Ser. Anal. **1**, 15–30 (1980)
3. Hauhs, M., Trancón y Widemann, B.: Applications of algebra and coalgebra in scientific modelling, illustrated with the logistic map. Electron. Notes Theoret. Comput. Sci. **264**(2), 105–123 (2010)
4. Hipel, K.W., McLeod, A.I.: Time Series Modelling of Water Resources and Environmental Systems. Elsevier, Amsterdam (1994)
5. Montanari, A., Rosso, R., Taqqu, M.S.: Fractionally differenced ARIMA models applied to hydrologic time series: identification, estimation, and simulation. Water Resour. Res. **33**(5), 1035–1044 (1997)
6. Montanari, A., Rosso, R., Taqqu, M.S.: A seasonal fractional ARIMA model applied to the Nile river monthly flows at Aswan. Water Resour. Res. **36**(5), 1249–1259 (2000)
7. Peters, R.H.: A Critique for Ecology. Cambridge University Press, Cambridge (1991)
8. Quinn, C., Vilkomir, S.A., Parnas, D.L., Kostic, S.: Specification of software component requirements using the trace function method. In: Proceedings of ICSEA, p. 50. IEEE Computer Society (2006)
9. Rosen, R.: Life Itself: a Comprehensive Inquiry into the Nature, Origin, and Fabrication of Life. Columbia University Press, New York (1991)
10. Rutten, J.: Universal coalgebra: a theory of systems. Theoret. Comput. Sci. **249**(1), 3–80 (2000)
11. The on-line encyclopedia of integer sequences: A000110. http://oeis.org/A000110 (2013)
12. Smolin, L.: Time Reborn: From the Crisis in Physics to the Future of the Universe. Houghton Mifflin Harcourt, Boston (2013)
13. Trancón y Widemann, B.: The recursion scheme of the trace function method. In: Filipe, J., Maciaszek, L.A. (eds.) Proceedings of ENASE, pp. 146–155 (2012)
14. Trnková, V.: Some properties of set functors. Comment. Math. Univ. Carol. **10**(2), 323–352 (1969)
15. Uustalu, T., Vene, V.: Primitive (co)recursion and course-of-value (co)iteration, categorically. Informatica **10**(1), 5–26 (1999)

Author Index

Printed in the United States
By Bookmasters